Soul, Psyche, Brain: New Directions in the Study of Religion and Brain–Mind Science

Edited by
Kelly Bulkeley

SOUL, PSYCHE, BRAIN
© Kelly Bulkeley, 2005.

All rights reserved. No part of this book may be used or reproduced in any manner whatsoever without written permission except in the case of brief quotations embodied in critical articles or reviews.

First published in 2005 by
PALGRAVE MACMILLAN™
175 Fifth Avenue, New York, N.Y. 10010 and
Houndmills, Basingstoke, Hampshire, England RG21 6XS
Companies and representatives throughout the world.

PALGRAVE MACMILLAN is the global academic imprint of the Palgrave Macmillan division of St. Martin's Press, LLC and of Palgrave Macmillan Ltd. Macmillan® is a registered trademark in the United States, United Kingdom and other countries. Palgrave is a registered trademark in the European Union and other countries.

ISBN 1–4039–6508–0
ISBN 1–4039–6509–9

Library of Congress Cataloging-in-Publication Data

    Soul, psyche, brain: new directions in the study of religion and brain-mind science / [edited by] Kelly Bulkeley.
      p. cm.
    Includes bibliographical references and index.
    ISBN 1–4039–6508–0—ISBN 1–4039–6509–9 (pbk.)
      1. Cognitive neuroscience. 2. Brain—Religion aspects.
  I. Bulkeley, Kelly, 1962–

QP360.5.S68 2005
612.8′233—dc22   2005040552

A catalogue record for this book is available from the British Library.

Design by Newgen Imaging Systems (P) Ltd., Chennai, India.

First edition: November 2005

10 9 8 7 6 5 4 3 2 1

Printed in the United States of America.

For Those Who Build Bridges

CONTENTS

Acknowledgments		vii
List of Contributors		viii
	Introduction Kelly Bulkeley	1
Chapter One	Genes, Brains, Minds: The Human Complex Holmes Rolston III	10
Chapter Two	Brain, Mind, and Spirit—A Clinician's Perspective, or Why I Am Not Afraid of Dualism James W. Jones	36
Chapter Three	Psychoneurological Dimensions of Anomalous Experience in Relation to Religious Belief and Spiritual Practice Stanley Krippner	61
Chapter Four	Sacred Emotions Robert A. Emmons	93
Chapter Five	Where Neurocognition Meets the Master: Attention and Metacognition in Zen Tracey L. Kahan and Patricia M. Simone	113
Chapter Six	From Chaos to Self-Organization: The Brain, Dreaming, and Religious Experience David Kahn	138
Chapter Seven	Converting: Toward a Cognitive Theory of Religious Change Patricia M. Davis and Lewis R. Rambo	159
Chapter Eight	Cognitive Science and Christian Theology Charlene P.E. Burns	174
Chapter Nine	Overcoming an Impoverished Ontology: Candrakīrti and the Mind–Brain Problem Richard K. Payne	197

Chapter Ten	Religion and Brain–Mind Science: Dreaming the Future *Kelly Bulkeley*	219
Chapter Eleven	Religion Out of Mind: The Ideology of Cognitive Science and Religion *Jeremy Carrette*	242
Chapter Twelve	Brain Science on Ethics: The Neurobiology of Making Choices *Walter J. Freeman*	262
Index		265

ACKNOWLEDGMENTS

An edited book is, by definition, a collaborative project. My goal in *Soul, Psyche, Brain* has been to create a platform for the outstanding group of scholars whose works you will read in the coming pages. My thanks, then, go first of all to the friends and colleagues who have contributed chapters to this book. Each of them has crystallized a career's worth of thought, research, and reflection into their chapters, and to these wonderful people—Holmes Rolston III, James W. Jones, Stanley Krippner, Robert Emmons, Tracey Kahan, Patricia Simone, David Kahn, Patricia Davis, Lewis Rambo, Charlene Burns, Richard Payne, Jeremy Carrette, and Walter Freeman—I express my deepest gratitude for their wisdom, intellectual passion, and critical acuity. No less crucial to the process of creating this book, the editorial staff of Palgrave Macmillan also deserves my abundant praise and thanks. Amanda Johnson, Matthew Ashford, Eva Talmadge, Yasmin Mathew, and Maran Elancheran have been steady, reliable companions in bringing *SPB* to fruition. Several schools, institutions, and audiences have provided venues for valuable conversations about religion and brain–mind science, and for that I thank the following: The International Association for the Study of Dreams, the Person, Culture, and Religion Group and the Religion and the Social Sciences Section of the American Academy of Religion, the Center for Theology and the Natural Sciences, the Graduate Theological Union, John F. Kennedy University, Santa Clara University, Harvard Divinity School, and St. Lawrence University. I am especially grateful to the many individuals who have generously shared their insights on the topics covered in *SPB*, including Nina Azari, Nancy Grace, Carol Rausch Albright, Don Browning, Bill Domhoff, Allan Hobson, Ed Pace-Schott, Roger Lohmann, Steven Bauman, and Ryan Hurd. Finally, over the past year I have been meeting regularly with three fellow alumnae from the University of Chicago Divinity School—Diane Jonte-Pace, Ann Taves, and Catherine Bell—to discuss the impact of recent developments in brain–mind science on the study of religion. Though we approach this subject from different directions, we agree that something very significant is happening here, and Diane, Ann, and Catherine have helped open my eyes to the exciting potentials, and formidable challenges, in this new era of religion–science dialogue.

LIST OF CONTRIBUTORS

Kelly Bulkeley is a Visiting Scholar at the Graduate Theological Union and author of several books on religion, psychology, and dreaming, including *The Wondering Brain*.

Charlene P.E. Burns is Associate Professor of Religious Studies at the University of Wisconsin, Eau Claire, and author of many works on Christian theology and psychology, including *Divine Becoming*.

Jeremy Carrette is Senior Lecturer in Theology and Religious Studies at the University of Kent, UK, and author of several books on religion, psychology, and politics, including *Michel Foucault and Theology*.

Patricia M. Davis is a doctoral student at the Graduate Theological Union and author of articles on Christian spirituality, metaphor theory, and dreaming.

Robert A. Emmons is Professor of Psychology at the University of California, Davis and author of numerous works on psychology, religion, and personality, including *The Psychology of Ultimate Concerns*.

Walter J. Freeman is Professor of Neurobiology at the University of California, Berkeley and the author of many books on neural interaction, perception, and chaos theory, including *How Brains Make Up Their Minds*.

James W. Jones is Professor of Psychology of Religion at Rutgers University and author of many books on psychoanalysis, psychotherapy, and religion, including *The Mirror of God*.

Tracey L. Kahan is Associate Professor of Psychology at Santa Clara University and author of several articles on consciousness, metacognition, and dreaming.

David Kahn is a Physicist at the Neurophysiology Laboratory at Harvard Medical School and author of multiple articles on brain development, self-organization, and dreaming.

Stanley Krippner is Professor of Psychology at the Saybrook Graduate School and author of numerous books on hypnosis, dreaming, and consciousness, including *The Varieties of Anomalous Experience*.

Richard K. Payne is Dean of the Institute of Buddhist Studies and editor of several books on Buddhism, including *Approaching the Land of Bliss*.

Lewis R. Rambo is Professor of Pastoral Psychology at San Francisco Theological Seminary and author of many books on the psychology of religion, including *Understanding Religious Conversion*.

Holmes Rolston III is Professor of Philosophy at Colorado State University and author of several books on religion, science, and the environment, including *Genes, Genesis, and God*.

Patricia M. Simone is Associate Professor of Psychology at Santa Clara University and author of a number of articles on cognitive neuropsychology, memory, attention, and gerontology.

Introduction

KELLY BULKELEY

Do prayer and meditation really "work" in changing the way the mind functions?

Is there a "God spot" in the brain, where religious experience originates?

Are humans genetically hardwired to be aggressively violent, or morally altruistic, or both?

Is philosophical dualism dead and monism triumphant? Can all religious and psychological experiences be explained in terms of chemical and electrical activities in the brain? Is the soul, finally, a scientifically outmoded concept?

Yes, no, both, and no, no, no.

Now that we have these simple-minded questions out of the way, we can get on with the task of exploring the new frontiers of religion and brain–mind science. There has been enough research in both religious studies and cognitive neuroscience to make this much clear: prayer and meditation do change the way the mind functions. There is no one specific neural region that triggers all religious experience. Humans are genetically predisposed toward both aggression and altruism. Philosophical monism is, at present, incapable of reducing all religious and psychological experiences to material brain functioning, and the soul remains a viable concept for understanding the fullness of what it means to be human.

   I am not saying that people no longer argue about these questions. They do, and will continue to do so for many years to come. What I am saying (along with all the other contributors to this book) is that the time has arrived for us to move *beyond* these rudimentary questions and investigate the more complex and more interesting issues that have emerged in the dialogue between religion and brain–mind science. In just the past few years, cognitive neuroscientists have made several remarkable discoveries about the development and functioning of the brain–mind system. These findings raise fascinating questions about theological and philosophical conceptions of human nature. At the same time, recent investigations in religious studies (in coordination with anthropology, history, and critical theory) have disclosed new ways of understanding the complex, multidimensional qualities of human religiosity. This work has tremendous significance for cognitive neuroscientific theories about selfhood, agency, and consciousness. *Soul, Psyche, Brain* brings together these two realms of research, offering a new introduction to a dynamic and growing area of study.

It should be emphasized at the outset that this collection of essays is part of a long tradition of Western scholarship investigating the psychophysiological aspects of religious experience. David Wulff's *Psychology of Religion: Classic and Contemporary* (1997) documents the efforts made throughout the twentieth century to analyze religion in scientific terms and explain it in relation to natural biological processes. All four of the major pioneers Wulff identifies in the field of psychology of religion—William James, Sigmund Freud, C.G. Jung, and G. Stanley Hall—were dedicated to the goal of discovering the deepest possible correlations between religious experience and brain–mind functioning. Psychology of religion researchers have scientifically examined the effects of particular behaviors long associated with religion, including fasting, sleep deprivation, sensory withdrawal, breath control, dancing, meditation, prayer, and the ingestion of various psychoactive substances. Researchers have also investigated certain brain phenomena (endorphin release, temporal lobe epilepsy, hemispheric specialization) in connection with subjective reports of spiritual experience. The results of these studies have been impressive insofar as they show that (1) religious experience is indeed rooted in the body, specifically in the psychophysiology of the brain–mind system, and (2) that humans have devised a wide variety of highly effective practices for altering that system in religiously significant ways. However, Wulff points out that something important is still missing in this research: "At issue is not whether neurophysiology plays a role in religious experiences—for presumably all experience is represented somewhere in the brain—but whether referral to brain and other bodily processes is *the most appropriate way* by which to comprehend them" (1997, p. 112, italics added).

This is precisely what every chapter in this book seeks—a more appropriate way of comprehending religion and spirituality in connection with the biological nature of our species. Each of the contributors is convinced, as am I, that religion and science can learn much from each other by combining their resources to explore the religiously activated brain. Every chapter offers a creative means of overcoming the conflict between absolutist positions at both extremes—the pro-religion advocates who reject evolutionary science and the pro-science advocates who reject all forms of religiosity. Although the contributors use quite different approaches (more on the details of their chapters in a moment), they all agree that the present moment offers a particularly auspicious time for developing new integrations of religious studies and psychological science. Wulff seems to have seen this coming, as he makes the following prediction in the 1997 edition of his book: "No other approach in the psychology of religion promises as revolutionary a future as the biological one" (p. 112). As you will see in the following chapters, that future is well nigh upon us.

In addition to its psychology of religion context, this book should also be seen as the continuation of a longer history of Western reflection on the evolutionary basis of religion. Charles Darwin himself was the first to speculate on this subject, beginning in the mid-1800s. From the start of his career, Darwin recognized the significance of evolution for everything having to do with human mental life. "The mind is function of body" he wrote in an early notebook, and foresaw, with a mix of gentlemanly trepidation and revolutionary excitement, that such a radical idea, if proven, would force a violent overthrow of many theological and metaphysical beliefs long cherished by the British upper class.

It is worth dwelling for a moment on Darwin's experiences with religion. He was born in 1839 and raised in a well-to-do English family with a tendency toward freethinking atheism (something rather unusual for members of their elevated social class). Having failed at medical school, Darwin made a half-hearted attempt at becoming a minister with the Church of England. He was saved from that fate by the glorious voyage of the H.M.S. Beagle, on which he served as the ship's official naturalist and upper-class companion for the captain. During his five-year circumambulation of the globe, Darwin came face to face with the *mysterium tremendum* of Nature. His eyes were opened to the incredibly diverse and interconnected phenomena of the natural world, and he experienced what were perhaps the most authentically spiritual sentiments of his life. The wild, verdant rain forests of South America inspired the twenty-seven-year-old Darwin to exclaim, "No one can stand unmoved in these solitudes, without feeling that there is more in man than the mere breath of his body." Elsewhere, he spoke of the jungles as "temples filled with the varied productions of the God of Nature" (Desmond and Moore 1991, p. 191).

Alas, this transcendent experience did not have a lasting spiritual impact. The trajectory of Darwin's intellectual development was decidedly away from religion in any form whatsoever. His experiences aboard the Beagle had shown him how puny humans are in the grand scheme of Nature, and he rejected Christian teachings about the special supremacy of humans in the created world. He felt that all the waste and violence in the world made a mockery of belief in a benevolent God, especially so after the deaths of his father and his beloved daughter Emma. These agonizing losses prompted Darwin to give up any pretense of Christian faith. He did not, however, join with those of his scientific comrades who were using evolutionary theory as a rhetorical weapon against organized religion. Darwin ended his days as a private but resolute agnostic, with no personal belief in God or an afterlife and yet no desire to force other people to believe as he did.

Still, Darwin continued thinking about religion and evolution, and two of his speculations are direct precursors to the major topics of this book. One is his idea that religious faith is not necessary for moral development or psychological maturity, because evolution has endowed humans with social instincts that naturally incline us to form and maintain bonds of friendship with other people. The Golden Rule, in Darwin's view, is a product not of divine decree but of evolved instinct. Second is the admittedly heretical idea that perhaps the experience of God can be explained as nothing more than the effect of a particular state of brain organization. Darwin saw quite clearly how materialist implications could be derived from evolutionary theory, implications that would be deeply disturbing to religious believers and would undermine the legitimacy (and political power) of church teachings.

Darwin's influence on current brain–mind research cannot be overestimated. The evolutionary processes he identified remain the primary framework used by cognitive neuroscientists to explain their findings. Specifically, Darwin's views on religion and other aspects of human psychology have spawned a growing literature in the evolutionary analysis of religious beliefs, rituals, and experiences. Particularly noteworthy in this regard are Pascal Boyer's *Religion Explained* (2001), Ilkka Pyysiäinen's *How Religion Works* (2001), and Thomas Lawson and Robert McCauley's *Rethinking Religion* (2002). These works carry out an essentially Darwinian project of reducing religious phenomena to their material basis in the biology of human evolution. By contrast, the chapters in *Soul, Psyche, Brain*, though deeply informed by Darwin's thought, do not stop with the materialist level of explanation. They take the materialist findings of brain–mind science and use them as a platform to ask *new* questions—about the future potential of our still evolving nature, about our capacity for creative imagination and spiritual growth, and about our understanding of what it means to lead a good, fulfilling, fully realized human life. If anything, this book is a call to return to Darwin's rapturous experiences in the Brazilian rain forests. His own theory makes no place for such experiences of "unchurched spirituality," but the contributors to this book show that the fullest extension of Darwin's evolutionary thought must include an openness to these religiously charged dimensions of human existence.

A third context for this book is the one that reveals its greatest limitations. The title *Soul, Psyche, Brain* is intended to highlight the multiplicity of terms and concepts used in human efforts to know ourselves. *Soul* is the term favored by many religious believers, *psyche* is the defining concept of the discipline of psychology, and *brain* is the central focus of cognitive neuroscience. There are obviously great differences among these terms, and much of this book is devoted to exploring their various meanings. However, it should be just as obvious that all three terms share a common cultural foundation in Western civilization. To the extent that almost all the chapters of this book work within the conceptual universe defined by the trio of terms soul, psyche, and brain, the result is that many *other* ways of thinking about these issues will be neglected. The only exceptions are the essays by Kahan and Simone (chapter 5) and Payne (chapter 9), both of which discuss Western psychology in connection with Buddhist points of view. This is actually a fair reflection of the current dialogue between religion and brain–mind science, at least in the United States. Most of the discussion involves Christian theology in connection with Western psychology and neuroscience, with a small but growing interest in Buddhism. How does this Western discussion relate to the rich traditions of self-knowledge in Islam, Hinduism, and the indigenous cultures of Africa, Australia, and the Americas? Unfortunately, you won't get much of an answer to that question in the present book. But at least you won't get a *wrong* answer, which is what happens too often when researchers use brain–mind science as an exhaustive (and dismissive) explanation for all of the world's religious and spiritual traditions. If nothing else, the contributors to *Soul, Psyche, Brain* agree that future progress in this field depends on greater humility, open-mindedness, and willingness to learn from others.

Let me say a few words about each of the chapters and their authors. The first is "Genes, Brains, Minds: The Human Complex," by Holmes Rolston III, a venerable professor of philosophy at Colorado State University and for many years a leading voice in the study of religion and science. Rolston begins with a wide-angle consideration of how our brains are shaped by the genetic inheritance of the human species. He shows how the emergence of culture allowed for tremendous advances in human psychological development, to the point where we now have a capacity for spiritual experience and self-transcendence. Just as he argued in his 1997 Gifford Lectures (later published as *Genes, Genesis, and God*), Rolston says the amazing new discoveries of evolutionary biology do not disprove religion in any simplistic way, but rather enrich our understanding of moral goodness, creative genius, and existential self-awareness.

James Jones of Rutgers University brings his experience as both a clinical psychologist and a religious studies scholar (along with training in the philosophy of science) to bear on the subject of his chapter, "Brain, Mind, and Spirit—A Clinician's Perspective, or Why I Am Not Afraid of Dualism." As already mentioned, the mainstream consensus among neuroscientists is that consciousness is a by-product of physical activities in the brain. There is no disembodied soul or purely rational mind—everything we feel, think, and experience can be explained in terms of brain neurophysiology. Many theologians and religious studies scholars have already challenged this materialist approach, though, as Jones shows in his careful analysis of Nancey Murphy's recent work, these religious responses are themselves inadequate in helping us understand the complex realities of human consciousness and spiritual experience. Jones pushes back hard against the neuroscientific claim that the mind–body problem has been solved in favor of monism over dualism, and argues that a brain-centered approach cannot account for the counter-monistic findings of research in behavioral medicine, meditation, hypnosis, and other fields of psychophysiology.

Stanley Krippner's long career as a globe-trotting, anthropologically informed psychologist is the foundation for the third chapter, "Psychoneurological Dimensions of Anomalous Experience in Relation to Religious Belief and Spiritual Practice." The psychology of religion has always taken an interest in unusual modes of awareness, knowledge, and power. Both Freud and Jung studied the precognitive dimensions of dreaming, whereas James examined people who claimed to be mediums. Recent findings in the neurosciences are adding new pieces of information to our understanding of such extraordinary psychological phenomena, and Krippner (of the Saybrook Institute) provides a concise survey of what is currently known about rare but emotionally and physiologically charged occurrences such as telepathy, mysticism, meditation, intensified dreaming, and near-death experience. He emphasizes that anomalous experiences such as these are not innately pathological or disordered; rather, they reflect the unusual activation of brain–mind processes, which are, in their ordinary condition, increasingly well understood by modern psychology.

Chapter 4, "Sacred Emotions" by Robert Emmons, considers the implications of perhaps the biggest change produced by cognitive neuroscience in our understanding of human nature—the discovery that reason cannot function without emotion. The psychological ideal of a purely rational mind, which goes back to Enlightenment philosophers like Descartes and Kant, has been exploded by neuroscientific research showing that human reasoning abilities suffer terribly if we lose our capacity for emotional experience. *We cannot be healthy and whole without emotions.* Emmons, a psychologist of religion at the University of California, Davis, argues that in light of these findings, we should reconsider the role of emotions in religion, particularly the way religions provide a context and direction for emotional experience and expression. He points to the considerable number of studies on "positive" emotions such as gratitude, awe, reverence, wonder, hope, forgiveness, and joy, all of which are regularly associated with a spiritual orientation toward life. For Emmons, the recent findings of psychological science are vitally important because they refute a simplistic, unidirectional brain→mind view of causality, and reveal instead a complex and dynamic interplay among the body, the mind, culture, and religion.

The practice of Zen Buddhist meditation is the subject of chapter 5, "Where Neurocognition Meets the Master: Attention and Metacognition in Zen." Tracey Kahan and Patricia Simone, a psychologist and a neuroscientist, respectively, at Santa Clara University, bring together a wealth of new evidence demonstrating the extraordinary qualities of brain–mind functioning during Zen meditation. Many psychological studies have shown that the human capacity for "metacognition," that is, thinking about thinking, is basic to our self-awareness, emotional regulation, and long-term planning. Of special interest to Kahan and Simone is the capacity for selective attention, involving the metacognitive process of deciding which perceptions, feelings, and ideas to attend to and which to ignore. What Zen meditation is able to do, according to Kahan and Simone, is discipline people's attention and sharpen their metacognitive focus so they can achieve and then sustain a present-centered awareness. Carrying on James Austin's project in *Zen and the Brain* (1998), Kahan and Simone further enrich our understanding of the way certain spiritual practices can dramatically transform brain–mind functioning.

David Kahn's "From Chaos to Self-Organization: The Brain, Dreaming, and Religious Experience," offers a state-of-the-art report on the neuroscience of brain development. Kahn's work at Harvard Medical School's Department of Psychiatry has focused on the neural and psychological dimensions of dream experience, and in this chapter he uses dreaming as an illustration of a crucial insight about the way the brain functions. The brain, he argues, is a self-organizing system whose healthy and creative development depends on a constant, lively tension between structure and chaos. Kahn's argument may be discomforting for religious believers insofar as he claims no special creator is necessary to account for the emergence of human intelligence. But scientific materialists may be equally disturbed by Kahn's evidence showing the inherently free, unpredictable, open-ended nature of human consciousness.

Kahn's interest in the neuroscience of self-organization is, despite its very different academic perspective, quite similar to the main topic discussed by psychology of religion scholars Patricia Davis and Lewis Rambo (of the Graduate Theological Union) in their chapter, "Converting: Toward a Cognitive Theory of Religious Change." The religious phenomenon of conversion, which Rambo has studied extensively, involves varying degrees of individual choice, along with multiple influences at the sociological, cultural, and psychological levels. By using the metaphor theory of cognitive linguist George Lakoff to analyze the language used by Christian converts as they describe their experiences, Davis and Rambo develop a new way of understanding the complex interplay of religious meanings, psychological functioning, and individual choice in experiences of conversion. What comes of Davis and Rambo's analysis is the recognition that, at least in the case of Christian conversion, the process of religious change is characterized by unpredictable bursts of growth in cognitive complexity and self-awareness.

As illustrated by the Davis and Rambo chapter, much work has been done in exploring the connection between specifically Christian religious traditions and brain–mind science. Charlene Burns' chapter "Cognitive Science and Christian Theology" gives a masterful overview of this particular area of religion–science dialogue. Burns, who teaches philosophy and religion at the University of Wisconsin, Eau Claire, gives special attention to the implications of brain–mind science for Christian theological claims about the soul. She critically reviews the ideas of the major researchers who have tried over the past several decades to correlate Christian belief with cognitive science, and, much like James Jones in chapter 2, rejects the "nonreductive physicalism" proposed by some contemporary theologians, even though that theory does mark an advance over the materialist reductionism of scientists who believe consciousness is a mere epiphenomenon of brain functioning. As an alternative to these unsuccessful theories, Burns points to resources in the Christian tradition that conceive of the human soul as a psychosomatic unity emerging in relation to a broader cultural community.

One does not have to be a Christian to appreciate the contemporary significance of these historical teachings about the embodied soul.

All of these issues look different when considered from the perspective of a religion other than Christianity. Richard Payne, dean of the Institute of Buddhist Studies at the Graduate Theological Union, explores in chapter 9 the connection between Western psychology and Buddhist teachings on the nature (and nonexistence) of the self. "Overcoming an Impoverished Ontology: Candrakirti and the Mind–Brain Problem" is devoted to the work of medieval Indian philosopher Candrakirti, who provides an especially lucid expression of Buddhist approaches to psychological self-awareness. In addition to providing a detailed portrait of Candrakirti's prescient ideas, Payne's chapter describes the long Buddhist history of careful philosophical analysis of the mind–brain question. He compares these teachings to recent Western psychological and anthropological work on the constructive nature of human perception, cognition, and selfhood. A new Western appreciation for the self as a social construct, combined with the ancient Buddhist spiritual quest for release from the illusion of the self—this is the possibility Payne wants us to consider. His chapter, along with chapter 5 by Kahan and Simone, points to an important (and non-Christian) direction for future investigation.

Chapter 10 presents my work on religious and psychological approaches to dreaming. "Religion and Brain–Mind Science: Dreaming the Future" brings together the leading findings about dreams and dreaming from both sides of the dialogue—historical and anthropological studies on the one hand, psychological and neuroscientific research on the other. We have learned a great deal in recent years about the many roles dreams have played in religious beliefs, practices, and experiences from cultures all around the world. We have also learned much about the basic neurocognitive processes that are and are not activated during REM dreaming. The best and most fruitful way of integrating these two areas of research (so I argue) is to study the phenomenology of what C.G. Jung called "big dreams," that is, dreams that are extraordinarily intense and vivid, with striking images, physiological carry-over effects, and a high degree of memorability. The cross-cultural occurrence of big dreams, combined with their rootedness in the brain, strongly suggest the possibility that such dreams serve powerful adaptive functions, which can be explained and understood in evolutionary terms, the *sine qua non* of Western psychological science.

The penultimate chapter, "Religion Out of Mind: The Ideology of Cognitive Science and Religion," is by Jeremy Carrette, a psychologist of religion at the University of Kent, whose work centers on a critical reappraisal of the social, economic, and political factors that have shaped, and continue to shape, the psychological study of religion. Carrette examines the recent work of evolutionary psychologists and cognitive scientists (particularly that of Lawson and McCauley) who claim to have identified the fundamental and universal mental processes that give rise to religion. He forcefully challenges the unspoken assumptions and biases that pervade

Lawson and McCauley's assertions. Without dismissing scientific research in its entirety, Carrette calls into question the automatic authority that cognitive scientists are granted in Western society, and makes us more aware of the subtle but powerful ideological influences shaping everyone's work in this field of study, including our own.

Walter Freeman, a neuroscientist at the University of California, Berkeley, reflects on the broader social implications of brain–mind research in the last chapter, "Brain Science on Ethics: The Neurobiology of Making Choices." Originally presented as an invited address at a high-school graduation in Italy, Freeman's brief chapter will hopefully encourage readers to think carefully about what moral, political, and spiritual lessons they draw from the latest findings of brain–mind research. Like David Kahn, Freeman appeals to research on chaos, complexity, and nonlinear systems in arguing that the human mind is fundamentally free and has the capacity to create its own future. If we accept Freeman's claim that the scientific materialists are wrong about psychological determinism, and if humans are indeed blessed with the capacity for free moral choice, then the ethical and spiritual teachings of the world's religious traditions become valuable resources in the future scientific study of the brain–mind system.

It will, I hope, come as no surprise that the book ends without a formal conclusion. There is no need to impose an artificial sense of closure on these issues—the future of religion and brain–mind science is truly wide open. We are living at a time when our sources of information about both religion and brain–mind science have far outstripped our theoretical understanding of how the two areas relate to one another. This gap is likely to widen in coming years, as religious studies scholars continue to analyze and evaluate religion's increasingly significant role in global life and conflict, while cognitive neuroscientists discover ever more detailed features of brain–mind functioning. The only thing we know right now is that the traditional frameworks used by both religion and science are, *by themselves*, inadequate to the task of making sense of this surging cascade of new information.

References

Austin, J.H. 1998. *Zen and the Brain*. Cambridge, MA: MIT Press.
Boyer, P. 2001. *Religion Explained: The Evolutionary Origins of Religious Thought*. New York: Basic Books.
Desmond, A. and James, M. 1991. *Darwin: The Life of a Tormented Evolutionist*. New York: W.W. Norton.
Lawson, E.T. and McCauley, R.N. 2002. *Bringing Ritual to Mind: Psychological Foundations of Cultural Forms*. Cambridge: Cambridge University Press.
Pyysiäinen, I. 2001. *How Religion Works: Towards a New Cognitive Science of Religion*. Leiden: Brill, the Netherlands.
Rolston, H. 1999. *Genes, Genesis, and God: Values and their Origins in Natural and Human History*. Cambridge: Cambridge University Press.
Wulff, D. 1997. *Psychology of Religion: Classic and Contemporary*. New York: John Wiley & Sons.

CHAPTER ONE

Genes, Brains, Minds: The Human Complex

HOLMES ROLSTON III

Earth is the planet where the most complex creativity of which we are aware has taken place; and on this Earth, the most complex creative thing known to us is the human mind. John Maynard Smith and Eörs Szathmáry analyze "the major transitions in evolution" with the resulting complexity, asking, "how and why this complexity has increased in the course of evolution." "Our thesis is that the increase has depended on a small number of major transitions in the way in which genetic information is transmitted between generations." Critical innovations have included "the origin of the genetic code itself," "the origin of eukaryotes from prokaryotes," "meiotic sex," "multicellular life," "animal societies," and especially "the emergence of human language with a universal grammar and unlimited semantic representation," this last innovation making possible human culture (1995, pp. 3, 14).

Maynard Smith, the dean of theoretical biologists, finds that each of these innovative levels is surprising, not scientifically predictable on the basis of the biological precedents. He and his colleague are deeply impressed with the cybernetic and, eventually, cognitive character of what has taken place in natural history, expressed so strikingly in the human mind. What makes the critical difference in evolutionary history is increase in the information possibility space, which is not something inherent in the precursor materials, nor in the evolutionary system, nor something for which biology has an evident explanation, although all these events, when they happen, are retrospectively interpretable in biological categories—at least all except perhaps culture are. The biological explanation is modestly incomplete, recognizing the importance of the genesis of new information channels.

Since we humans find ourselves at the apex of these complex events, it becomes us, as far as we can, to figure out what to make of ourselves, both who we are and where we are. We proceed with an analysis of nature and culture, adapted versus adaptable minds, genes making human brains, human minds making brains, and the spirited human self and our self-transcendence.

At such levels of complexity, we will often be in "over our heads"; but one conclusion is inescapable: what is in our heads is as startling as anything else yet known in the universe. We will be left wondering how far what is going on in our heads is a key, at cosmological and metaphysical levels, to what is going on over our heads.

Nature and Culture

Both "nature" and "culture" have multiple layers of meaning. If one is a metaphysical naturalist, nature is all that there is, and so all things in culture—computers, artificial limbs, or presidential elections—are natural. Nature has no contrast class. At another level, however, culture contrasts with nature; and we need to be adequately discriminating about the real differences between them. Animals, much less plants, do not form cumulative transmissible cultures. Information in wild nature travels intergenerationally largely on genes; information in human culture travels neurally as persons are educated into transmissible cultures.

The determinants of animal and plant behavior are never anthropological, political, economic, technological, scientific, philosophical, ethical, or religious. The intellectual and social heritage of past generations, lived out in the present, re-formed and transmitted to the next generation, is regularly decisive in culture. Culture, by Margaret Mead's account, is "the systematic body of learned behavior which is transmitted from parents to children" (1989, p. xi).[1] Culture, according to Edward B. Tylor's classic definition, is "that complex whole which includes knowledge, belief, art, morals, law, custom, and any other capabilities and habits acquired by man as a member of society" (1903, p. 1).

Animal ethologists have complained that such accounts of culture are too anthropocentric (indeed chauvinistic!) and need to be more inclusive of animals (de Waal, 1999). Partly because of new animal behaviors observed, but mostly by enlarging (or, if you like, shrinking) the definition, it has become fashionable to claim that animals have culture. Robert Boyd and Peter J. Richerson revise the definition: "Culture is information capable of affecting individuals' phenotypes which they acquire from other conspecifics by teaching or imitation" (1985, p. 33). The addition of "imitation" greatly expands and simultaneously dilutes what counts as culture. By this account, there is culture when apes "ape" each other, but also culture in horses and dogs, beavers, rats—wherever animals imitate the behaviors of parents and conspecifics. Geese, with a genetic tendency to migrate, learn the route by following others; warblers, with a tendency to sing, learn to sing better when they hear others. Whales and dolphins communicate by copying the noises they hear from others; this vocal imitation constitutes culture at sea (Rendell and Whitehead 2001).

But with culture extending from people to warblers, it has become a nondiscriminating category for the concerns we wish to analyze here.

One finds widespread animal cultures by lowering the standards of evidence. Critical to a more discriminating analysis is the difference between mind–mind interactions, sharing ideas, pervasive in human cultures, and not mere behavioral imitation, copying what another does, which is widespread among animals, that can acquire information. If we are going to call what warblers and geese do culture, then we will need to invent another word "super-culture"—to describe what humans do, which is indeed "super" to these animal capacities.

Opening an anthology on *Chimpanzee Cultures*, Wrangham et al. doubt, interestingly, whether there is much of such a thing: "Cultural transmission among chimpanzees is, at best, inefficient, and possibly absent" (1994, p. 2). There is scant and in some cases negative evidence for active imitation or teaching of the likeliest features to be transmitted, such as tool-using techniques. Chimpanzees clearly influence each other's behavior, and seem to intend to do that; they copy the behavior of others. Chimps do seem to know when another chimp has seen something (e.g., where food is). But they do not differentiate between those who know and those who do not when they communicate with other chimps. The chimp world is local. In terms of acquired information, if a chimp doesn't see it (or hear, taste, smell it), he doesn't know it. If a brother chimp departs and disperses to another troop for a year and then returns, he does not remember and recognize (re-cognize) his brother; they take their family and troop cues from whoever is nearby and do not have the concept of "brother."

There is no clear evidence that chimps attribute mental states to others. They seem, conclude these authors, "restricted to private conceptual worlds." In the technical vocabulary, the chimps have little or no "theory of mind"; they do not know of other minds' being there with whom they might communicate to learn what they know. Without some concept of teaching, of ideas' moving from mind to mind, from parent to child, from teacher to pupil, a cumulative transmissible culture is impossible. Humans learn what they realize others know; they employ these ideas and resulting behaviors; they test and modify them, and, in turn, teach others what they know, including the next generation. So human cultures cumulate, but with animals there is no such cultural "ratchet" effect.

In a lead article in *Behavioral and Brain Sciences*, Michael Tomasello, Ann Cale Kruger, and Hiliary Horn Ratner pinpoint this difference:

> Simply put, human beings learn from one another in ways that nonhuman animals do not . . . . Human beings are able to learn from one another in this way because they have very powerful, perhaps uniquely powerful forms of social cognition. Human beings understand and take the perspective of others in a manner and to a degree that allows them to participate more intimately than nonhuman animals in the knowledge and skills of conspecifics." (1993, p. 495)

Bennett G. Galef, Jr. concludes: "As far as is known, no nonhuman animal teaches" (1992, p. 161).

We can better dissect nature, culture, and cumulative transmissible cultures with degrees of intentionality (Dennett 1987). Animals are variously socialized, and become what they become due to interactions with their surroundings, which include the groups in which they live. But there is little or no evidence of any higher-order intentionality, even among primates that are highly social. Organisms with zero-order intentionality have no beliefs or desires at all. Animals, such as vervet monkeys, intend to change the behavior of other animals—this represents first-order intentionality. Second-order intentionality would involve intent to change the mind, as distinguished from the behavior (though perhaps the behavior as well), of another animal, that is, to teach by passing ideas from mind to mind. Third-order intentionality involves knowledge that another, a teacher, intends to change one's mind. Human language is in this sense recursive; animal communication is not. Primates do not seem to realize that there are minds in others to teach, although they often imitate each other's behavior, as when adults are imitated by their offspring.

In this higher-order sense of communication, conclude Dorothy L. Cheney and Robert M. Seyfarth, "signaler and recipient take into account each others' states of mind. By this criterion, it is highly doubtful that *any* animal signals could ever be described as truly communicative" (1990, pp. 142–143). They continue:

> It is far from clear whether any nonhuman primates ever communicate with the intent to inform in the sense that they recognize that they have information that others do not possess . . . There is as yet little evidence of any higher-order intentionality among nonhuman species . . . Teaching would seem to demand some ability to attribute states of mind to others . . . Even in the most well documented cases, however, active instruction by adults seem to be absent . . . The social environment in most primate species is probably too simple to require higher-order intentionality. (pp. 209, 223, 252)

David Premack finds that humans are quite unique in their capacity to teach: "Teaching, which is strictly human, reverses the flow of information found in imitation. Unlike imitation, in which the novice observes the expert, the teacher observes the novice—and not only observes, but also judges and modifies" (2004, p. 318). In due course, in human societies, the pupil likewise judges and modifies what the teacher teaches. In such recursive loops, cumulative transmissible cultures can be endlessly generated and regenerated.

Cumulative transmissible cultures are made possible by the distinctive human capacities for language. Language "comes naturally" to us, in the sense that humans everywhere have it. The child picks up speech during normal development with marvelous rapidity; language acquisition is only more or less intentional. The mind of a child is innately prepared for such learning (Chomsky 1986). Human language, when it comes, is elevated remarkably above anything known in nonhuman nature. The capacities for symbolization, abstraction, vocabulary development, teaching, literary expression, argument are quite advanced; they do not come naturally as an inheritance from other primates, whatever may otherwise be our genetic similarity with them. Though language comes naturally to humans, what is learned has been culturally transmitted; the specific language and content of childhood education is that of an acquired, nongenetic culture. The development, transmission, and criticism of culture depends on this capacity for language.

In a major recent study to determine whether animals have language, the authors Hauser et al. conclude: "It seems relatively clear, after nearly a century of intensive research on animal communication, that no species other than humans has a comparable capacity to recombine meaningful units into an unlimited variety of larger structures, each differing systemically in meaning" (2002, p. 1576). The primate communication "system apparently never takes on the open-ended generative properties of human language" (p. 1577).

After 30 years of study of communication in mountain gorillas, the researchers Harcourt and Stewart conclude:

> Gorilla close-calls [those made within the group] are very far from being language-like, they seem to be of the order of complexity of threat displays, as indeed do chimpanzee calls. That simplicity raises the question of why apes, popularly considered more intelligent than monkeys, have apparently a simpler mode of communication, in the sense that they apparently do not label the environment by association of specific calls with specific contexts . . . We have no answer for the contrast. (2001, pp. 257–258)

Cheney and Seyfarth (1990) found that vervet monkeys give different alarm signals for snakes, leopards, and eagles; other monkeys hear these alarms and take cover appropriately to differing predators. Hence, it seemed that the calling monkey intended to refer and communicate its knowledge to others. But the most recent evidence raises doubt about whether the seeming "callers" intend to inform. Rather, these differing noises appear to be spontaneous response grunts in alarm, although other monkeys can learn from such grunts and respond appropriately to the predator that is present. Such signals cannot "be considered as precursors for, or homologs of, human words." "There is no evidence that calling is intentional in the sense of taking into account what other individuals believe or want" (Hauser et al. 2002, p. 1576).

What is missing in the primates is precisely what makes a human cumulative transmissible culture possible. The central idea is that acquired knowledge and behavior is learned and transmitted from person to person, by one generation teaching another and ideas passing from mind to mind, in large part through the medium of language, with such knowledge and behavior resulting in a greatly rebuilt, or cultured, environment. Humans have genes, of course; but humans live under what Boyd and Richerson call "a dual inheritance system" (1985; Durham 1991). They live both in nature and in culture. Discovery of the nature and origins of human language, making possible this emergence of culture, is quite possibly "the hardest problem in science" (Christiansen and Kirby 2003, p. 1).

Adapted versus Adaptable Minds

In nature, in the lives of animals, the microscopic determinants are coded in the genes, but the macroscopic determinants are found in the ecological niches these animals inhabit, in their need to cope, to survive, as this has been honed by natural selection. We next need to place the mind, which makes culture possible, in an evolutionary context. Mind is at once a survival tool in both nature and in culture. But this evolutionary past, while necessary for explaining our mental powers, may not be sufficient for a complete explanation.

Biologists distinguish between proximate and ultimate explanations (Mayr 1988, p. 28). Why does a plant turn toward light? Cells on the darker side of a stem elongate faster than cells on the brighter side because of an asymmetric distribution of auxin moving down from the shoot tip. But the ultimate explanation is that, over evolutionary time, in the competition for sunlight, there were suitable mutations, and such phototropism increases photosynthesis. Analogously, in the developing infant, genes produce a brain, which sponsors a mind. But the developing infant also inherits a long evolutionary past. The results of this ancient history are delivered biologically at birth to (all normal) members of *Homo sapiens*. These past evolutionary events (phylogenesis) are recapitulated (more or less) and generate a contemporary brain (ontogenesis), sponsoring a mind. What was achieved in millions of years (even billions if one includes all the biochemistries) is, via DNA suitably emplaced in a zygote in the womb, coded and copied, reenacted in the few natal/childhood months and years.

Therefore, whatever the proximate explanations about how an infant develops a brain and a mind, a more comprehensive explanatory framework is the evolutionary success; brains must have been good for something. Fish have fins, birds have wings, humans have brains—all for adaptive success. Fish must swim, birds must fly, and humans must be cultured. That seems obviously what the distinctive human brain is for. The infant, coming of age, needs to inherit a long cultural past. But there is a vital disanalogy. The information fish need to swim is in their genes, inborn and with some cutting and splicing of this information in the developing embryo; likewise with the birds who fly. The cultural information the infant needs, however, is not in his or her genes. It must be acquired by cultural learning. The previously solitary mind is able to import the acquired knowledge of others and to export its own acquired knowledge. So minds become ideationally webworked where previously only bodies were genetically and ecosystemically webworked.

One might first think that genes and culture coevolve, and on some scales that can seem reasonable. Humans have lived in cultures for perhaps a million years, during which time they have reproduced across thousands of generations. There is every reason to expect that over these millennia, those humans who do best culturally will do best reproductively also, and vice versa that a genotype will be selected to produce a culturally congenial phenotype.

As cultures become more fluid and complex, however, any tight co-evolutionary connections become problematic. The genes need to produce a keen, critical, open mind, which can evaluate cultural options for their functional usefulness and for their contribution to a meaningful life. The direction of selection in humans, as evidenced by their enormous potential for diverse cultures, would then select for an unspecialized intellect with open educable capacity—from those of the Neanderthals to our high-tech computer age—all of which require intelligence in various roles.

When we try to map the evolution of the brain onto the mind's acquisition of cultures, we immediately confront a time-joint problem. Evolution proceeds slowly over geologic timescapes; cultural changes can be quite rapid, especially in these modern times. The result is something like linking a horse and buggy with a jet plane. Information transfer in culture can be several orders of magnitude faster and overleap genetic lines. There is a radically accelerated transmission speed. Evolving genes shift in ecosystemic webs and this takes centuries and millennia. Passing ideas around takes minutes, hours, days, though these ideas do accumulate over millennia. The shift is something like that from snail mail delivered on horseback to e-mail on the Internet. The best strategy for slow-paced genes that need to succeed in fast-paced culture is not to build a relatively inflexible mind whose pace and preferences are genetically biased toward one culture or another, since these biases could misdirect persons in the rapidly shifting vicissitudes of culture. Rather, the genes will need to build a flexible mind, which can make preferences independently of any genetic/cultural biases.

When there emerges a later-evolved method of communication at the neural past the genetic level, the genes will subsequently need to develop so as to favor *teachability* above all. What will get selected is not so much specific gene traits coevolving lockstep with matching cultural behaviors as open teachability, which is to say that the genes will have to abandon tight control of behavior and cast their luck with launching a human organism whose behavior results from an education beyond their control. As more and more knowledge is loaded into the tradition (fire-building, agriculture, writing, weaponry, industrial processes, ethical codes, electronic technology, legal history), the genome selected will be the set that is maximally instructible by the increasingly knowledgeable tradition. This will require that the genes produce a flexible and open intellect, which is generalized and unspecialized, able to accommodate lots of learning and to do so speedily, able to adopt behaviors that are functional in, or conform to, whatever cultures they find themselves in. Perhaps the owners of these genes may choose another culture and migrate there. Perhaps soldiers or traders from a variant culture will invade their territory and force their culture upon them.

Theodosius Dobzhansky, a principal founder of modern genetics, reached this conclusion: "A genetically fixed capacity to acquire only a certain culture, or only a certain role within a culture, would however be perilous; cultures and roles change too rapidly . . . Human genes insure that a culture can be acquired, they do not ordain which particular culture this will be" (1963, p. 146). Boyd and Richerson, wondering whether genetics might bias our cultural dispositions in our dual inheritance system, conclude: "Genetic differentiation between human populations for determinants of biases is unlikely" (1985, pp. 284–285). It is better to be able to learn any of the myriad human languages than to be genetically dispositioned to learn French, better to eat a cosmopolitan fare than to like only Italian food, better to be able to use any of the various cultural ideas than to be genetically inclined to use only Polynesian-originated ones.

Intelligence, based on neurology, allows an organism to make an appropriate, rapid response to an environmental opportunity or threat, protecting it against the necessity of making slower, less reversible responses at the genetic level. If the genes supply intelligence in sufficient amounts, they need not themselves be closely tuned to directing behavior that can track environmental changes; they turn this over to the general intelligence they have created.

But, reply the evolutionary psychologists, this idea of a "global learning capacity" can be exaggerated. The genes do not build a *tabula rasa* mind; humans do need behavioral dispositions of some kinds, such as to fear snakes or spiders, to seek mates, to avoid incest, to protect their children, to reciprocate for mutual benefits, to obey parents, or follow leaders. Every earthbound culture must provide for persons to be washed, sheltered, go to the toilet, mate, and so on. Every culture must express and control the human emotions—love, fear, joy, grief, guilt, anxiety—and allow artistic, musical, religious expression, protect property and privacy, and provide for various activities to which they are "by nature" inclined. Perhaps humans could be genetically disposed toward religious beliefs or ethical practices, because of cultural group selection; those in such cultures prosper (Wilson 1978). So a genetic bias toward ideas useful in various cultures can be expected, and welcomed.

This account of evolutionary psychology can become too restrictive, however, with the claim that humans have more of an evolutionary adapted mind than a culturally adaptable one. John Tooby and Leda Cosmides, denying any all-purpose mind, claim that humans have what they call an "adapted mind." The mind is made up of "a complex pluralism of mechanisms," "a bag of tricks," a set of "complex adaptations" that, over our evolutionary history, have promoted survival. "What is special about the human mind is not that it gave up 'instinct' in order to become flexible, but that it proliferated 'instincts'—that is, content-specific problem-solving specializations" (1992, pp. 61, 69, 113). "These evolved psychological mechanisms are adaptations, constructed by natural selection over evolutionary time" (Cosmides et al. 1992, p. 5). These form a set of behavioral subroutines, selected for coping in culture, by which humans maximize their offspring. The human mind is "an integrated bundle of complex mechanisms (*adaptations*)" (Symons 1992, p. 138). The mind is, says Cosmides, more like a Swiss army knife, tools for this and that, rather than a general purpose learning device.[2]

Humans have needed teachability; but they have also needed channeled reaction patterns. The adapted mind evolved a complex of behavior-disposition "modules," "Darwinian algorithms," each dedicated to task-specific functions in one or the other dimension of life, such as picking mates, or helping family, or obeying parents, or being suspicious of strangers, or dealing with noncooperators by ostracizing them, or preferring savannah-type landscapes. In picking mates, for example, men are disposed to select younger women, likely to be fertile. Women are disposed to select men of social status, likely to be good providers (Buss 1989; Buss et al. 1990; Symons 1992). Further, these dispositions to behavior, still present in any contemporary culture, are those that meant survival in a Pleistocene environment (such as fear of strangers, or desiring many children); and this may mean that they are neither optimal nor altogether desirable dispositions in a modern environment (where people may need to cooperate with strangers, have fewer children, and live in cities) (Cosmides et al. 1992, p. 5).

The human mind is indeed complex, and various subroutines to which we are genetically programmed (e.g., caring for children, obeying parents, and even ostracizing noncooperators or being suspicious of strangers) may indeed be convenient shortcuts to survival—reliable modes of operating whether or not we have reflected rationally over these behaviors. It seems plausible that humans are disposed to see colors in certain ways, or to like sweets and fats, or use nouns and verbs in our languages. Some more or less "automatic" behavior is desirable. It is hardly surprising that males look for females likely to be good mothers (able to bear children and care for them) and females look for males likely to be good fathers (able and likely to provide resources and care about the family). It would be surprising if evolution had selected any other dispositions.

It is also possible that selective forces in earlier cultures (for men with strength enough to hunt or plow) differ from those of later cultures (for persons who can read, write, and do arithmetic). We should probably not assume, however, that there was some one kind of Pleistocene environment, either in the various kinds of landscapes on which humans lived or in the various cultures that they developed. The Pleistocene environment too demanded multiple skills, and an adaptable mind that could integrate them well. Many of the successful behaviors (recognizing faces, planning for tomorrow, being resolute in difficult times, cooperating with others, learning from mistakes, using appropriate caution, controlling jealousy, or lust, or forgiving others) were just as relevant then as they are now. There is much evidence, for example, that humans now taken as infants out of aboriginal cultures can do quite well when educated into a modern European culture.

The mind is not overly compartmentalized, because behaviors interconnect. Behavioral and genetic psychologists are fond of speaking of mental "mechanisms," and any machine-like function, working instinctively, diminishes the cognitive reflection required. But if women are prone to choose men of status, that requires considerable capacity to make judgments about what counts as status—economically, politically, religiously. They will have to judge which one from among their suitors, often still relatively young, is most likely to attain it in the decades of their child rearing. If men are to be good providers, that requires judgments about cooperation, and if one is operating in a barter or market culture, judgments will be needed about trading with strangers, or ostracizing merchants who renege on their promises. Men need to judge potential mates not just on their likely fertility, but also on whether they too are likely to be good providers, able and willing to care for offspring, and to educate them successfully into their culture, until these offspring reach childbearing age.

Any such articulated behavioral mode needs to be figured back into a more generalized intelligence (Sterelny 1995). Genetically programmed algorithms seem unlikely for the detail of such decisions under changing cultural conditions. Such decisions are difficult even for well-educated persons; they may require insight into character and evaluation based on intuition, additionally to conscious, explicit calculations; decisions at this level take considerable capacity for judgment, not simply mental mechanisms. The strongest finding by far in the cross-cultural study of mate preference is that both sexes from cultures around the globe consistently agree on the most promising characteristics they look for in a mate: kindness, understanding, and intelligence (Buss 1989, p. 13; Buss et al. 1990, pp. 18–20). Capacities to select such a mate are perhaps somewhat "instinctive," but they are unlikely to be an adaptive mechanism isolated from general intelligence and moral sensitivity.

Apparently, the mind is not so compartmentalized that humans—modern ones who read this literature at least—cannot make a critical appraisal of what behavioral subroutines they do inherit by genetic disposition, and choose, if they wish, to offset these "Stone Age" dispositions in their evolutionary psychology. Cosmides and Tooby are doing just that—if we may be permitted an *ad hominem* argument. They themselves illustrate that the human mind is more than a patchwork of naturally selected response routines when they call for "conceptual integration" of the diverse academic disciplines studying humans, their behavior, and their minds. These include "evolutionary biology, cognitive science, behavioral ecology, psychology, hunter-gatherer studies, social anthropology, biological anthropology, primatology, and neurobiology," among others (Cosmides et al. 1992, pp. 4, 23–24).

These are not disciplines in which one becomes an expert by behavioral mechanisms in a Swiss-army-knife mind adapted for Pleistocene environment.

At least they and their readers must have quite broadly analytical and synoptic minds.[3] The mind is fully capable of evaluating any such behavioral modules, and of recommending appropriate education so as to reshape these dispositions in result. These psychologists seem to be quite able to re-adapt by critical thought their own adapted minds; nor is there any reason to think that they and their colleagues in evolutionary psychology are alone in this capacity. Neuroscientist Beatriz Luna and her colleagues (2004) have found that the brain switches from relying heavily on local regions in childhood to more distributive and collaborative interactions among distant brain regions on becoming an adult.

All sorts of cultures demand all sorts of capacities and skills, and nearly all humans have sufficiently rich talents to find a niche in their culture. If so, there might not be any differential selection pressures when cultural patterns differ across place and time. On statistical average, different human populations in different cultures might not be detectably different genetically so far as their capacities for either culture in general or this or that culture are involved. S.L. Washburn, surveying the archaeological record, concludes "that there has been no important change in human abilities in the last 30,000 years" (1978, p. 57). If so, then all the changes are technological, historical, political, religious, or some other form of cultural change.

In present human populations, it seems that a baby taken from any race on Earth, appropriately reared, can receive almost any sort of general education. This does not mean that any baby can become a mathematician, or a musician, or a professional basketball player. But different babies can be found in any particular race who can do all these things well, and any normal baby can learn enough of these things to function more or less normally in any culture. Geneticists find that the vastest part of human variation is not across races or continents but within local populations (Lewontin 1972, p. 397; 1982).

Culture is quite a diverse affair, and it might be culture that reinforces genetic disposition for some practices (incest avoidance), but not for others (learning nuclear physics), with interaction sometimes and independence at other times. Whether or not adults have enzymes for digesting fresh milk will determine their pastoral practices. But, the differences, say, between the Druids of ancient Britain and the Maoists in modern China, would be non-genetic and have to be sought in the historical courses peculiar to these cumulative transmissible cultures. Such cultures catch their member humans up into an ongoing tradition, give them their identity, and radically differentiate persons historically, even though Druids and Chinese have a biochemistry and a biological nature largely held in common (though there can be differences in skin color or in blood groups).

Genes Making Human Brains

Genes make such varied cultures possible by making up each human brain with one trillion neurons, each with hundreds and sometimes thousands of possible synaptic connections, providing virtually endless opportunities for encoding ideas. These hookups code cumulative cultural discoveries and transmit them in new networks of information transfer (language and books, and, more recently, telephones, television, and the Internet). When this has gone on for a hundred thousand years and more, one can expect some startling outcomes. In fact, we have recently experienced such a startling outcome: we humans have decoded our own genome. That simultaneously impresses us with the marvel of these genes that encode and transmit millennia of evolutionary discoveries, and the still greater marvel of the powers of the brain that the genes make, which can decode its own genome.

To the marvelous discoveries in genetics we now have to add equally stunning progress in the neurosciences, again simultaneously impressing us with the powers of the brain. We humans are beginning to decode our own brains. Neuroscience is, at present, less accomplished than genetic biosciences; and this is to be expected since its focus is orders of magnitude more complex than is the genome. What we do not know vastly exceeds what we know. Neuroscientists and psychologists face a conceptual problem, since scientists are using their brains to understand their brains, and while we can well suppose that the brain might understand itself in part and in outline, can any logical system transcend itself exhaustively to critique its own structures?

All other sciences study a simpler other, while in psychological science and neuroscience, mind tackles itself. That may imply limits to the possibility of a human science. We may run afoul of a limit to our resolving power, namely, that a system of great complexity can perhaps not be wholly understood, predicted, or controlled either by itself or by some observer of the same type and complexity. Meanwhile, what we do know leaves us impressed, and puzzled.

Here we find some "cognitive dissonance." The information in the human genome is quite impressive. If the DNA in the myriad cells of the human body were uncoiled and stretched out end to end, that microscopically slender thread would reach to the sun and back over half a dozen times.[4] But this is far too little information with which to build a functioning human brain. The number of neurons and their possible connections is far more vast than the number of genes coding for the neural system, and so it is impossible for the genes to specify all the needed neural connections. We already knew that when we thought the human genome would contain 100,000 genes, but a further recent surprise is the finding that we humans do not have as many genes as we thought, only some 25,000. Humans have 100 trillion cells in their body, one trillion in their brains, but only half again as many genes as the roundworm, with a body of 959 cells of which 302 are its "brain" (Venter et al. 2001; Wade 2001).

Nevertheless, there is this enormous amount of information in human genes, and the genes in the fetus and the womb seem to have learned how to generate, by repeated algorithms, a dynamic and open-ended neural network, which, in due course, makes itself. Brain-forming genes do not specify some product with stereotyped function; rather, by splicing and re-splicing, cutting and shuffling, the brain genes proliferate cascading neurons with almost endless possibilities of organization, depending on how they synaptically connect themselves up. Genes create the instruments, but the orchestration is cerebral. Our fewer genes does not mean that we have less intelligence than before; rather, it means that the secret of our advanced information lies somewhere else, resulting from genetic flexibility that opens up cerebral capacity. In generating the human brain, Barry J. Dickson concludes: "The ultimate challenge, after all, is to find out how a comparatively small number of guidance molecules generate such astonishingly complex patterns of neuronal wiring" (2002, p. 1963).

Richard Lewontin puts it this way:

> Our DNA is a powerful influence on our anatomies and physiologies. In particular, it makes possible the complex brain that characterizes human beings. But having made that brain possible, the genes have made possible human nature, a social nature whose limitations and possible shapes we do not know except insofar as we know what human consciousness has already made possible . . . History far transcends any narrow limitations that are claimed for either the power of the genes or the power of the environment to circumscribe us . . . The genes, in making possible the development of human consciousness, have surrendered their power both to determine the individual and its environment. They have been replaced by an entirely new level of causation, that of social interaction with its own laws and its own nature. (1991, p. 123)

The genes outdo themselves.

Theodosius Dobzhansky, realizing that genes underdetermine culture, had already anticipated this. Culture takes on a life of its own.

> Human genes have accomplished what no other genes succeeded in doing. They formed the biological basis for a superorganic culture, which proved to be the most powerful method of adaptation to the environment ever developed by any species . . . The development of culture shows regularities *sui generis*, not found in biological nature, just as biological phenomena are subject to biological laws which are different from, without being contrary to, the laws of inorganic nature. (1956, pp. 121–122)

Animal brains are already impressive. According to an estimate, in a cubic millimeter (about a pinhead) of mouse cortex, there are 450 meters of dendrites and 1–2 kilometers of axons; each neuron can synapse on thousands of others (Braitenberg and Schüz 1998). But this cognitive development has come to a striking expression point in the hominid lines leading to *Homo sapiens*, going from about 300 to 1,400 cubic centimeters of cranial capacity in a few million years. The human brain has a cortex 3,000 times larger than that of the mouse. The genes keep building a bigger and bigger brain. E.O. Wilson, Harvard sociobiologist, emphasizes: "No organ in the history of life has grown faster" (1978, p. 87).[5] The connecting fibers in a human brain, when extended, can wrap around the Earth 40 times. This line seems "headed for more head," so to speak.

Generally, in body structures such as blood or liver, humans and chimpanzees are 95–98 percent identical in their genomic DNA sequences and the resulting proteins, but this is not true of their brains. "Changes in protein and gene expression have been particularly pronounced in the human brain. Striking differences exist in morphology and cognitive abilities between humans and their closest evolutionary relatives, the chimpanzees." So conclude a team of molecular biologists and evolutionary anthropologists from the Max–Planck Institutes in Germany (Enard et al. 2002). The puzzle is how so little genetic difference can make such an enormous brainpower difference. "This is one of the major questions that those of us interested in our own biology would like to ask. What does that 1.5% difference look like?" asks Francis Collins, director of the National Human Genome Research Institute (in Gibbons 1998). Some threshold seems to have been crossed, a trans-genetic crossing, a quantum leap, a change of state of order of magnitude similar to that when life once originated, or when previously instinctively stereotyped organisms gained the capacity to acquire new, nongenetic information during their lifetimes.

Biologists sometimes make claims like this based on the 95–98 percent protein identity: "DNA evidence provides an objective non-anthropocentric view of the place of humans in evolution. We humans appear as only slightly remodeled chimpanzee-like apes" (Wildman et al. 2003, p. 7181). But humans have over three times the brain size of chimps, so that 3 percent, or whatever, in protein structures makes 300 percent bigger brains. Cognitively, we are not 3 percent but 300 percent different (Marks 2002, p. 23). A few percent different may be the way we humans appear from the perspective of DNA but appearances are often deceiving; when you compare Einstein with a chimp, it does not appear that Einstein is only slightly remodeled; nor do we wonder whether an atomic bomb built with his theory that $E = mc^2$ is a slightly remodeled ant-fishing stick.

An information explosion gets pinpointed in humans, an event otherwise unknown, but undoubtedly present in us. Perhaps only one line leads to persons, but in that line, at least, the steady growth of cranial capacity makes it difficult to think that intelligence is not being selected for and conserved when it is achieved. This know-how for building bigger brains is genetically coded, of course, but here genetic history transcends itself and passes over into something else. Chimps do not attempt to construct persuasive arguments. I am not a chimp because I do. You are not a chimp either, because you are reading this book and looking for such arguments. Such arguments require language with its advanced conceptual and symbolic powers enabling abstraction, analysis, evaluation, which is present in humans but unprecedented in animals. "All the odd elaborations of human life, socially and individually, including the heights of imagination, the depths of depravity, moral abstraction, and a sense of God, depend on this *symbolic coding of the nonvisible*" (Potts 2004, p. 263). In that capacity, humans are not a few percent different; they differ by a thousand orders of magnitude.

The human brain is of such complexity that descriptive numbers are astronomical and difficult to fathom.[6] A typical estimate is 10^{12} neurons, each with several thousand synapses (possibly tens of thousands), a flexible neural network, more complex by far than anything else known in the universe. Each neuron can "talk" to many others. This network can be formed and reformed, making possible virtually endless mental activity (Braitenberg and Schüz 1998). The result of such combinatorial explosion is that the 1,500 cubic centimeters of a human brain is capable of forming more possible thoughts than there are atoms in the universe (Flanagan 1992, p. 37). Compare how many sentences can be composed rearranging the 26 letters of the English alphabet. The most startling phenomenon yet found in the universe is right behind the eyes we are looking with. We noted earlier a marvelous information in genetic nature; but now, in the human brain, the combinatorial cybernetic explosion is recompounded.

Genes repeatedly make animal brains. But does evolution repeatedly produce this ideational intelligence characteristic of humans? Increasing diversity and complexity appear repeatedly in evolutionary history. In the animal world, eyes evolved many different times, and similarly with muscles, with organs of hearing, taste, smell. Legs, fins, and wings evolved several times. Genetically based skills are widely distributed and shared. Much of this increased complexity depends on neural development, allowing, from the skin in, centered identity and integrated control of animal life, and, from the skin out, cognitive powers for information perception and processing important for survival. On the one hand, such mental powers evidently have survival value; on the other, most species (plants, insects, crustaceans) survive quite well with little intelligence and develop no more over the millennia.

So one cannot claim that all animals, much less organisms in general, evolve steadily toward higher intelligence. Only some do. But perhaps it is highly likely that some will. Christian de Duve, a Nobel laureate, concludes that neural power, where it luckily arises, has such "decisive selective advantage" that there is high probability of its increase:

> The direction leading toward polyneuronal circuit formation is likely to be specially privileged in this respect, so great are the advantages linked with it. Let something like a neuron once emerge, and neuronal networks of increasing complexity are almost bound to arise. The drive toward larger brains and, therefore, toward more consciousness, intelligence, and communication ability dominates the animal limb of the tree of life on Earth. (1995, p. 297)

Perhaps that is so with certain kinds of intelligence, but still it is rather surprising that of the 5–10 million species on Earth at present, of the perhaps 5–10 billion species that have come and gone over evolutionary time, only one has reached self-conscious personality sufficient to build cumulative transmissible cultures. Ernst Mayr, despite finding other kinds of progress undeniable in the evolutionary record, reflects on the evolution of intelligence with conclusions opposite from those of de Duve:

> We know that the particular kind of life (system of macromolecules) that exists on Earth can produce intelligence ... We can now ask what was the probability of this system producing intelligence (remembering that the same system was able to produce eyes no less than 40 times). We have two large super-kingdoms of life on Earth, the prokaryote evolutionary lines each of which could lead theoretically to intelligence. In actual fact none of the thousands of lines among the prokaryotes came anywhere near it.
>
> There are 4 kingdoms among the eukaryotes, each again with thousands or ten thousands of evolutionary lineages. But in three of these kingdoms, the protists, fungi, and plants, no trace of intelligence evolved. This leaves the kingdom of Animalia to which we belong. It consists of about 25 major branches, the so-called phyla, indeed if we include extinct phyla, more than 30 of them. Again, only one of them developed real intelligence, the chordates. There are numerous Classes in the chordates, I would guess more than 50 of them, but only one of them (the mammals) developed real intelligence, as in Man. The mammals consist of 20-odd orders, only one of them, the primates, acquiring intelligence, and among the well over 100 species of primates only one, Man, has the kind of intelligence that would permit [the development of advanced culture]. Hence, in contrast to eyes, an evolution of intelligence is not probable. (Quoted in Barrow and Tipler 1986, pp. 132–133)

Repeatedly, Mayr concludes: "An evolutionist is impressed by the incredible improbability of intelligent life ever to have evolved" (1988, p. 69; 1994). Mind of the human kind is unusual, even on this unusual Earth.

What is surprising in humans is not so much that they have intelligence generically, for many other animals have specific forms of a generic intelligence; nor is it that humans have intelligence with subjectivity, for there are precursors of this too in the primates. The surprise is that this intelligence becomes reflectively self-conscious and builds cumulative transmissible cultures. *Homo sapiens*, as we have named ourselves, is the "wise" species, and some of this is "wisdom" programmed into our genes, universal to all. Still, the specific reference largely denotes the wisdom achieved during human historical careers, and passed on culturally to generations to come. The wisdom peculiar to humans lies in the powers of their self-conscious minds and builds in their cumulatively transmissible cultures.

J. Craig Venter and over 200 coauthors, reporting on the completion of the Celera Genomics version of the human genome project, caution in their concluding paragraph:

> In organisms with complex nervous systems, neither gene number, neuron number, nor number of cell types correlates in any meaningful manner with even simplistic measures of structural or behavioral complexity . . . Between humans and chimpanzees, the gene number, gene structures and functions, chromosomal and genomic organizations, and cell types and neuroanatomies are almost indistinguishable, yet the development modifications that predisposed human lineages to cortical expansion and development of the larynx, giving rise to language culminated in a massive singularity that by even the simplest of criteria made humans more complex in a behavioral sense . . .
>
> There are two fallacies to be avoided: determinism, the idea that all characteristics of the person are "hard-wired" by the genome; and reductionism, the view that with complete knowledge of the human genome sequence, it is only a matter of time before our understanding of gene functions and interactions will provide a complete causal description of the human variability. The real challenge of human biology, beyond the task of finding out how genes orchestrate the construction and maintenance of the miraculous mechanism of our bodies, will lie ahead as we seek to explain how our minds have come to organize thoughts sufficiently well to investigate our own existence. (2001, pp. 1347–1348)

Human Minds Making Brains

Genes make the kind of human brains that facilitate an open mind possible. But when that happens, these processes can also work the other way around. Minds employ and reshape their brains to facilitate their chosen ideologies and lifestyles. Our ideas and practices configure and reconfigure our own sponsoring brain structures. Michael Merzenich, a neuroscientist, reports his increasing appreciation of "what is the most remarkable quality of our brain: its capacity to develop and to specialize its own processing machinery, to shape its own abilities, and to enable, through hard brainwork, its own achievements" (Merzenich 2001, p. 418).

In the vocabulary of neuroscience, we have "mutable maps" in our cortical representations, formed and re-formed by our deliberated changes in thinking and resulting behaviors. For example, with the decision to play a violin well and resolute practice, string musicians alter the structural configuration of their brains to facilitate the differential use of left and right arms—fingering the strings with one and drawing the bow with the other (Elbert et al. 1995). Likewise, musicians enhance their hearing sensitivity to tones, enlarging the relevant auditory cortex by 25 percent compared with nonmusicians (Pantev et al. 1998).

So our minds shape our brains. The authors of a leading neuroscience text conclude: "The amount of cortex devoted to the fingers of the left hand is greatly enlarged in string musicians. It is likely that this is an exaggerated version of a continuous mapping process that goes on in everyone's brain as their life experiences vary" (Bear et al. 2001, p. 418). With the decision to become a taxi driver in London, and long experience driving about in the city, drivers likewise alter their brain structures, devoting more space to navigation-related skills than non-taxi drivers have. "There is a capacity for local plastic change in the structure of the healthy adult human brain in response to environmental demands" (Maguire et al. 2000, p. 4398). Similarly, researchers have found that "the structure of the human brain is altered by the experience of acquiring a second language" (Mechelli et al. 2004), or by learning to juggle (Draganski et al. 2004).

One can say that finding differing locations in the brain where differing kinds of mental activities takes place is evidence for the physical basis of our mental activities. This is true. But another way to interpret the same evidence is that our mental decisions to become a violin player, taxi driver, or learn a second language reallocate brain locations to new functions in support of these decisions. Violin players, taxi drivers, jugglers use highly localized areas of their brains. But other skills, such as gaining a higher education, are more pervasively distributed. We have no apparatus to measure such more global synaptic changes, but every reason to think they are there.

This brain is as open as it is wired up; the self we become is registered by its synaptic configurations, which is to say that the information from personal experience, both explicit and implicit, goes to pattern the brain. The informing of the mind, our psychological experiences, reconfigure brain process, and there are no known limits to this global flexibility and interactivity. This is what philosophers call "top down" causation (an emergent phenomenon reshaping and controlling its precedents), as contrasted with "bottom up" causation (precedent, simpler causes fully determinative of more complex outcomes). Quantitative genetic differences add into qualitative differences in capacity, an emerging cognitive possibility that exceeds previous evolutionary achievements.

So the genes-producing-brains-producing-behavior model, always too simplistic, has now been quite replaced by a dual model, where genes produce neural networks with open possibilities, and the awakening person dynamically self-organizes a brain interactively with complex environmental influences in both nature and culture. Dean Hamer (2002) models the alternatives as shown in figure 1.1.

Indeed, strange though it may seem at first, and despite the astronomical numbers of neurons in the adult brain, in the early generation of the brain

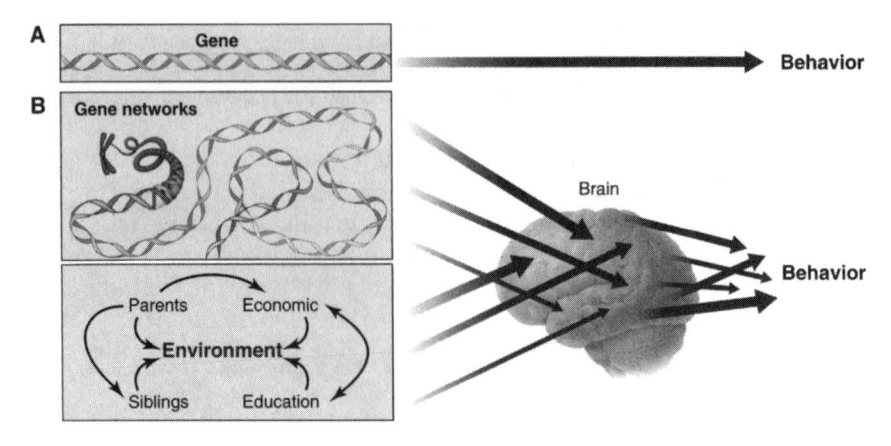

Figure 1.1 Two views of behavior genetics. (**A**) A simplified model underlying much behavior genetics research envisages a direct linear relationship between individual genes and behavior. (**B**) The reality is likely to be far more complex with gene networks and multiple environmental factors impacting brain development and fuction, which, in turn, will influence behavior. From Dean Hamer, "Rethinking Behavioral Genetics." *Science* 298 (October 4, 2002), 71–72.

during the first years of life, there are made far more neurons than the maturing brain needs. The awakening mind organizes itself by pruning away neurons that it is not using, as well as by facilitating new synaptic connections that it comes to need to support its developing lifestyle (Bear et al. 2001, Chapter 22, especially pp. 719–722). Neuroscientists may speak of the "death" of such brain cells. A more comprehensive perspective interprets this as further evidence of the excessively huge possibility of space open to the developing brain, its potential freedom and openness, coupled with the reduction of such possibilities required when some possibilities and not others are actualized.

In philosophical circles (more than among neuroscientists), it is currently fashionable to envision the brain as a kind of computer. The computational mind is the model for much cognitive science. But there is an important disanalogy with computers. Christof Koch and Gilles Laurent caution us about this:

> Software and hardware, which can be easily separated in a computer, are completely interwoven in brains . . . Brains wire themselves up during development as well as during adult life, by modifying, updating, replacing connections, and even in some circuits by generating new neurons. While brains do indeed perform something akin to information processing, they differ profoundly from any existing computer in the scale of their intrinsic structural and dynamic complexity. (1999, p. 98)

Computers, of course, do not have minds with which to reconfigure themselves. Minds, everyone knows, can do some computing.

In evolutionary history, with the coming of humans, there appears the genesis of ideas; and in culture thereafter, ideas are perennially generated and regenerated. This phenomenon too has to be incorporated into any unified worldview. But only in the human world does consciousness become recompounded though the compounding of transmissible cultures; that is the peculiar genius of the human "spirit." Superposed on biology, we become, so to speak, "free spirits," not free from either the worlds of nature or culture, but free in those environments. That humans are embodied spirits, bodies with self-reflective psychological experience, capable of thinking about themselves and what they can and ought to do is really beyond dispute. The act of disputing it, verifies it.

Humans must mate, their genes degenerate unless they outbreed; and so, perhaps, biology shapes marriage customs, or what humans think about incest. But consider what educated people think about polygamy, or abortion, or birth control—or disarmament, or evolutionary theory—all done on circuits in the brains that the genes have made possible. What is happening when a developed nation sends food to those underfed in a developing nation? Such beliefs and events are the results of decisions, perhaps individual, perhaps corporate, but it no longer seems plausible to hold that the principal determinant is something basically biological, such as producing more offspring in the next generation, or that the decision is only the resultant of some complex of basically instinctive, adaptive behavioral subroutines, more or less stereotyped by the genetics. Culture relaxes the pressures of natural selection, and the genetically constructed but experientially completed mind opens up new levels of freedom.

Spirited Self and Self-Transcendence

What is really exciting is that human intelligence is now "spirited," an ego with felt, self-reflective psychological inwardness. In the most organized structure in the universe, so far as is known, molecules, trillions of them, spin around in this astronomically complex webwork and generate the unified, centrally focused experience of mind. For this process, neuroscience can as yet scarcely imagine a theory. A multiple net of billions of neurons objectively supports one unified mental subject—a singular center of experience. Synapses, neurotransmitters, axon growth—all these can be and must be viewed as objects from the "outside" when neuroscience studies them. But what we also know, immediately, is that these events have "insides" to them, subjective experience. There is "somebody there."

The self-actualizing and self-organizing characteristic of all living organisms now in humans doubles back on itself in this reflexive animal with the qualitative emergence of what the Germans call "Geist," what existentialists call "Existenz," what philosophers and theologians often call spirit. (Like nature and culture, spirit too has multiple layers of meaning.) This sense of the existential self, the Cartesian "I think, therefore I am," which is present in all normal persons, remains at once our central certainty and the great unknown. An object, the brained body, becomes a spirited subject. A team of neuroscientists (Bear et al.) concludes: "It is difficult to study the brain without developing a sense of awe about how well it works." They also concede: "Exactly how the parallel streams of sensory data are melded into perception, images, and ideas remains the Holy Grail of neuroscience" (2001, pp. 434, 740).

We must further recompound this complexity when we look forward in the directions in which contemporary evolutionary biology, molecular biology, neuroscience, and psychology are all pointing. In nature, once there were two metaphysical fundamentals: matter and energy. The physicists reduced these two to one: matter–energy; the biologists afterward discovered that there are still two metaphysical fundamentals: matter–energy and information. At the start of the cybernetic age, Norbert Wiener insisted: "Information is information, not matter or energy" (1948, p. 155). What is already spectacular in biology on Earth, differing from the physics and chemistry of the stars, is an information explosion. Biological information is actively agential, self-actualizing. Only on Earth (so far as we yet know) can anything be learned, and the first secret of such animated life is genetic coding of an organismic self enabling coping in an environment.

But there are multiple orders of magnitude change with the coming of humans. This cybernetic or cognitive tendency does not "reduce" well; rather, it tends to "expand." This seems especially true when nature goes on information searches, and generates human brains with almost unlimited searching capacity. Yes, the evolutionary cognitive trajectory continues, but the past is not a good guide to what the future holds when there is this discontinuous "massive singularity" (Venter, 2001) at the coming of the human brain. Perhaps the most we can conclude is that the secret of such creativity lies in new domains of information searched and gained, in new information possibilities opening up. If so, the kind of ultimate destiny we now must envision, and perhaps also the kind of ultimate explanation needed, can as plausibly be said to be mind-like as mindless mechanicity. In this sense, the evolutionary, the genetic, the neurological, and the psychological sciences suggest that we inhabit a "spiritual" universe. We can wonder if there is a "Logos" in, with, and under the logic of such nature.

Alone among the other species on Earth, *Homo sapiens* is cognitively remarkable for being a spirited self and for self-transcendence. We humans are at once "spirited selves," enjoying our incarnation in flesh and blood, empowered for survival by our brain/minds, defending our personal selves, and yet transcending ourselves and our local concerns. *Homo sapiens* is the only part of the world free to orient itself with a view of the whole. That makes us, if you like, free spirits; it also makes us self-transcending spirits.

Consider this self-transcendence first in the sciences—and now it is revealing to look beyond genetics and neuroscience, beyond the sciences where we study ourselves. Physics and astronomy are within our scientific cultures, and yet with these disciplines, we transcend our cultures. With our instrumented intelligences and constructed theories, we now know of phenomena at structural levels from quarks to quasars. We measure distances from picometers to the extent of the visible universe in light years, across 40 orders of magnitude. We measure the strengths of the four major binding forces in nature (gravity, electromagnetism, the strong and weak nuclear forces), again across 40 orders of magnitude. We measure time at ranges across 34 orders of magnitude, from attoseconds to the billions-of-years age of the universe. Nature gave us our mind-sponsoring brains; nature gave us our hands. Nature did not give us radiotelescopes with which to "see" pulsars, or relativity theory with which to compute time dilation. These come from human genius cumulated in our transmissible cultures (though we do not forget that nature supplies these marvelous processes analyzed by radiotelemetry and relativity theory).

These extremes are beyond our embodied experience. No one experiences a light year or a picosecond. But they are not beyond our comprehension entirely; else, we could not use such concepts so effectively in science. The instrumentation is a construction (radio telescopes and mathematics), a cultural invention, a "social construct," if you must. But precisely this construction enables us dramatically to extend our native ranges of perception. The construction disembodies us. It distances us from our embodiment. No one has an everyday "picture" of a quark or a pulsar. But we have good theory about why nothing can be "seen" at such ranges in the ordinary sense of see, which requires light in the wavelength range of 400–700 nanometers, with quarks and pulsars far outside that range. We can ask whether a molecule is too small to be colored, or whether an electron, in its superposition states, is so radically different as to have no position, no "place" in the native range sense, but only a probabilistic location.

Owing to their linguistic abilities, humans have enormous powers of symbolic thought and abstraction, of extrapolation and theory construction, of hypothesis testing and paradigm evaluation. We often attempt understanding by analogy. Metaphor makes initial contact, and then we critique the imagery with counter-imagery, with more precision in analysis, with measurement, further imagination. We may decide to prefer the account that mathematics suggests, even if this seems counterintuitive to analogies drawn from native range experience (as in quantum mechanics). Science involves a long history of breaking up commonsense understandings with more sophisticated ones. We greatly extrapolate and radically transform any such originating metaphor. We get loose enough from our positions and places to consider other time–space scales. Our bodies with our perceptions, our brains with their concepts, which are figured out on synaptic circuits, and our spirited selves expand our location and build up overviews of the global and astronomical whole, ranging from subatomic levels through organismic, evolutionary, and ecosystemic levels.

That transcends startpoint location enabling us to reach standpoint location greater than ourselves. No animal, humans included, knows everything going on at all levels, quarks to cosmos ("the God's eye view").

Some animals, sometimes humans, know little of what is going on at any level; they have only functional behaviors, genetically coded or behaviorally acquired, that work, more or less, for survival. They have, we might say, limited know-how and no know-that. But humans can sometimes enjoy an epistemic genius transcending their own sector, and take an overview (Earth seen from space, the planet's hydrologic cycles), or take in particulars outside their embodiment (sonar in bats, low-frequency elephant communication).

Humans find themselves uniquely emplaced on a unique planet—in their world—cognitively and critically, as no other species is. Our bodily incarnation embeds us in this biospheric community; we are Earthlings. Our mental genius, our spirited self, enables us to rise to transcending overview. Eugene P. Wigner, a mathematical physicist, calls the mathematical facility humans have achieved a "miracle in itself," and comments, "Certainly it is hard to believe that our reasoning power was brought, by Darwin's process of natural selection, to the perfection which it seems to possess" (1960, p. 3).

Max Delbrück, the father of molecular genetics and a Nobel laureate, finds deeply puzzling the fact that human rationality has evolved out of natural history, selected for better survival in the jungle, for producing more offspring, yet providing an exodus by which we transcend our origins to probe the depths of the universe:

> Evolutionary thinking ... suggests, in fact it demands, that our concrete mental operations are indeed adaptations to the mode of life in which we had to compete for survival a long, long time before science. As such we are saddled with them, just as we are with our organs of locomotion and our eyes and ears. But in science we can transcend them, as electronics transcends our sense organs.
>
> Why, then, do the formal operations of the mind carry us so much further? Were those abilities not also matters of biological evolution? If they, too, evolved to let us get along in the cave, how can it be that they permit us to obtain deep insights into cosmology, elementary particles, molecular genetics, number theory? To this question I have no answer. (1978, p. 353; cf. 1986, p. 280).

Science is rooted in human nature, employs biologically evolved perceptual and conceptual faculties, and is a social construct; but, for all that, it sometimes flowers to discover objective truths—such as the relativity theory or the atomic table—which are true universally, that is, all over our universe.

If human capacities in the sciences are so startling, what does this suggest for human capacities in the arts, in ethics, in religion? On the one hand, these too evolved in the jungle and helped us survive, and, like science, continue to do so. But here too we may well transcend our local selves, our local presence, by multiple orders of magnitude. We humans live on Earth; the spiritual formation required must be of earthly use and globally inclusive. Beyond that, it does not follow that nothing universally true can appear in human morality because it emerges while humans are in residence on Earth.

Some insights in our human moral systems may be transhuman. Keep promises. Tell the truth. Do not steal. Respect property. There is nothing particularly earthbound about "Do to others as you would have them do to you." Love your enemies; do good to those who hate you. Such commandments may be imperatives on other planets where there are no humans, but rather where alien species of moral agents inhabit inertial reference frames that have no contact with ours. Wherever there are moral agents living in a culture that has been elevated above natural selection, one can hope that there is love, justice, and freedom, although we cannot specify what content these activities will take in their forms of life. The miracle of the mind is as much its capacity for seeking righteousness as its capacity for figuring mathematics. Nothing known in genetics or neuroscience prevents our claiming that humans are spirited selves who can transcend themselves in their spiritual life.

Once critics might have said that mind is rare, and drawn the conclusion that mind is an epiphenomenon, a freakish accident, that reveals nothing about the nature of nature or about forces superintending or transcending nature. But scientists now realize that anomalous events can be quite revelatory of deep-down truths. Scientists look for places where some phenomenon in nature has come to an unusually intense expression in order to study it more carefully there. Our human minds are a phenomenon of that intense kind. If so, what we humans have cognitively become, and what we morally ought to be, our trajectory, reveal a great deal more than our origins in the matter out of which we were launched and have been assembled. Perhaps after all, this primate rising from the dust of the Earth, on becoming so remarkably spiritually informed, bears the image of God.

Notes

1. For 164 definitions of culture, see Kroeber and Kluckhohn, 1963.
2. Cosmides once started a lecture by holding up a Swiss army knife as a model of the mind. This was at a joint meeting of the Royal Society of London and the British Academy, April 4–6, 1995, London, and the proceedings titled "Evolution of Social Behavior Patterns in Primates and Man" published were in a volume by Runciman et al. (1996).
3. As Cosmides must have believed while speaking at a joint meeting of the Royal Society of London, dealing with the sciences, and the British Academy, dealing with arts, asking the cross-disciplinary arts–sciences audience to evaluate the model of a Swiss-army-knife mind.
4. Estimated from data in Orten and Neuhaus (1982, pp. 8, 154).
5. Some early humans had slightly larger brains than modern humans, though a smaller brain to body ratio, but modern brains are more convoluted and complex. Brain size is only an approximate index of intelligence; some individuals with quite small brains have been fully human.
6. Nor are the estimates always consistent; they can differ by an order of magnitude, partly owing to their astronomical nature, partly due to our ignorance of neuroscience.

References

Barrow, J.D. and Tipler, F.J. 1986. *The Anthropic Cosmological Principle*. New York: Oxford University Press.
Bear, M.F., Connors, B.W., and Paradiso, M.A. 2001. *Neuroscience: Exploring the Brain*, 2nd ed. Baltimore: Lippincott Williams and Wilkins.
Boyd, R. and Richerson, P.J. 1985. *Culture and the Evolutionary Process*. Chicago: University of Chicago Press.
Braitenberg, V. and Schüz, A. 1998. *Cortex: Statistics and Geometry of Neuronal Connectivity*, 2nd ed. New York: Springer.
Buss, D. 1989. Sex differences in human mate preferences: evolutionary hypotheses tested in 37 cultures. *Behavioral and Brain Sciences*, 12, 1–49.
Buss, D. et al. 1990. International preferences in selecting mates: a study of 37 cultures. *Journal of Cross-Cultural Psychology*, 21, 5–47.
Cheney, D.L. and Seyfarth, R.M. 1990. *How Monkeys See the World*. Chicago: University of Chicago Press.
Chomsky, N. 1986. *Knowledge of Language: Its Nature, Origin, and Use*. New York: Praeger Scientific.
Christiansen, M.H. and Kirby, S. 2003. Language evolution: the hardest problem in science?, in M.H. Christiansen and S. Kirby (eds.), *Language Evolution*. New York: Oxford University Press, pp. 1–15.
Cosmides, L., Tooby, J., and Barkow, J.H. 1992. Introduction: evolutionary psychology and conceptual integration, in Jerome H. Barkow, Leda Cosmides, and John Tooby (eds.), *The Adapted Mind: Evolutionary Psychology and the Generation of Culture*. New York: Oxford University Press, pp. 3–15.
de Duve, Christian, 1995. *Vital Dust: Life as a Cosmic Imperative*. New York: Basic Books.
Delbrück, M. 1978. Mind from Matter? *American Scholar*, 47, 339–353.
———. 1986. *Mind from Matter: An Essay on Evolutionary Epistemology*. Palo Alto, CA: Blackwell Scientific.
Dennett, D.C. 1987. *The Intentional Stance*. Cambridge, MA: The MIT Press.
de Waal, F.B.M. 1999. Cultural primatology comes of age. *Nature* 399 (June 17), 635–636.
Dickson, B.J. 2002. Molecular mechanisms of axon guidance. *Science*, 298, 1959–1964.
Dobzhansky, T. 1956. *The Biological Basis of Human Freedom*. New York: Columbia University Press.
Dobzhansky, T., 1963, Anthropology and the natural sciences: the problem of human evolution. *Current Anthropology*, 4, 138, 146–148.
Draganski, B., Gaser, C., Busch, V., Schuierer, G., Bogdahn, U., and May, A. 2004. Changes in grey matter induced by training. *Nature*, 427 (January 22), 311–312.
Durham, W.H. 1991. *Coevolution: Genes, Culture, and Human Diversity*. Stanford, CA: Stanford University Press.
Elbert, T., Pantev, C., Wienbruch, C., Rockstroh, B., and Taub, E. 1995. Increased cortical representation of the fingers of the left hand in string players. *Science*, 270 (October 13), 305–307.
Enard, W. et al. 2002. Intra- and interspecific variation in primate gene expression patterns. *Science*, 296 (April 12), 340–343.
Flanagan, O. 1992. *Consciousness Reconsidered*. Cambridge, MA: The MIT Press.
Galef, B.G., Jr. 1992. The question of animal culture. *Human Nature*, 3 (no. 2), 157–178.
Gibbons, A. 1998. Which of our genes make us human? *Science*, 281 (September 4), 1432–1434.
Hamer, D. 2002. Rethinking behavior genetics. *Science*, 298 (October 4), 71–72.
Harcourt, A.H. and Stewart, K.J. 2001. Vocal relationships of wild mountain gorillas, in Martha M. Robbins, Pascale Sicotte, and Kelly J. Stewart (eds.), *Mountain Gorillas: Three Decades of Research at Karisoke*. Cambridge: Cambridge University Press, pp. 241–262.
Hauser, M.D., Chomsky, N., and Fitch, W.T. 2002. The faculty of language: What is it, who has it, and how did it evolve? *Science*, 298 (November 22), 1569–1579.
Koch, C. and Laurent, G. 1999. Complexity and the nervous system. *Science*, 284 (April 2), 96–98.
Kroeber, A.L. and Kluckhohn, C. 1963. *Culture: A Critical Review of Concepts and Definitions*. New York: Vintage Books, Random House.
Lewontin, R.C. 1972. The apportionment of human diversity. *Evolutionary Biology*, 6, 381–398.
———. 1982. *Human Diversity*. San Francisco: W. H. Freeman.
———. 1991. *Biology as Ideology: The Doctrine of DNA*. New York: HarperCollins Publishers.

Luna, B., Krista E. Garver, Trinity A. Urban, Nicole A. Lazar, and John A. Sweeney. 2004. Maturation of cognitive processes from late childhood to adulthood. *Child Development*, 75, 1357–1372.

Maguire, E.A. et al. 2000. Navigation-related structural change in the hippocampi of taxi drivers. *Proceedings of the National Academy of Sciences*, USA, 97 (no. 8), 4398–4403.

Marks, J. 2002. *What It Means to be 98% Chimpanzee: Apes, People, and their Genes.* Berkeley: University of California Press.

Maynard Smith, J. and Szathmáry, E. 1995. *The Major Transitions in Evolution.* New York: W. H. Freeman.

Mayr, E. 1988. *Toward a New Philosophy of Biology.* Cambridge, MA: Harvard University Press.

———. 1994. Does it pay to acquire high intelligence? *Perspectives in Biology and Medicine*, 37, 337–338.

Mead, M. 1959, 1989. Preface, in Ruth Benedict, *Patterns of Culture.* Boston: Houghton Mifflin, pp. xi–xiv.

Mechelli, A. et al. 2004. Structural plasticity in the bilingual brain. *Nature*, 431 (October 14), 757.

Merzenich, M. 2001. The power of mutable maps, box essay, p. 418, in Bear, Connors, and Paradiso.

Orten, J.M. and Neuhaus, O.W. 1982. *Human Biochemistry*, 10th ed. St. Louis: C.V. Mosby Co.

Pantev, C., Oostenveld, R., Engellien, A., Ross, B., Roberts, L.E., and Hoke, M. 1998. Increased auditory cortical representation in musicians. *Nature*, 392 (April 23), 811–814.

Potts, R. 2004. Sociality and the concept of culture in human origins, in Robert W. Sussman and Audrey R. Chapman (eds.), *The Origins and Nature of Sociality.* New York: Aldine de Gruyter, pp. 249–269.

Premack, D. 2004. Is language the key to human intelligence?. *Science*, 303 (January 16), 318–320.

Rendell, L. and Whitehead, H. 2001. Culture in whales and dolphins. *Behavioral and Brain Sciences*, 24, 309–382.

Runciman, W.G., Maynard Smith, J., and Dunbar, R.I.M. 1996. *Evolution of Social Behaviour Patterns in Primates and Man.* Oxford: Oxford University Press.

Sterelny, K. 1995. The adapted mind. *Biology and Philosophy*, 10, 365–380.

Symons, D. 1992. On the use and misuse of Darwinism in the study of human behavior, in Jerome H. Barkow, Leda Cosmides, and John Tooby (eds.), *The Adapted Mind: Evolutionary Psychology and the Generation of Culture.* New York: Oxford University Press, pp. 137–159.

Tomasello, M., Kruger, A.C., and Ratner, H.H. 1993. Cultural learning. *Behavioral and Brain Sciences*, 16, 495–552.

Tooby, J. and Cosmides, L. 1992. The psychological foundations of culture, in Jerome H. Barkow, Leda Cosmides, and John Tooby (eds.), *The Adapted Mind: Evolutionary Psychology and the Generation of Culture.* New York: Oxford University Press, pp. 19–136.

Tylor, E.B. 1903. *Primitive Cultures*, 4th ed., 2 vols. London: John Murray.

Venter, J.C. et al. 2001. The sequence of the human genome. *Science*, 291 (February 16), 1304–1351.

Wade, N. 2001. Genome's riddle: few genes, much complexity. *New York Times* (February 13), D1, D4.

Washburn, S.L. 1978. Animal behavior and social anthropology, in Michael S. Gregory, Anita Silvers, and Diane Sutch (eds.), *Sociobiology and Human Nature.* San Francisco: Jossey-Bass Publishers, pp. 53–74.

Wiener, N. 1948. *Cybernetics.* New York: John Wiley.

Wigner, E.P. 1960. The unreasonable effectiveness of mathematics in the natural sciences. *Communications on Pure and Applied Mathematics*, 13, 1–14.

Wildman, D.E., Uddin, M., Liu, G., Grossman, L.I., and Goodman, M. 2003. Implications of natural selection in shaping 99.4% nonsynonymous DNA identity between humans and chimpanzees: enlarging genus Homo. *Proceedings of the National Academy of Sciences*, USA, 100, 7181–7188.

Wilson, D.S. 2002. *Darwin's Cathedral: Evolution, Religion, and the Nature of Society.* Chicago: University of Chicago Press.

Wilson, E.O., 1978. *On Human Nature.* Cambridge, MA: Harvard University Press.

Wrangham, R.W., McGrew, W.C., de Waal, F.B.M., and Heltne, P.G. (eds.) 1994. *Chimpanzee Cultures.* Cambridge, MA: Harvard University Press.

CHAPTER TWO

Brain, Mind, and Spirit—A Clinician's Perspective, or Why I Am Not Afraid of Dualism

JAMES W. JONES

Physicalism reigns supreme. Even theologians have rushed to embrace it. Not the "greedy" (Dennett) reductionistic physicalism of yore but a new, kindler, gentler physicalism called (of course!) "nonreductive physicalism." Rather than the militantly antireligious drive of earlier reductive materialists, the new nonreductive types insist that theirs is a position fundamentally affirming of the religious and moral life (Brown et al. 1998). Rather than eliminating the features of human life on which religion depends, this form of physicalism affirms that "consciousness and religious awareness are emergent properties and they have top–down causal influence on the body" (Murphy 1998, p. 131). "The long-banned subjective states and qualities are now put up front—in the driver's seat as it were . . .," enthuses one of this position's most famous spokesman, the Nobel laureate Roger Sperry (1991, p. 244).

I fail to share this enthusiasm. Not because I am an eliminative materialist or a Cartesian dualist. Far from it. Rather because I am, among other things, a clinical psychologist practicing and teaching in the area of psychoneuroimmunology (or behavioral medicine), as well as teaching religious studies. Working directly at the interface of the body, the mind, and the spirit, with suffering patients, gives one another perspective on the neuroscience and religion discussion. In brief, my argument will be that the nonreductive physicalists' account of consciousness and the spirit as emergent or supervenient properties is not adequate to the data of psychoneuroimmunology. Advocates of this position make a point of affirming the reality of "top–down" causation from mind to body and not just "bottom–up" causation from brain to mind. An affirmation at the heart of mind–body medicine. However, my contention will be that nonreductive physicalism, as currently formulated, cannot account for such top–down activities of the mind.

What Needs to be Explained

Hypnosis

For several years I have practiced clinical hypnosis as part of my behavioral medicine work. I have found it particularly effective in the treatment of anxiety, chronic pain, stress-related disorders, and smoking cessation. My claims in this chapter go beyond clinical anecdotes. Hypnotic interventions have been extensively documented to be effective in these areas (Brown and Fromm 1987). Central to my own practice has been the use of imagery. For example, having patients imagine they are warming their hands over a fire has been shown to increase the blood flow to their hands and so dilate blood vessels. This may help in relieving vascular headaches. Or, in reverse, having patients imagine putting their hands in a bucket of cold water can induce a numbness in the hands, which can be transferred to other parts of the body and so serve to relieve chronic pain (Barber 1996). It is common to remove warts by having the patient imagine them gone (DuBreuil and Spanos 1993). Research has also documented that by using imagery under hypnosis, a person can impact the functioning of his or her immune system (Rurzyla-Smith et al. 1995; Wood et al. 2003). Brain scans of subjects undergoing hypnotic interventions for pain modulation and emotional arousal reveal consistent effects of hypnotic suggestions on the relevant brain centers (Feldman 2004).

It is hard to interpret such findings in any other way than as illustrating the power of mental imagery to affect the body. A person forms a purely inner, mental act (an image) and the *following* result is that the blood pressure changes, or pain sensations decrease, or other physiological processes alter (Sheikh et al. 1996). In light of such practices, it is hard for me to deny that inner, mental activities can control physiological processes. The question for this chapter is the extent to which nonreductive physicalism can account for this.

Biofeedback

Biofeedback often uses interventions similar to those used in hypnosis. But biofeedback goes beyond clinical hypnosis in documenting the effects on the body. Employing imagery, direct instructions for calmness, and various relaxation techniques, biofeedback demonstrates under laboratory conditions that imagining a relaxed state, or heaviness in the limbs, or images of light or color, or prescinding from active thought can reduce heart rate, change skin conductance, relax musculoskeletal tension, and even shift brain wave patterns (Basmajian 1983; Green and Green 1977; Schwartz and Beatty 1977).

The demonstrated capacity to control one's brain waves is most philosophically interesting. Reductive physicalism, and perhaps all forms of physicalism, attribute primary causation to physical factors, that is brain activity. It is certainly true that changes in electrical activity in the brain correlate with and may be said to cause mental activity in many circumstances. EEG biofeedback of electrical activity in the brain implies that under other conditions, understanding a set of instructions or forming a mental image comes first and is reliably *followed* by changes in patterns of brain activity (Green and Green 1977).

In addition, brain scans comparing subjects visualizing an object with subjects actually seeing the object show differential blood flow to the visual cortex (Kosslyn et al. 1993). Likewise, brain scans comparing hearing music played with a hypnotic hallucination of hearing music and simply imagining hearing the music show that imagining an experience produces a different neuronal pattern than actually having the experience (Woody and Szechtman 2000). In all these cases, it would appear that mental activity (imagining a sound or image) is the primary cause of changes in brain activity and that one can learn to intentionally control his or her brain waves and other neuronal activities.

Other clinical interventions also reveal the possibilities of consciously affecting one's neurophysiology. For example, studies have shown that consciously choosing to redirect attention and act against powerful compulsive urges not only effectively treats obsessive–compulsive disorder (OCD) but also modifies the underlying neuronal circuitry. Brain scans of patients successfully treated for OCD by such cognitive–behavioral treatments reveal significant changes in their cerebral physiology (Schwartz 1999). Similar results have been shown in the treatment of depressed patients. Here too, active psychological interventions have produced measurable and significant alterations in cerebral activity directly attributable to intentional cognitive changes and reliably associated with relief from depression (Goldapple et al. 2004).

While reductive physicalists insist that consciousness is but the result of cerebral functioning, the results of biofeedback, hypnosis, and brain scans of patients treated with active psychological interventions demonstrate that consciously choosing to form an image, redirect attention, refocus thoughts, act differently can directly affect basic cerebral activity. What is the cause and what is the effect here?

Meditation Research

Meditation-derived techniques have been increasingly deployed in the practice of behavioral medicine. The last two decades have witnessed an exponential increase in the number of articles detailing the psychophysiological effects of meditation (reviews by Marlett and Kristeller 1999 and Andresen 2000). For some time now, the clinical literature has described the effectiveness of meditation-derived techniques for the treatment of anxiety disorders (Kabat-Zinn 1990; Kabat-Zinn et al. 1992), stress (Carlson et al. 2003; Kabat-Zinn 1990; Shapiro et al. 1998), and, more recently, eating disorders (Kristeller and Hallet 1999), depression (Segal et al. 2002), and personality disorders (Linehan 1993). More recent psychophysiological research has demonstrated the impact of meditation on such basic physiological functions as brain hemispheric lateralization, immune system functioning, and emotional processing. Even short-term meditation practice has been shown to increase activity in the left cerebral hemisphere (a result associated with an increase in positive emotional responses) and improve immune functioning (Davidson et al. 2003; Goleman 2003). More advanced meditators have demonstrated, under laboratory conditions, the ability to control fundamental physiological processes, such as basic reflexes, formerly thought to be beyond conscious control (Goleman 2003). Studies have also shown that a variety of cognitive processes can be altered through regular meditation practice (Deikman 2000; Goleman 2003). Meditation has been shown to dramatically improve the mind's ability to focus and maintain attention, and to develop the capacity to detach from engrained emotional and cognitive reactions to familiar thoughts and feelings. This has been demonstrated to be important clinically in weakening and modifying long-standing patterns of anxious rumination, depressive thinking, addictive attachment, or reactive anger. Such meditation-based cognitive changes facilitate the emergence of self-regulatory functions that are experienced as healthier, saner, and more balanced (Austin 1998). Thus, the conscious choice to undertake a meditative discipline impacts a variety of physical and psychological domains.

What does psychoneuroimmunology contribute to the brain, mind, spirit discussion? It would appear to rule out a Cartesian dualism in which the mind or spirit are seen as disconnected from the body. It also seems to rule out an eliminativist physicalism in which mental activity is regarded as epiphenomenal and irrelevant to neurological and physiological functioning. At minimum, it also sharpens the idea of "downward causation," and suggests that a rather strong notion of mental causation is essential to a complete understanding of the role of the mind. The kind of self-regulation currently being demonstrated in psychophysiological laboratories and clinical practice, involving hypnosis, biofeedback, and meditation, demands a robust account of mental causation. The question for this chapter is whether the model of nonreductive physicalism can provide a strong enough account of mental causation.

Nonreductive Physicalism

Arguments in support of nonreductive physicalism must go beyond simply describing the functioning of neural organizations or pointing out correlations between conscious events and neuronal activity. Reductive physicalists, nonreductive physicalists, and dualists, all agree about the functioning of various neurotransmitters, the growth and decay of neuronal cells, and (because of sophisticated EEG monitoring and brain scans) which parts of the brain are more active or more quiescent during various mental activities.

There is little dispute these days about these findings. They are a major part of the data of contemporary cognitive neuroscience and psychophysiology.

Nonreductive physicalism, like its cousin, reductive physicalism, and its antagonist, dualism, is not simply a set of experimental findings. It is, rather, the interpretations of these findings. And, as the history and philosophy of science amply demonstrates, most significant scientific disputes (as well as most contemporary disputes between science and religion) are not about the data but rather about how the data are to be interpreted. For the most part reductive and nonreductive physicalists, and even today's dualists (Eccles 1982; Penfield 1975; Popper and Eccles 1977), agree on the results of current neuroscience experiments. Their disagreements are about the interpretation of these findings. So these disagreements will probably not be settled by appeals to experimental data (on which they all virtually agree) but rather to which position gives the most complete, coherent, and compelling account of that data. This chapter will argue that nonreductive physicalism does not appear successful on that score.

Two constructs have been recently added to the discourse of physicalism to make it nonreductive: emergence and/or supervenience. The category of "emergence" has spawned a metaphysical vision of a hierarchical universe with the higher levels "emerging" from of the lower ones (for a recent review of this discussion, see the *Journal of Consciousness Studies* 2001, devoted to the topic of emergence and Russell et al. 1999). Such a metaphysical position has obvious applications to the problem of consciousness. Roger Sperry writes that the central nervous system is "governed by novel emergent properties of its own" (1991, p. 246). Consciousness is "no longer a mere impotent epiphenomenon of brain activity. It becomes a powerful impelling force in its own right" (p. 239). Conscious agency emerges from neuronal organization and then exercises control over it.

This emergent power of consciousness necessitates a new model of causality, which "combines traditional bottom–up with emergent top–down causation," and in which mental activities "exert a concomitant supervenient form of downward control over their constituent neurocellular activities" (p. 239).[1]

Like Sperry, Philip Clayton also draws on the metaphor of emergence to conceptualize the relationship of the mind and the brain. For Clayton, emergence is less about novel properties and more about new levels of explanation.

> A property is thus emergent only if laws cannot be formulated at the lower level that predict its occurrence and subsequent behavior ... A set of phenomena is designated as emergentist only when an exhaustive description of the underlying state of affairs, although necessary, is not sufficient for explaining emergent properties. (1999, p. 201)

Labeling consciousness "emergent" means that it cannot be completely explained in terms of neuronal processes. Clayton calls this "the insufficiency thesis," which "predicts that neuroscience will *not* be sufficient to explain all that we come to know about the human person" (p. 188). Restricted to explanations using only the categories of neurophysiology, the neurosciences can never completely explain domains that emerge out of, and therefore go beyond, sheer physiology. "Without questioning the dependence on the physical," Clayton writes, his position "understands mental properties to be different in kind from those observed at lower levels and to exercise a type of causal influence unique to this new emergent level" (p. 203). Emergence means that something genuinely new, unique, and unpredictable can arise from within the natural order.

Like Sperry, Clayton affirms the reality of downward causation as an emergent property. But neither thinker specifies in detail how such causality might work. In a footnote (p. 195), at least, Clayton recognizes the problem. There he suggests that accounting for the causal influence of ideas would require "nothing less than a new theory of causality," because such a claim "diverges from the standard use of the term causality in science."

In the first instance, the category of emergence appears concerned with hierarchies of explanation since the strongest arguments in its favor appear to say that at each level, new categories of explanation are required. This is part of a much larger discussion in the philosophy of science about whether the theories of particular sciences can all be reduced to or derived from fundamental theories of physics. This seems highly unlikely. Chemistry requires categories beyond those of physics; biology requires categories beyond those of chemistry; and the study of consciousness (Clayton maintains) requires categories that are "irreducibly psychological" (p. 205).

But Clayton is not satisfied with a purely epistemological argument about the irreducibility of explanatory theories. He moves directly to ground this emergent hierarchy of explanations in ontology. Clayton advocates a "pluralistic ontology" in which the physical domain consists of a variety of entities, some of which may be rather different from what we think of as physical in our ordinary sense. He calls this "property pluralism," but this remains a form of "monism," Clayton writes,

> One's overall ontology should be monist. There is only one natural order, although it includes many different types of things. Mental causation is not supernatural; it is natural. It is thus amenable to explanation in this-worldly terms, although at least part of the explanation will need to employ irreducibly psychological concepts. (p. 205)

In other places, he refers to his position as "emergentist monism" because "monism asserts that only one kind of thing exists . . . monism is a necessary assumption for those who wish to do science" (p. 209). Despite this commitment to monism, Clayton rejects the appellation of physicalism, for "human persons, correctly and fully understood, include a spiritual dimension which, whatever else it is, is more than physical" (p. 212). He is clearly struggling with the problem of continuity and discontinuity within nature.

Science requires continuity and the category of monism supplies that. But theology and spirituality require a degree of discontinuity in order to avoid a hard reductionism. The category of emergence is relied on to provide the necessary degree of discontinuity in order to make such realities as downward causation comprehensible.

Both Sperry and Clayton clearly and forcefully wish to maintain the reality of the spiritual domain. Both recognize that some notion of mental causation is a necessary component of that domain. While Sperry appears to rely mainly on rhetoric, Clayton attempts to work out a more complete metaphysical picture with emergence at the center. The question here is whether their category of emergence can do all the work required of it in order to affirm a robust model of mental causation. This essay will respond with a reluctant "no."

Is Consciousness an Emergent Property?

Clearly, complex systems possess properties that their component parts do not: words, cells, and water have properties that emerge out of the organization of their letters, their macromolecules, their atomic constitution. What, then, are some of the characteristics of an emergent property? At least three minimal conditions must be present. A1 can be said to be an emergent property of system A if:

A1 cannot exist without A.
A1 has constituent elements in common with A.
A1 has characteristics not possessed by the individual components of A.

This would clearly describe the relation between words and letters: a word is a system of letters and it cannot exist without the letters, and both the letters and the word are linguistic, often written, forms. Or the relation between a cell and its chemicals: a cell is a system of chemicals and it cannot exist apart from the chemicals, and the cell and the chemicals that make it up are both composed of atoms and molecules. Or that between molecules of water and the atoms of hydrogen and oxygen; water cannot exist apart from hydrogen and oxygen and both are composed on subatomic elements. But note that all these examples involve properties emerging from systems composed of similar entities (letters, chemicals, atoms).

However, if we say that consciousness is an emergent property of a system of neurons, we run into immediate problems.

1. The claim that consciousness cannot exist apart from the brain is one of the things that such a model was supposed to demonstrate. An argument that begins by assuming this tenet may be simply circular and may end up concluding what it has already taken for granted. However, we might grant that consciousness may not exist apart from

the brain in order to go on and explore the logic of this model. We must beware of using this model, however, to argue that consciousness cannot be separate from the brain since this model of emergent properties seems to depend on precisely this claim.
2. A more serious problem exists: the second assumption points out that this emergent model of the mind depends on the assumption that minds and brains are at least partially similar. Calling this position a form of physicalism underscores this assumption. If all that is real about human nature is physical, and consciousness is real, it too must be in some sense physical, that is, it must in some sense be not just correlated with but rather similar to physiological activity in the brain.[2]

Such a claim has serious logical difficulties. In what sense can thoughts and neurons be said to be similar enough to be parts of the same system? Practically none. Consider:

(a) Neurons and other components of the central nervous system, like all physical entities, are always described in the categories of space and time. Thoughts and images are never described, except perhaps under poetic license, in terms of their mass, energy coefficient, or width.
(b) I may make a claim about the neurons in my brain—their number, density, organization, or development and be mistaken about it. As philosophers have pointed out for centuries, I cannot be mistaken about the ideas or sensations I have in my mind. If I say I feel a pain in my foot, I cannot be mistaken about feeling such a sensation, even if I do not have a foot.

All of this is so obvious that it is a little silly to repeat it except that it seems to be a fatal blow to the emergent model of consciousness, and any physicalist position, no matter how "nonreductive." If thoughts and neurons are neither described in the same categories nor governed by the same logic of explanation, in what sense can they possibly be even partially similar? And if thoughts and neurons are not at least basically similar, in what sense can thoughts be understood as a property of a system of neurons? Certainly not in the same sense that a word can be understood as a system of letters or a cell as a system of chemicals. (An oft-cited critique of this theory on which the system's model appears to depend can be found in Nagel 1974; Poulton 1973; see also Watkins 1982.)

Put most starkly, a thought is not a thing. As philosophers have noted for centuries, the sensation of seeing red is not reducible to or translatable into statements about wavelengths, rods and cones, or neuronal processing (Chalmers 1995; Robinson 1976; Velmans 2000). No description of physics or neurology can lead from there to a description of the experience of redness. They are simply two separate and distinct linguistic systems. One of the claimed advantages of the emergent model in contrast to dualism is that it removes the dilemma of specifying how mind and brain, spirit and matter, are connected. Renaming consciousness as an emergent property may not account for the emergence of consciousness without some way of specifying how two such different things as thoughts and brains can be aspects of a single system. Of course, the nonreductive physicalist wants to claim that both thoughts and brains are, in some sense, physical. But my point is that specifying in exactly what sense images, thoughts, intentions are themselves physical (as distinct from simply possessing physical correlates) is far from clear. We will return to this point again and again in the coming pages.

The model of an emergent system's properties is supposed to be simpler than its competitors, but it is not clear in what sense this simplicity is a virtue if it provides no explanation of the process that most needs explaining—the connection of neuronal states and conscious states. As fervently as the proponents of this model might hope otherwise, it is not clear that just calling consciousness an emergent system's property removes the need (which dualism also has) to provide a theoretical bridge between brains and thoughts.

Supervenience

Nancey Murphy is less happy with the metaphor of emergence; she prefers the category of "supervenience" (1999b, 1999c). Supervenience defines a dependent but nonreductive relationship between properties: Property G in Domain A is said to supervene on Property F in Domain B, if an x instantiates G is in virtue of x also instantiating F under circumstances c (Murphy 1999a, p. 150). For example, the property of being a penny supervenes on being a copper disk with Lincoln's head under the circumstances of being minted by a legitimate U.S. mint (p. 150). Thus, there is a "*codetermination* of the supervenient property by the subvenient property or properties and the circumstances" (p. 152). The property of being a penny is codetermined by the circular, copper disk and the U.S. currency laws. Neither the physical properties nor the legal context by themselves make something a penny. Thus, context or circumstances can be genuine determining factors for a given property.

Since supervenience "reflects both dependence and nonreducibility," it "gives us a way of talking about the genuine dependence of human characteristics on the brain, but leaves room for the codetermination of *some* of those characteristics by the external world, especially by culture" (p. 151). On this definition, downward causation exists when a more encompassing set of circumstances effect lower level processes. For example, macroevolutionary processes select for some DNA sequences and not others. Thus, macroevolutionary processes cause some DNA sequences to survive and not others. Selection produces termites with strong jaws and therefore causes their DNA sequences to survive and reproduce (Murphy's example, p. 155).

This position is clearly dependent on that of the philosopher Donald Davidson who she quotes with approval. Davidson argues that the following are observed in downward causation:

1. All higher level processes are restrained by and act in conformity to the laws of the lower levels.
2. Higher levels require lower level processes and structures for their implementation.
3. The emergentist principle. Higher level processes cannot be completely described by the laws and terms of the lower levels. This is a principle of nonreducibility.
4. Laws operating at the higher, organism level, like natural selection, determine events at the lower level, for example, which DNA sequences survive (pp. 155–156).

Applying this to the issue of mental causation, Murphy argues that the macro context of culture provides us with reasons to think, or conditions us to think, that, say, $7 + 7 = 14$ as part of a larger systems of arithmetic. This learning gets instantiated in our brains and in our cognitive processing systems according to the laws of learning and memory, which every beginner in psychology learns. So, when asked what $7 + 7$ is, I immediately think of 14. Hence, my thinking processes supervene on my neurological processes, but it is not sufficient to answer the question of why $7 + 7 = 14$ by simply describing the neuronal activity in my brain. I must also give reasons in terms of the rules of arithmetic. Thus, it is permissible to claim that reasons and reasoning are not simply illusory but have a real effect on our lives in the world, in this case on how we perform addition. The same reasoning, Murphy argues, can apply to moral reasoning. The moral principles and reasons we have been taught, likewise, can be said to affect our lives. These reasons are causally effective not because they directly act on the neurological architecture (Murphy explicitly rejects any such claim!), but rather because culture and learning have linked these cognitions to certain neural–physiological activities.

So supervenience captures a certain relationship between properties, and the way in which a larger context impacts upon lower level processes. Since the lower level physical processes may underdetermine the macro properties (not every copper disk with Lincoln's head is a penny—the one I stamped out in my basement clearly isn't), the subvenient property can cause the supervenient property without determining it. But such underdetermination is not the same as top–down causation, a fact that Murphy herself acknowledges when she says directly "I reject all moves to make supervenience or realization a causal relationship!" (p. 154).

Where, then, does this leave the issue of mental causation? Focused primarily on relationships between properties, the category of supervenience alone cannot speak to that. Murphy attempts to answer this question directly in an article entitled "Downward Causation and Why the Mental Matters." She rejects Sperry's claim that higher level, emergent mental processes "overpower" or "over rule" lower level physiological processes (1999b, p. 15). Instead, she begins with a crucial distinction between structuring and triggering causes. Triggering causes are the brute physical processes; structural causes direct, channel, and structure the physical processes. Electrical impulses travel through my computer according to the general laws of physics, but the software directs those electrons in certain patterns. Impulses travel between the neurons according to the laws governing the movement of ions through the neurons. But these impulses are structured by the density of the neurons, the amount of neurotransmitters, the strength of the impulses (p. 15).

Mental causation is an example of structuring causation. "Downward causation is not overpowering but selective activation of lower-level processes" (p. 17). A similar position can be found in the works of Meyering (1999) and Van Gulick (1993). This argument depends on an interesting definition of causality. We usually think of causality as a relationship between events. The event of my striking the ball with a cue causes a second event of the ball moving across the billiard table. Murphy proposes "that we enrich our resources for understanding causation by countenancing the causal role of properties of *entities* or *objects*, along with the causal role of events" (1999b, p. 14).[3] This allows her to argue that the property of the brain (an object) having a certain structure can be a genuinely causal factor. So the property of a set of neurons being linked in a certain pattern can be said to cause the electrical activity in my brain to be channeled in a specific way, thereby causing me to think and act in a specific way. But how does this set of neurons acquire this property of being organized in this certain way? In some instances, it may be from the top–down effect of the environment on the brain. Remember the earlier discussion of supervenience in which the larger context (e.g., culture or the environment) structures our neurological activity. "The neural system—because of its plasticity—can be *trained* to perform various operations. Much of this training is thought to happen by means of feedback from the environment" (p.18). As time goes on, this training establishes certain cognitive structures (language, rules of calculation, cognitive schema, and so on). Thus,

> A cognitive process supervenes on a *pattern* of neural activation. These patterns also act in a downward way. Over time the pattern itself activates or deactivates component causal capacities . . . No laws governing operation of the system at the micro level need to be violated; rather, some of the original causal pathways are simply disused. (p. 17)

This appears to be essentially a presentation of a theory of how learning and memory get encoded neurologically. It is not clear how this is really a theory of intentionality and mental causation. Basically, this sounds like a conditioning model of learning applied to the brain. But Murphy goes beyond simple conditioning by arguing for a kind of self-regulation—a "supervisory system" or a "meta-organizing system," which is a higher level in a hierarchical model of cognitive processing. She seems to be making the traditional argument in cognitive psychology that intentionality and consciousness can be understood as increasingly complex systems of feedback [this appears to me to be the obvious conclusion of her increasingly complex diagrams (p. 20)]. But the claim that consciousness and intentionality can be completely described with the metaphor of feedback loops is far from settled (Jones 1992).

Murphy's idea of structuring causation (which is shared by Meyering and Van Gulick) may be a possible account of the physiology of downward or mental causation. But it is not compelling if the very concept of downward causation itself is flawed, at least when positioned in a nonreductive physicalists framework. The ordinary account (some would say "folk psychology account") of mental causation again suggests that two *events* are causally related: a mental event (the thought "It is raining, I best take my umbrella") and a physical event (I pick up my umbrella). This implicit dualism is said to shipwreck on the lack of an explanation of how the mental can causally impact on the physical; how a mental idea can cause my muscles to contract and my arm to move. The physicalist does not have that problem. Since the mental event is a physical event, there is no problem explaining how one physical event can cause another. The problem is accounting for the consciously experienced connection between the mental event and the physical act. For Murphy, that connection appears to be explained by the laws of learning and conditioning. We learn to associate perceptions of rain with reaching for an umbrella. This conditioning sets up *two* sorts of associations or connections: *first*, between certain neuronal connections and the learned conjunction of rain and reaching for umbrellas *and second*, simultaneously, between those neuronal connections (that connect the two physical events of perceiving rain and reaching for an umbrella) and the thought "it is raining" and the intention to reach for an umbrella. Thus there is, in reality, a rational, learned conjunction between the mental event "it is raining" and the act of reaching for an umbrella. But the thought does *not* cause the action, rather it is conjoined to it or, as Murphy would say, the thought supervenes on the neural physiology of perceiving rain and picking up an umbrella. But, and this is the crucial point here, the mental events as mental are not causative; rather, it is only their physiological substrata that has direct causal connections with the larger neural–physiological domain. As a supervenient (or emergent) property, the mental cannot be exhaustively described in the language of neural physiology alone. And so it acquires an explanatory autonomy while the causal processes are entirely neural physiological (this is based on Murphy 1999a, p. 14). The experience of mental causation is thus artfully defended while the reality of mental causation as mental realities exerting a causal influence is denied. The discourse of thoughts and intentions is explanatorily relevant to human behavior but mental properties or entities are not causally efficacious on human behavior.

Murphy rejects Sperry's assertion that mental processes can overrule physiological ones. In her discussion of supervenience, she explicitly denies that supervenient properties are causal. Supervenience is a conjunctive relationship, not a causal one. That would seem to eliminate mental causation entirely. Mental processes are important because they are conjoined to physical processes and because they provide "reasons" for our actions. But they do no causal "work" in the physical world. [A similar critique is often made of Davidson's position, see Heil and Mele (1993).]

Of course, to the physicalist, the "folk psychological" account of mental causation is wrong to begin with. It depends on an ordinary model of "event" or "entity" being applied to the mental realm. And, for the physicalist, there are no mental entities or domains. There is only the physical domain that gives rise to the experience of a mental life. But that experience of a mental life is really a physical reality, even though it cannot be exhaustively described in physical categories, even though it requires subjective language to communicate itself. But then again we seem to be back to the earlier discussion of the sense in which thoughts and intentions can be said to be physical—something nonreductive physicalism demands—given their obvious and irreducible differences.

In these discussions of downward causation in relation to the categories of emergence and supervenience, there appears to be a confusion between categories of explanation and causal agents. Most philosophers and neuroscientists agree that higher-order cognitive processes can only be described in categories that go beyond simple accounts of neuronal firings and neurotransmitter releases or synaptic organization. These higher-order cognitive and psychological processes require meta-level categories of explanation and description. I agree that is a sufficient reason for adopting metaphors of emergence and supervenience in our description of mental processes as opposed to a hard reductionist stand. But, as Dennis Bielfeldt writes, "Semantic irreducibility does not entail causal autonomy" (2001, p. 170). Bielfeldt goes on to quote Murphy's own acknowledgment that the language used "at each level cannot be reduced to that of a lower level, even though what happens at each level is uniquely determined by the coordinated action taking place at the lower levels, where it is fully described in terms of the lower level language" (Murphy and Ellis 1996, p. 28). This along with Murphy's rejection of supervenience as a "causal relationship" and Clayton's (correct, in my estimation) claim that mental causation requires a "new form of causation" that "diverges from the standard use of the term causality in science" suggest that supervenience and emergence, while adequate accounts of the relationship among *levels of explanation* required to account for conscious mental life, are not adequate to a robust account of mental causation.

Awareness

Murphy's theory of supervenience focuses primarily on the contents of consciousness—thoughts, reasons, intentions—some of which might readily become linked to neurological processes in the quasi-conditioning way that she describes. Many authors (Chalmers, Hutto, Nagel, Velmans) suggest that the real problem of consciousness involves not simply its contents but rather the brute fact of awareness itself. And contrary to the physicalist's account, it is not so easy to see how awareness itself can be completely mapped neurophysiologically.

Consider the following thought experiment. It is probably not possible in practice, but it is easy enough to visualize. Suppose you are on an operating table with your brain exposed and a series of cameras and screens allow you to observe your own brain functioning. Since the brain itself carries little sensation, neurosurgery can be done with the patient awake. You notice the color red in the corner of the room, and at the same time you become aware of the neuronal discharge that represents the visual experience of seeing red. And you realize that the neuronal activity in the visual cortex is connected to the experience of seeing red. And, simultaneously, you notice the neuronal discharge that represents drawing the connection between the previous occipital activity and the experience of redness. And then—or simultaneously?—you see the neuronal correlate of drawing the conclusion that the previous neuronal activity represents drawing the conclusion about the experience of redness. And, of course, there would have to be a neuronal correlate of that conclusion, but again, where in the sequence would you see it? And where would you see the neuronal correlate of seeing that?

Why is this so confusing? Because you are watching your brain record the experience of watching your brain record the experience of watching your brain, ad infinitum. You see the brain configuration change as you think new thoughts, but what do you see that goes with the recognition that you are watching the brain configuration change as you think new thoughts? What neuronal activity would you observe that goes with the awareness of your awareness?

The sequence of observing one's own brain might be diagramed as follows, where N.S. stands for a neuronal state and C.E. for a conscious experience:

N.S.1 > C.E.1 (I see my brain)
[N S.1 > C.E.1 (I see my brain)] > [N.S.2 > C.E.2 (I am aware that I am seeing my brain)]
{[N.S.1 > C.E.1 (I see my brain)] > [N.S.2 > C.E.2 (I am aware that I am seeing my brain)]} > [N.S.3 > C.E.3 (I am aware that I am seeing my brain and the connection of that awareness to my brain)]
{[N.S.1 > C.E.1 (I see my brain)] > [N.S.2 > C.E.2 (I am aware that I am seeing my brain)]} > [N.S.3 > C.E.3 (I am aware that I am seeing *my* brain and the connection of that awareness to my brain)] >
[N.S.4 > C.E.4 (I am aware that I am seeing my brain and seeing the connection of that awareness to my brain and seeing the connection of seeing that awareness of my brain to my brain)]

It is harder to imagine mapping an increasing (hypothetically infinite) series of hierarchies onto the shifting linear configurations of neuronal activities when one of those hierarchies represents an awareness of those shifting configurations of neuronal activities and another hierarchy represents an awareness of that awareness of those shifting configurations. What is the state of the system that goes with observing that state of the system? The nonreductive physicalist's model of emergent or supervenient properties, in fact, may not do away with the paradoxical relation between cortical states and conscious experiences, especially when the conscious experience in question is of the cortical state that goes with that conscious experience of that cortical state.

Can Nonreductive Physicalism Explain Top–Down Causality?

Obviously, I agree with the nonreductive physicalists' assertion of top–down causality from mind to brain and then from brain to the other physiological systems that comprise the human being. This is a fundamental assumption of psychoneuroimmunology. And it is required to distinguish nonreductive from reductive physicalism. My concern is that even the strongest doctrine of emergence or supervenience cannot really provide a sufficiently powerful model of top–down causality to account for the findings of psychophysiology.

As noted before, the term supervenience denotes a conjunction between two sets of properties without having to specify the exact nature of these properties, except to say they can only occur in conjunction with one another. And, in addition, that one level is not easily (or at all) reducible to the other. The characteristics of water—its fluidity, ability to freeze or boil—supervene on its molecular structure. Presumably, any element that has that exact chemical composition (H_2O) would have the same properties. But these characteristics cannot be described by descriptions of oxygen or hydrogen alone. Or the meaning of a sentence supervenes on the sounds of its words. Presumably, any two sentences that sounded alike would have the same meaning. But the meaning cannot be described simply by descriptions of the phonetics of the sound. Or the beauty of the Mona Lisa supervenes on the arrangements of the pigments that compose it. And any similarly arranged set of pigments would be as beautiful. But that beauty cannot be described in terms of the chemistry of pigments alone. Note that so defined, supervenience simply requires conjunction and not causality. The supervening properties cannot be descriptively reduced to the categories applied to the micro properties. But that does not give them any ontological priority or causal efficacy. In this sense, supervenience is compatible with a diversity of positions from the interactive dualism of Popper and Eccles (1977), to the panpsychism of the Whiteheadeans (Griffin 1998, 2002), as well as nonreductive physicalism. By itself, the notion of supervenience does not address the central issue of this chapter—mental causation.

A sufficiently strong doctrine of top–down causation must go beyond simply describing the functioning of neural systems or finding correlations between conscious events of neuronal activity. It must assert that emergent properties now exert direct causal power over the lower levels—something Sperry (1991) does in fact assert very forcefully. In her discussion of supervenience, Murphy directly denies that supervenient properties are causal. However, in her piece on "Why the Mental Matters," she seems to suggest that mental properties can be downwardly causal (and not just conjoined with physical processes) but in a special sense—they provide structuring, not triggering causes. However, if the higher level properties can exert any kind of causality (triggering or structuring) over its constituent parts, this implies that the larger system has causal properties not derived from or controlled by the causal properties of the parts. In this case, that the mind has causal powers not derived from the causal properties of the neurons.

There are at least two questions to be raised about any claim of downward causation (Clayton's, Sperry's, or Murphy's). First, if these powers of causality are not entirely determined by the causal processes in the brain, where do they arise from? From where does the mind acquire the property of downward causation? It is certainly the case that the meaning of a word determines the order of the letters in that word, and that function of a cell determines the behavior of the macromolecules that make it up. In that sense, they are exerting a kind of downward causality. Is this the same as the kind of downward causality the mind demonstrates in biofeedback and the placebo effect? A stronger example might be the ways in which a society regulates the behavior of its members. But this would require seeing the brain as a society of neurons in a very strong sense and not just as a convenient metaphor. For the analogy of brain and society to really work, the neurons would have to be given a certain degree of autonomy and agency (or, perhaps, the mind–brain is a strictly totalitarian state). And that would just push the question down a level to the concern of where the individual neurons acquire this semi-autonomy from. Escaping the Scylla of reductionism only to come close to the Charybdis of panpsychism.

Physical science has assumed the macro features of a system are determined by the causal properties of its parts. The causal processes going on among its macromolecules govern what a cell can and cannot do. The meaning of the words govern what a sentence can and cannot mean. In most, if not all, cases of emergent properties, any causation at the macro level is derived from causation at the micro level. In none of these cases can the macro processes overrule or alter or even "structure" micro level causal activity. But in the case of consciousness, the nonreductive physicalist says that a new principle of causation, "top–down causation," suddenly appears and influences, if not overrules, the micro level processes.

On the issue of consciousness as a cause, the nonreductive physicalist appears to be in a no-win situation. He can maintain the common scientific position that all causality arises from fundamental micro level processes. But then, he would be practically indistinguishable from the reductive physicalist. And then, mental causality becomes simply conjunction (between neuronal and mental events). Sometimes, when describing supervenience, Murphy sounds like this is her position. Thus, mental causation is effectively denied. Too weak a model of causation for mind–body medicine. Or the nonreductive physicalist can affirm a strong causal power of consciousness to overrule, or at least redirect, those micro level causal properties, but at the cost of leaving inexplicable the origin of this top–down causality.

And by claiming that macro level causality can act on micro level processes in ways at least semi-independent of micro level deterministic laws, he seems to be implying a violation of basic natural law. Of course, he might reply that these macro level causal powers are limited by the micro level properties, as the meaning of a sentence is limited by the meanings of the words that make it up. But that is exactly what a strong model of downward causation must deny. Hence, either downward causation must be weakened into insignificance or basic natural laws must be violated.

The second concern is that if brain processes can be overruled by a higher-order mental causation (as Sperry and perhaps Clayton suggest), then it would appear that the central nervous system is not really a closed, physical system. Again, Murphy denies that mental processes overrule basic physiology but at the cost of weakening, if not eliminating, any doctrine of mental causation. But the principle of the physical world as a closed system, not amenable to intrusions from beyond, is a major assumption of scientific physicalism. Of course, the nonreductive physicalist can assert that the mind too is physical, operating within the constraints of the physical world. But that brings us back to the earlier problems associated with any such "identity" theory. If you simply say that everything that is real is physical, and that consciousness is real, then consciousness becomes physical by definition. A tautology is all that has been produced here: that mental entities are real entails that mental entities are physical, because real is equivalent to physical. The problem has been solved by definition.

But a new problem has been created: what exactly is meant by physical? What are the limits of the physical in the nonreductive physicalist account? It would seem that the domain of the physical is without clear boundaries here, that there are no real criteria for what are genuinely, authentically physical.

The reductionist says simply that the physical is what is described by the physical sciences. Period. Here, the reductive physicalist has the virtue of simplicity. The nonreductive physicalist, on the other hand, needs to assent that mental properties cannot be completely described in terms of physics and chemistry. Otherwise, they would be reductive physicalists. Yet they also want to say that mental properties are physical? In what sense?

Once again, on the issue of consciousness as a cause, the nonreducive physicalist appears to be in a no-win situation. She can insist that mental processes are really physical and so the closure of the system of nature is not violated. But that claim borders on an identity theory that appears problematic for reasons discussed previously and undercuts any real difference between reductive and nonreductive physicalism. Or she can reject the identity theory and stress the difference between mental and neuronal domains, and hence maintain her nonreductive stance. But then it becomes less clear in what sense her position is really one of physicalism.

These considerations leave me wondering whether nonreductive physicalism is really a coherent position. I'm not sure the nonreductive physicalist can have it both ways: trying the maintain both the reductive physicalist's tie to current natural science and the dualists' affirmation of conscious causality without either vicious reductionism or scientific incompatibility.[4]

In addition, it is clearly one thing to simply assert the arising of consciousness from neuronal activity, and something else to specify the actual processes by which that happens. Virtually all writers agree that no such account is currently available (e.g. Chalmers 1995; Hutto 1993; Libet 1982, 1996; McGinn 1989; Velmans 2000). Some go as far as to suggest that we cannot even conceive of what such a count might hypothetically look like. All attempts to do that based in contemporary science have had serious problems. Quantum theories have trouble finding places in which quantum events immediately appear in the ordinary world of brains and choices. Theories drawing on nonlinear dynamics and the emergence of complexity have trouble locating such processes in ordinary neurophysiology. Contrary to both quantum indeterminacy and chaos theory, the neurons in brains seem to obey deterministic biological laws. And, more to the point, advocates of quantum theories or nonlinear dynamics agree that such processes by themselves probably could not give rise to a strong version of downward causation (Scott 2004; Silberstein 2001). I do not want to push this point too hard. It is, after all, something of an argument from silence. The future may well produce a compelling scientific model of how neuronal processes give rise to conscious experience. But it should, at least, suggest a more humble and nuanced position than a plain assertion that consciousness is simply produced by the brain.

The Problem of Incompleteness

Before we consider the implications of this discussion for religion, we must note something about the nature of scientific theorizing, especially as applied to consciousness. Any present or future neurological theory of consciousness, including various forms of physicalism, will be incomplete in at least three senses.

Goedel's Incompleteness

My thought experiment involving the possible neurology of our awareness of our awareness points to another aspect of this problem. Essential to human consciousness is the self-reflexiveness that makes conscious accounts of consciousness possible. The mathematician and philosopher Kurt Goedel demonstrated that formal systems of reasonable complexity cannot validate all their assumptions and claims and are "incomplete" in that sense. His investigations included claims within a formal system that referred to that system itself, thus involving him in the problem of self-reflexiveness and leading him to conclude that there is an inherent incompleteness in any account involving self-reflexiveness. In diagramming my thought experiment regarding the observation of the brain state that goes with observing that brain state, I created a set of Goedelian sentences (Findlay 1952), suggesting a possible incompleteness in every description of consciousness.

The application of Goedel's theorem to the problem of consciousness has a controversial history, much of it centering around a paper by J. Lucas (1964) in which he argues that Goedel's theorem makes a consistent materialistic philosophy of mind impossible (a position Goedel himself may have held) and thus supports a kind of dualism, almost by default [a discussion of the controversy and critique of Lucas can be found in Hofstadter and Dennett (1981), especially pp. 276–283]. While covering much of the same ground as Lucas, my discussion is not necessarily designed to argue for dualism, but rather only for incompleteness.

Experimental Incompleteness

Contemporary neuroscience has uncovered areas of incompleteness in the investigation of brain functioning—Penfield's failure to access or localize self-awareness, for example, or Libet's experiments, which appear to demonstrate a serious disjunction between sufficient neuronal activity and the correlated conscious experience (at least in the area of somatic sensations) (Libet 1967, 1978, 1982), or the lack of an agreed-upon theory of the transformation of neuronal processes into conscious experience (Chalmers 1995; Libet 1996, 1982). These may be resolved by further investigation. They may also reflect an incompleteness inherent in the subject under study.

There is a paradox in neuroscience: the primary instrument for studying the mind–brain is the mind–brain. Does that make neuroscience different from, say, physics or chemistry? It would probably be misleading to say that physics consists of electrons studying electrons or chemistry consists of chemicals studying chemicals, but it is not misleading to say that neuroscience consists of the brain studying the brain. The study of consciousness may contain a limitation that can never be completely resolved, since we are using the brain to study the brain and using the categories of cognitive processing to study the categories of cognitive processing.

This may parallel the dispute in physics about the "collapse of the wave function" in which experimental phenomena set limits on our knowledge of the subatomic domain in Heisenberg's "uncertainty principle." Schrodinger, Wigner, Jeans, and others suggest that the uncertainty principle not only puts an inevitable limitation on our knowledge of the physical world, but also points to the irreducible nature of consciousness, which has become an indispensable component in the experiments of quantum mechanics (Jones

1984; Morowitz 1981). Likewise, some gaps in our current neurological knowledge of consciousness may well be filled by further investigation; others may reflect intrinsic and abiding limitations on the field.

Put another way, in our investigations of consciousness, we never stand outside the domain of consciousness. Even the latest and most sophisticated brain scanning technologies still take place within the field of consciousness. Only a conscious and intentional agent can invent such machines, design experiments using them, gather the results, and interpret the data. Consciousness is presupposed in every experiment. It is never studied entirely from the outside; rather, all experiments and model building take place within the field of consciousness.

Theoretical Incompleteness

The issue of incompleteness is as much a philosophy-of-science issue as a neurological one. I have argued elsewhere (Jones 1981) that, as a matter of logic, no scientific theory can or will ever be complete. It is not a criticism of any neuroscientific model to say that it is not a complete account, for all theories are incomplete in several senses—for example, selectivity must limit a theory's range and scope.

Using the analogy of a painting: I can give a complete description of the chemistry of the pigments, but is that a complete account of Picasso's *Guernica*? Obviously not. Many aspects of the work are not touched by such a discussion. Each field-dependent analysis may be complete on its own terms but cover only certain aspects of the painting. [The analogy is discussed in more depth and detail elsewhere (Jones 1981).]

What are the implications of these different types of incompleteness for the model of nonreductive physicalism? As we have seen, the need to be in continuity with the worldview of mainstream science has created serious problems for nonreductive physicalism, especially because of mainstream science's commitment to the causal closure of the physical world. From a pragmatic perspective in the philosophy of science, such postulates as the causal closure of the physical world can be seen as heuristic instruments, not as inviolable natural "laws" (Jones 1981; Toulmin 1960). If all scientific claims are incomplete or limited in the senses just mentioned, this would apply to the model of nature as a causally closed system. For purposes of scientific investigation, the natural world is framed as a closed causal system. Part of the motivation of science is to see how much heuristic gain can be obtained from investigating the world on the assumption. Obviously a lot. But the explanatory successes of science may have blinded us to the inherent limitations of all human systems of knowledge and led us to regard such principles as the causal closure of nature as absolute truths rather than as exceedingly fruitful heuristic tools. If nonreductive physicalists could loosen the grip of the principle of causal closure on their thinking, they might be able to fashion a more coherent position. But, to be sure, one less in continuity with current scientific models.

Put more bluntly, science does not say that science is the only valid way to envision the world. Some scientists may hold that as an article of faith. However, such a claim is hardly an empirical one; no experiment could demonstrate its truth. Nor is such a claim necessary for the conduct of science. Many brilliant scientists have also held various, nonempirical, religious, and metaphysical views of reality, which have not interfered with their scientific work. The standing incompleteness—in the senses previously discussed—within all current (and I think future) neurological theories leaves room for multiple models of consciousness. No neurological account of the human person can be used to preclude all theological ones (Jones 1992).

Implications for Religion

The stark truth seems to be that natural science as currently conceived cannot provide a robust enough account of mental causation to account for the findings of research in behavioral medicine, meditation, hypnosis, and other fields of psychophysiology. Others have reached this same conclusion: Bielfeldt (1999b); Khilstrom (2002); Velmans (1996, 2000, 2002); as well as those philosophers who find nonreductive physicalism wanting on philosophical grounds. And Clayton seems to share it too when suggesting the need for a new theory of causation. If self-regulation research continues to be borne out (and I see no indications in the literature that it will not be), we may well have to revise our scientific consensus. We may be at one of those historical points where scientific research is uncovering data that cannot be adequately explained in the terms of the reigning consensus of what is "scientific." This might serve as a warning to those engaged in the science–religion discussion not to base all their theorizing on a model of physicalism that may be empirically fraying at the edges.

It could also mean that the future may hold a rethinking of what constitutes the physical. This is the stand of those who argue that only an expansion of what constitutes the boundaries of "the physical" can solve the problem of mental causation on physicalist terms. This is the position of the Whiteheadians such as David Griffin and those like Chalmers and Velmans who want to say that consciousness is simply another irreducible dimension of the universe. Another more radical option is to abandon physicalism and the doctrine of the universe as a closed system as metaphysical commitments (Bielfeldt 2001).

Religious people have additional intellectual resources to address the problem of consciousness. That is one of my assumptions here. For example, religious people for whom the system of nature is part of a larger and more encompassing reality need not, and probably should not, absolutize the metaphor of nature as a casually closed system.

In a more encompassing religious framework, proposed solutions to the problem of consciousness that make no sense in a more limited physicalist framework become coherent. For example, Chalmers's (1995) and Velmans's (2000, 2002) proposals that there is an *urgrund*, which subsumes both consciousness and physical reality, makes sense in the context of those religious philosophies that have always affirmed that the physical world has a spiritual dimension or is the expression of a spiritual *grund*.

Or, meditative practices can train the practitioner to experience the ways in which consciousness gives rise to the thoughts and the categories through which we experience the world, including the scientific models we use to study consciousness. Such experiential knowing makes it harder to lose sight of the fact that in all our studies of consciousness we never escape the domain of consciousness.

Consciousness is presupposed in every human method of understanding. It is the final basis of every claim we make. In that sense, it pervades every object we know. In meditation, this theoretical assertion is given experiential validation; we may become aware that central to mind or consciousness (as they are known experientially) is the activity of generating our awareness and the categories that shape that awareness. This insight into the creative power of consciousness and its inseparability from everything we know is a window on a reality beyond the subject–object duality, the radiant *grund* from which the world as we know it springs. Such is the testimony of generations of Buddhist and Christian contemplatives, as well as those from many other traditions (Jones 2003).

Since they can now be placed within an empirically derived framework, values and beliefs and moral choices are acceptable to the nonreductive physicalist. One result of this is a Kantian reduction of religion to ethical behavior (Sperry 1991), or ethical behavior plus belief in a revealed, transcendent God (Murphy 1998). And certainly, many religions, especially monotheistic ones, might subsist on such a primarily ethical model of what religion is. But of course, all religions contain other traditions besides this purely ethical self-definition. Such traditions, which might loosely be called "mystical" or "contemplative," claim that within the depths of human consciousness is a window on the universal and the divine. In all religions, such a claim is presented as a quasi-empirical one—one that can be demonstrated within experience by those willing to undertake the requisite spiritual disciplines (Jones 2003). Those whose religious practice involves the immediate experience of the divine *grund* as well as ethical behavior may require a rather different understanding of human nature than that offered by nonreductive physicalism. Or, to put it differently, nonreductive physicalism, like all strict physicalisms, may provide too narrow a definition of human nature to support the full range and richness of religious practices and experiences.

Notes

In 1992, in *Zygon*, I published an article "Can neuroscience provide a complete account of human nature," which covers some of the same ground as here. This chapter is an expansion, elaboration, and up-dating of some of the points made there.

1. I have described and discussed Sperry's position at length elsewhere (Jones 1992) and will not repeat that discussion here.
2. Clayton maintains that his position is not a form of physicalism. But he affirms a naturalistic monism. So, if consciousness is not physical, it must at least be similar enough to what is physical to be part of the same system. Clayton does not specify what that similarity is. Sperry and Murphy both call their positions forms of physicalism.
3. This question of what constitutes a "property" or an "event" is exceedingly controversial in the philosophy of mind and is far beyond the scope of this chapter. See Heil and Mele (1993).
4. In many ways, my argument here follows that of J. Kim (1998), where he repudiates his earlier advocacy of the position of nonreductive physicalism. Clearly, however, I am using it in the service of a radically different position. And I am arriving at it more from the standpoint of clinical and experimental evidence and less from a strictly logical analysis. Also, after completing this chapter, I came across a paper by Dennis Bielfeldt (1999a), which covers much of the same ground as this chapter. Bielfeldt draws on Kim's work more directly and his concerns are theologically focused on using downward causation to explain divine action rather than to account for research in self-regulation and behavioral medicine.

Kim's (1998) treatment illustrates the way in which the argument about physicalism and mental causation depends upon certain (rather robust) models of causation. This raises the further question of whether such strong (virtually classical) models of causation are compelling. Although he does not directly assert it, Silberstein (2001) implies that contemporary physics offers a rather different model of causation, which might be relevant to the issue at hand. In Jones (1984), in an analysis of the theories David Bohm, I also suggest a more open-textured model of causation. How such a new model of causality might impact our understanding of mental causation and the relationship of consciousness and the brain is way beyond the scope of this chapter. Sufficient to say, with as much caution as possible, that such more current models of causality will probably not produce a view of the physical universe as inimical to a religious vision as did classical models of causation (Jones 1984).

References

Andresen, J. 2000. Meditation meets behavioral medicine: The story of experimental research on meditation. *Journal of Consciousness Studies*, 7, 17–74.
Austin, J. 1998. *Zen and the Brain*. Cambridge, MA: MIT Press.
Barber, J. 1996, *Hypnosis and Suggestion in the Treatment of Pain*. New York: Norton.
Basmajian, J.V. 1983. *Biofeedback—Principles and Practices for Clinicians*. Baltimore: Williams & Wilkins.
Bielfeldt, D. 1999a. Can downward causation make the mental matter? A reply to Meyering and Murphy. *CTNS Bulletin*, 19/4, 11–21.
———. 1999b. Mancey Murphy's nonreductive physicalism. *Zygon*, 34/4, 619–628.
———. 2001. Can Western monotheism avoid substance dualism? *Zygon*, 36/1, 153–177.
Brown, D. and Fromm, E. 1987. *Hypnosis and Behavioral Medicine*. Hillsdale, NJ: Erlbaum.
Brown, W., Murphy, N., and Malony, N. (eds.) 1998. *Whatever Happened to the Soul?* Minneapolis: Fortress Press.
Carlson, L., Speca, M. Petal, K., and Goodey, E. 2003. Minfulness-based stress reduction in relation to quality of life, mood, symptoms of stress, and immune parameters in breast and prostate cancer outpatients. *Psychosomatic Medicine*, 65, 571–581.
Chalmers, D. 1995. Facing up to the problem of consciousness. *Journal of Consciousness Studies*, 2, 200–219.
Clayton, P. 1999. Neuroscience, the person, and God, in R. Russell, N. Murphy, T. Meyering, and M. Arbib (eds.), *Neuroscience and the Person*. Notre Dame: University of Notre Dame Press, pp. 181–214.
Davidson, R., Kabat-Zinn, J., Shumacher, J., Rosenkranz, M., Muller, D., Santorelli, S., Urbanowski, F., Harrington, A., Bonus, K., and Sheridan, J. 2003. Alterations in brain and immune function produced by mindfulness meditation. *Psychosomatic Medicine*, 65, 564–570.
Deikman, A. 2000. A functional approach to mysticism. *Journal of Consciousness Studies*. 7, 75–92.
DuBreuil, S. and Spanos, N. 1993. Psychological treatment of warts, in J. Rhue, S. Lynn, and I. Kirsh (eds.), *Handbook of Clinical Hypnosis*. Washington: American Psychological Association, pp. 623–648.

Eccles, J. 1982. *Mind and Brain*. New York: Paragon.
Feldman, J. 2004. The neurobiology of pain, affect, and hypnosis. *American Journal of Clinical Hypnosis*, 46, 187–200.
Findlay, J. 1952. Goedelian sentences. A non-numerical approach. *Mind,* 51, 259–265.
Goldapple, K., Segal, Z., Garson, C., Lau, M., Bieling, P., Kennedy, S., and Mayberg, H. 2004. Modulation of cortical–limbic pathways in major depression. *Archives of General Psychiatry*, 61, 34–41.
Goleman, D. 2003. *Destructive Emotions*. New York: Bantam.
Green, E. and Green, A. 1977. *Beyond Biofeedback*. New York: Delacorte.
Griffin, D. 1998. *Unsnarling the World Knot: Consciousness, Freedom and the Mind–Body Problem*. Berkeley: University of California Press.
———. 2002. Scientific naturalism, the mind–body relation, and religious experience. *Zygon*, 37, 361–380.
Heil, J. and Mele, A. 1993. *Mental Causation*. Oxford: Clarendon Press.
Hofstadter, D. R. and Dennett, D.C. (eds.) 1981. *The Mind's I*. New York. Basic Books.
Hutto, D. 1998. An ideal solution to the problem of consciousness. *Journal of Consciousness Studies*, 5/3, 328–341.
Jones, J. 1981. *The Texture of Knowledge*. Lanaham: University Press of America.
———. 1984. *The Redemption of Matter*. Lanaham: University Press of America.
———. 1992. Can neuroscience provide a complete account of human nature. *Zygon*, 27, 187–202.
———. 2003. *The Mirror of God: Christian Faith as Spiritual Practice—Lessons from Buddhism and Psychotherapy*. New York: Palgrave.
Journal of Consciousness Studies, 2001, 8/9–10.
Kabat-Zinn, J. 1990. *Full Catastrophe Living*. New York: Delacorte.
Kabat-Zinn, J., Massion, A.O., Kristeller, J., Peterson L.G., Fletcher, K.E., Pbert, L., Lenderking, W.R., and Santorelli, S.F. 1992. Effectiveness of a meditation-based stress reduction program in the treatment of anxiety disorders. *American Journal of Psychology*, 149, 936–943.
Kihlstrom, J. 2002. The seductions of materialism and the pleasures of dualism. *Journal of Conscious Studies*, 9, 30–34.
Kim, J. 1998. *Mind in a Physical World*. Cambridge, MA: MIT Press.
Kosslyn, S., Alpert, N., Thompson, S., Weise, C., Chambris, S., Hamilton, S., Rauch, S., and Buonanno, F. 1993. Visual mental imagery activates topographically organized visual cortex. *Journal of Cognitive Neuroscience*, 5, 263–287.
Kristeller, J.L. and Hallett, B. 1999. Effects of a meditation-based intervention in the treatment of binge eating. *Journal of Health Psychology*, 4(3), 357–363.
Libet, B. 1967. Responses of human somatosensory cortex to stimuli below threshold for conscious experience. *Science*, 158, 1597–1600.
———. 1978. Neuronal versus subjective timing for a conscious sensory experience, in P.A. Buser and A. Rougeul-Buser (eds.), *Cerebral Correlates of Conscious Experience*. Amsterdam: North Holland Press.
———. 1982. Subjective and neuronal time factors in conscious sensory experience and their implications for the mind–Brain relationship, in J. Eccles (ed.), *Mind and Brain*. New York: Paragon, pp. 99–102.
———. 1996. Neural processes in conscious experience, in M. Velmans, (ed.), *The Science of Consciousness*. London: Routledge, pp. 96–117.
Linehan, M. 1993. *Cognitive-behavioral treatment of borderline personality disorder*. New York: Guilford.
Lucas, J. 1964. Minds, machines, and goedel, in J. Anderson (ed.), *Minds and Machines*. Englewood Cliffs, NJ: Prentice Hall.
Marlett, G.A. and Kristeller, J.L. 1999. Mindfulness and meditation, in W.R. Miller (ed.), *Integrating Spirituality in Treatment*. American Psychological Association Books, pp. 67–84.
McGinn, C. 1989. Can we solve the mind–body problem? *Mind*, 88, 349–366.
Meyering, T. 1999, Mind matters, in R. Russell, N. Murphy, T. Meyering, and M. Arbib (eds.), *Neuroscience and the Person*. Notre Dame: University of Notre Dame Press, pp. 165–180.
Morowitz, H.J. 1981. Rediscovering the mind, in D.R. Hofstader and D.C. Dennett (eds.), *The Mind's I*. New York: Basic Books, pp. 34–42.
Murphy, N. 1998. Non-reductive physicalism: Philosophical issues, in W. Brown, and N. Murphy, and N. Malony (eds.), *Whatever Happened to the Soul?* Minneapolis: Fortress Press, pp. 127–148.

Murphy, N. 1999a. Supervenience and the downward efficacy of the mental: A nonreductive physicalist account of human action, in R. Russell, N. Murphy, T. Meyering, and M. Arbib (eds.), *Neuroscience and the Person*. Notre Dame: University of Notre Dame Press, pp. 147–164.

———. 1999b. Downward causation and why the mental matters. *CTNS Bulletin*, 19/1, 13–21.

———. 1999c. Physicalism without reductionism: Toward a scientifically, philosophically, and theologically sound portrait of human nature. *Zygon*, 34/4, 551–571.

Murphy, N and Ellis, G.F. 1996. *On the Moral Nature of the Universe*.

Nagel, T. 1974. What is it like to be a bat? *Philosophical Review*, 83, 435–450.

Penfield, W. 1975. *The Mystery of the Mind*. Princeton: Princeton University Press.

Popper, K. and Eccles, J. 1977. *The Self and its Brain*. London: Routledge and Kegan, Paul.

Poulton, E.P. 1973. *Critique of the Psycho-Physical Identity Theory*. The Hague: Mouton.

Robinson, D.M. 1982. Some thoughts on the matter of the mind/body problem, in J. Eccles (ed.), *Mind and Brain*. New York. Paragon, pp. 197–206.

Robinson, H.M. 1976. The mind–body problem in contemporary philosophy. *Zygon*, 11, 346–60.

Rurzyla-Smith, P., Barabasz, A., Barabasz, M., and Warner, D. 1995. Effects of hypnosis on immune response. *American Journal of Clinical Hypnosis*, 38, 71–79.

Russell, R., Murphy, N., Meyering, T., and Arbib, M. (eds.) 1999. *Neuroscience and the Person*. Berkeley, CA: Center for Theology and the Natural Sciences.

Schwartz, G. and Beatty, J. 1977. *Biofeedback Theory and Research*. New York: Academic Press.

Schwartz, J. 1999. A role for volition in the generation of new brain circuitry. *Journal of Consciousness Studies*, 6, 115–142.

Scott, A. 2004. Reductionism revisited. *Journal of Consciousness Studies*, 11, 51–68.

Segal, Z., Williams, J., and Teasdale, J. 2002. *Mindfulness-Based Cognitive Therapy for Depression*. New York: Guilford.

Shapiro, S., Schwartz, G., and Bonner, G. 1998. Effects of mindfulness-based stress reduction on medical and premedical students. *Journal of Behavioral Medicine*, 21, 581–599.

Sheikh, A., Kunzendorf, R., and and Sheikh, K. 1996. Somatic consequences, of consciousness, in M. Velmans (ed.), *The Science of Consciousness*. London: Routledge, pp. 140–180.

Silberstein, M. 2001. Converging on emergence: Consciousness, causation, and explanation. *Journal of Consciousness Studies*, 8, 61–98.

Sperry, R.W. 1991. Search for beliefs to live by consistent with science. *Zygon*, 26, 237–358.

Toulmin, S. 1960. *The Philosophy of Science*. New York: Harper & Row.

Van Gulick, R. 1993. Who's in charge here? Who's doing all the work? in J. Heil and A. Mele (eds.), *Mental Causation*. Oxford: Clarendon Press, pp. 223–256.

Velmans, M. (ed.) 1996. *The Science of Consciousness*. London: Routledge.

Velmans, M. 2000. *Understanding Consciousness*. London: Routledge.

———. 2002. How could conscious experiences effect brains? *Journal of Conscious Studies*, 9, 3–29.

Watkins, J. 1982. A basic difficulty in the mind–brain identity hypothesis, in J. Eccles (ed.), *Mind and Brain*. New York: Paragon, pp. 221–232.

Wood, G. Bughi, S. Morrison, J. Tanavoli, S. and Zadeh, H. 2003. Hypnosis, differential expression cytokines by T-cell subsets, and the hypothalamo-pituitary axis. *American Journal of Clinical Hypnosis*, 45, 179–193.

Woody, E. and Szechtman, H. 2000. Hypnotic hallucinations: Towards a biology of epistemology. *Contemporary Hypnosis*, 17, 4–14.

CHAPTER THREE

Psychoneurological Dimensions of Anomalous Experience in Relation to Religious Belief and Spiritual Practice

STANLEY KRIPPNER

For several years, I have been interested in the Brazilian churches that use a particular mind-altering brew as a sacrament. This brew is referred to as "ayahuasca," "yage," "hoasca," etc., depending on the part of the country in which one encounters it. I am often invited to participate in evening ceremonies by the church elders, and one visit was especially memorable.

A close friend Chris Ryan and I were on our way to an "ayahuasca church" located far from the metropolis where we had been staying. As we started on a dirt road, at about 10 kilometers per hour, passing small farms and energetic chickens, I was attempting to speak Portuguese with our hosts. Chris, not conversant in Portuguese, was looking out of the window. Suddenly, in his words, "I saw a beautiful woman walking along the side of the road in the other direction. She was wearing a green *sari* and a *bindi*, or ornamental mark, on her forehead, carefully carrying a shallow bowl which I assumed was filled with soup or some other food." Chris was sitting in the right rear seat; I was sitting directly in front of him but when he mentioned the lovely lady a few moments later I confessed that I had not noticed her.

We arrived at the church and were given a tour of the facilities. An hour or so later, entering the room where the ayahuasca ceremony would be held, Chris noticed a young woman in a green *sari*, and wearing a *bindi*, sitting in one of the front rows. He whispered to me, "That is the woman I saw on the road." I agreed with him that she was exceptionally lovely. Later, we introduced ourselves and Chris arranged to see her the following day. Her name was Rasilia; she had been born in India, and had been raised in both the United States and Brazil.

Toward the end of their final conversation, Chris remarked, "I knew when I first saw you in such a strange context that you must be very special." Rasilia paused for a moment, then replied, "I wouldn't normally tell anyone this but I think I know you well enough to know it won't freak you out too much. I was never out on the road that day. I was playing with the children behind the house and wouldn't have had any reason to go out on the road. I never leave the center when I go there for a ceremony." Chris' introduction to ayahuasca had been a dramatic event. But his encounter with Rasilia remained the highlight not only of his evening but also of his entire stay in Brazil.

Anomalous Experiences

This type of experience is often referred to as "anomalous," an English language word that derives from the Greek *anomalos*, meaning irregular, uneven, or unequal, in contrast to *homalos* meaning the same or common. Hence, an "anomalous experience" is irregular in that it differs from common experiences. It is uneven in that it is not the same as experiences that are even and ordinary. It is typically unequal in that it lacks the power to access the same attention given to regular experiences. Hence, an anomalous experience is uncommon and/or deviates from a dominant explanatory paradigm.

When two of my colleagues and I edited a book on anomalous experiences for the American Psychological Association, we used synesthesia (e.g., the experience of "hearing colors" or "tasting sounds") as an example of an uncommon experience that can be explained by a variety of mainstream psychoneurological models, and telepathy (i.e., the so-called mind-to-mind communication) as an example of a fairly common experience that deviates from the usually accepted explanations of reality (Cardeña et al. 2000, p. 4). In the words of the sociologist Marcello Truzzi, anomalous phenomena "contradict commonsense or institutionalized (scientific or religious) knowledge"; they are "anomalous to our generally accepted cultural storehouse of truths" (1971, p. 367). However, we did not assume that these experiences were necessarily symptomatic of mental or emotional disorders or that, when better understood, they would violate the paradigms of scientific psychology.

This perspective is in the tradition of the great American psychologist William James who explored anomalous experiences and wrote about them, most notably in his book *The Varieties of Religious Experience* (1902/1958). His friend Theodore Flournoy (1900/1994), a psychology professor at the University of Geneva, wrote an in-depth case study of a purported "medium" who spoke in different voices, wrote in different handwriting styles, and used different names. Rather than positing deception or accepting the medium's claim of contact with the "spirit world," Flournoy produced a sophisticated interpretation of the psychodynamic foundations of the imaginary languages involved. A friend of Flournoy's, Carl G. Jung (1902/1970), conducted a landmark study with another medium, using a word-association test he had developed to trace the origins of the names she gave him, not only of her own alleged spirit guides but of the "spiritual forces" that supposedly guide the universe. Jung terminated his work when the medium's performances took on fraudulent aspects. Both mediums, incidentally, reported visits to the planet Mars.

Notwithstanding the presence of anomalous experiences in case studies of disturbed individuals over the years, surveys of people reporting these events have found little relationship to any obvious form of psychopathology (e.g., Greeley 1975; Spanos et al. 1993). The attribution of personal meaning to anomalous experiences has been addressed by such writers as the sociologist James McClenon (1994b) who used the term "wondrous events" (suggesting that they stimulated the development of religious ideologies), the psychologist Daniel Helminiak (1984) who called them "extraordinary experiences" (focusing on whether they further the experient's "authentic growth"), and the parapsychologist Rhea A. White (1995) who referred to them as "exceptional human experiences" (noting their "transformational" potential in people's lives). McClenon's, Helminiak's, and White's phrases could be subsumed under the term "anomalous experiences" as they include conversion and "born again" experiences, claims of "kundalini energy," fire walking, and glossolalia ("speaking in tongues"), and memories of "former lives," "near-death" experiences (NDEs), "out-of-body" experiences (OBEs), "possession" experiences, and "visionary" experiences. McClenon's reviews of the altered states of consciousness literature, as well as cross-cultural surveys of anomalous experience, have persuaded him that such traits as absorption, dissociation, fantasy-proneness, and hypnotic susceptibility need to be added to the list. He considers all of these "normal human capacities which have not been thoroughly studied in non-clinical populations" (1994a, p. 129).

The Russian psychologist Boris Bratus (1988/1990) used the term "anomalies of personality" to describe personality characteristics and behaviors that deviate from a cultural norm. Rejecting the term "abnormal" and objecting to the use of either statistical or psychoanalytic criteria to determine what is "normal," Bratus replaced dehumanizing terminology and conceptualizations with a focus on uncommon activity. These activities often involve creativity, values, and quests for meaning, which, although anomalous within a given culture, appear to be adaptive for those individuals under question.

I agree with Bratus that what is anomalous in one culture may not be anomalous in another, and would add that what is anomalous within one paradigm need not be anomalous within another. Summarizing a number of surveys conducted in the United States, W.L. MacDonald concluded that age, education, gender, race, religion, and socioeconomic status influence the likelihood of reporting anomalous experiences, attributing the differences to different "shaping of individual realities." He conjectured that "the reality of human experience is socially constructed and is therefore subject to variation depending on the social context" (1994, p. 36).

Mystical Experiences

The claims of mystics to have an intuitive sense of reality that belies everyday assumptions have contributed to the origin of most religions and have, directly or indirectly, touched the lives of most of humanity (Cardeña et al. 2000, p. 15). The term mystic derives from the Latin word *mysticus* or mysteries and from the Greek *mystikos* from *mystes* or to initiate. David Wulff's scholarly assessment of these phenomena indicates that they are rare and fleeting, yet often stand out as defining moments in the lives of the experients (2000, pp. 397–398).

William James described several examples of these types of experiences and found that they shared several characteristics: ineffability, transiency, passivity, and a "noetic quality," that is, a type of knowledge unknown to the experient's discursive intellect (1902/1958, pp. 302–303). The philosopher W.T. Stace described the "core" of mystical experience (whether Christian, Judaic, Islamic, Buddhist, Hindu, Taoist, etc.) as the emergence of a "unitary" consciousness located neither in space nor in time, one marked by peace, bliss, joy, blessedness, and the feeling of having encountered something "divine." Stace also included "ineffability" in his list of characteristics, although he was careful to use the term "alleged." In addition to this "introvertive" mystical experience, Stace also described an "extraverted" experience, one that focuses on the external world and finds unity and life "in all things" (1960, p. 131).

Robert Masters and Jean Houston, with their backgrounds in both psychology and philosophy, differentiated between "mystical" and "religious" experiences, the former involving "fusion" and "union," and the latter marked by "confrontation" and "encounter" with God, the Ground of Being, or some other "Fundamental Reality" (1973, p. 100). For Masters and Houston, the aftereffects of these experiences were critically important, and they attempted to follow-up their research participants. In one instance, a theologian claimed to experience an ecstatic vision of unity while swinging in one of Masters and Houston's devices; he reported that a long-standing writer's block dissolved, that his teaching ability was enhanced, and that his relations with family members improved.

Perhaps the most systematic empirical study of a mystical experience's aftereffects was undertaken by the psychiatrist Walter Pahnke (1966), who enlisted 20 Protestant seminary students as his research participants. Pahnke gave half the group the psychedelic drug psilocybin and the other half a mild stimulant (a type of vitamin B) with no psychedelic properties; together, the seminarians listened to a radio broadcast of a Good Friday church service. At day's end, all participants completed a 147-item questionnaire based on Stace's descriptive elements of core mystical experience; those who had ingested psilocybin reported experiences significantly closer to Stace's description than those who received the vitamin concoction. Six months later, all 20 of the participants completed a questionnaire that focused on subsequent changes in their lives; in addition to confirming the mystical elements in their experience, the psilocybin group reported significantly more positive changes in their attitudes and behaviors. Seven participants of the psilocybin group and 9 members of the vitamin group were located a quarter of a century later. Responses to the same questionnaire indicated that the experience was one of the high points of their spiritual lives for the psilocybin group with persisting personal changes that were higher than those reported by the vitamin group. Specifically, the psilocybin participants reported that the experience had deepened their faith, broadened their understanding of Christianity, increased their identification with minorities, women, and the environment, heightened their appreciation of beauty, and had reduced their fear of death (Doblin 1991).

Mainstream neuroscience suspects that mystic experiences are correlated with certain brain activities, noting that several electroencephalographic (EEG) studies of meditation, a frequent precursor of mystical experiences, show shifts of neural activity while participants are meditating (see Murphy and Donovan 1996, for a summary of these studies). However, mystical experiences are not as easily generated as meditative states; hence, the neurological markers for the latter cannot be equated with those for the former (Wulff 2000, p. 405). In addition, many meditative studies have been conducted by partisan groups of investigators and need to be repeated by researchers without covert agendas (d'Aquili 1993, p. 260).

The initial clues as to the possible neural correlates of mystical experience came from research with epileptics, who often report spiritual experiences (Ramachandran and Blakeslee 1998; Wulff 2000, p. 407). Michael Persinger (1987a), a Canadian neuroscientist, postulated a continuum of temporal lobe lability representing the varying degrees to which individuals are predisposed to experience momentary foci of electrical activity, which he felt yielded mystical and related experiences. Makarec and Persinger (1985) found that the number of EEG spikes in the brain's temporal lobe, responding to rhythmic sounds and pulsating light, corresponded to measures of mystical experience and associated phenomena.

On the basis of several related studies, Persinger (1987a) distinguished between "god experiences" (emotionally charged transient phenomena typically characterized by a sense of profound meaningfulness, peacefulness, and serenity), "god concepts" (linguistic culturally shaped descriptions of god experiences), and "god beliefs" (a combination of god experiences with the culturally influenced god concepts). "Mystics" often harbor a range of god concepts that do not share the god concepts of their culture (Albright 2000), while "religious zealots" have shaped their god experiences into conventional cultural god concepts, strongly identifying with the latter. Many, if not most, religious "believers" have not had god experiences but subscribe to their cultural, subcultural, or familial god concepts; "atheists" and "agnostics" generally lack god experiences and do not subscribe to either conventional or unconventional god concepts.

Persinger notes that a singular god experience can change the life of an individual; a collection of these experiences can "form the dynamic core of a religious movement" (1987a, p. ix). Paul of Tarsus reported that the vision of Jesus on the road to Damascus stopped his persecution of Christians and inaugurated a series of missionary sojourns that, with their follow-up letters, shaped the new religious movement. Mohammed's reported "night journeys" produced insights that were transcribed by assistants, producing not only the *Holy Quran* but also a religious and political ideology, which spread throughout the world.

Over the centuries, there have been Christian and Muslim mystics (e.g., St. Francis of Assisi, various Sufi saints) whose phenomena (e.g., St. Francis' stigmata, Sufi dancing and whirling) did not represent mainstream Christianity and Islam, even leading to initial rejection and persecution. However, mainstream Christian and Islamic zealots have carried the teachings of their religions to every corner of the world, often establishing schools, hospitals, and social service centers along the way. At the same time, a few zealots have murdered "disbelievers" and "infidels," burned down birth control clinics, and have fought "holy wars" in the name of their faith.

Religious zealots have not been typical participants in psychoneurological studies, but Andrew Newberg and Eugene d'Aquili (2002) have presented data indicating that both Buddhist meditators and Franciscan nuns could, on occasion, evoke states of consciousness approximating god experiences through meditation, prayer, and/or ritual. Using single photon emission computed tomography (SPECT), they found that (when compared to the baseline waking record) specific regions of their participants' brains showed increased blood flow, while other regions showed marked decreases during a participant's experiential report. It is not unreasonable to suggest that these meditators and nuns permanently altered their brains in such a way to predispose themselves to god experiences through a neurological process referred to as "kindling" (Ramachandran and Blakeslee 1998, p. 180).

However, a group of experients identified by Persinger (1987a) and other investigators (e.g., Geschwind 1983; Mandel 1980) had unexpected god experiences without the practice of contemplation. Instead, Persinger suggests that their experiences were brought about not only by epileptic-type tendencies but by drugs, unusual diets, grief, fatigue, sensory deprivation, sensory overstimulation (e.g., piercing music), and various personal dilemmas and stressors, all of which seem to be associated with temporal lobe lability.

Rhawn Joseph (2002) asks whether one's neural network could be "tweaked" to provide a mystical experience for volunteer participants, and Persinger has designed such a program involving the controlled stimulation of neural circuitry through the external application of magnetic field patterns. In one of his reports, he stated that "with a single burst in the temporal lobe, people find structure and meaning in seconds" (1987a, p. 17). And if artificially generated magnetic field patterns can evoke mystical experiences, could naturally occurring patterns do the same? Derr and Persinger (1989) investigated the recurring appearance of apparitions over a church in Zeitoun, Egypt, discovering that the phenomenon corresponded to an unprecedented increase in seismic activity, and that the content of the apparitions usually reflected the religious background of the experients. Hearing of this study, one writer (J. Hitt) asked, "Might it surprise anyone to learn, in view of Persinger's theories, that when Joseph Smith was visited by the angel Moroni before founding Mormonism, and when Charles Taze Russell started the Jehovah's Witnesses, powerful Leonid meteor showers were occurring?" (1999, p. 11).

Long before the use of contemporary technology to evoke mystical experiences, indigenous cultural groups had developed a repertoire of drugs, chants, rhythmic music and movement, self-regulatory practices, and voluntary shifts in attentional states to produce what Persinger would call god experiences (e.g., Winkelman 1986). Although some traditions advocate celibacy as a route to higher consciousness, another venerable vehicle for attaining mystical insights has been sexual experience. Jenny Wade (2004), a transpersonal psychologist, interviewed nearly 100 people with various sexual orientations and from several religious backgrounds, constructing a typology of mystical, spiritual, or transcendent experiences during sex. Her typology was based on a model developed by the psychiatrist Stanislav Grof (1980, 1998), who analyzed reports from thousands of clients participating in LSD-facilitated psychotherapy and/or so-called holotropic breathwork. Wade reported that her research participants' transcendent experiences resembled those of Grof's group, especially in such areas as "ecstatic union" and the movement of kundalini energy.

Indeed, Newberg and d'Aquili posited that romantic love, as characterized by the phrase, "It's bigger than both of us," may be a transitional phase between aesthetic and religious experience, a signpost on the way to a state they call "absolute unitary being," a state in which the boundaries of entities within the world disappear and the self–other dichotomy is obliterated. It is equated with the "void," with "nirvana," and/or "the experience of God" (2001, p. 236).

For Newberg and d'Aquili, all spiritual and mystical experiences, at least those having a strong emotional component, are located somewhere along an aesthetic–religious continuum. Spiritual and mystical experiences share several characteristics: a progressive increase of unity over diversity, a progressive sense of transcendence or otherworldliness, progressive incorporation of an "observing Self" in the experience, and a progressive increase of certainty in the objective existence of what was experienced in the state (p. 243).

This body of research is sometimes referred to as "neurotheology" (e.g., Joseph 2002). However, I think this term is a misnomer. "Theology" is properly defined as the methodological study of the nature and properties of alleged deities (or of a single deity). Cognitive and affective neuroscience have not provided, and have not attempted to provide, this type of knowledge. The studies just cited have focused on peoples' experiential reports, the psychoneurological correlates of these reports, the way these reports can be categorized, and the manner in which these experiences might be evoked (through contemplation, drugs, electrical stimulation, and the like). No claim can be made that these correlates indicate causation; it would be both reductionistic and premature to claim that brain activity "causes" a god experience. The question can be asked, "Is a naturally occurring god experience more valid than an induced experience?" Such a question assumes that the "validity" of "naturally occurring" god experiences has been established. However, investigators have not even agreed on the questions that would need to be asked to establish such a validity.

On the other hand, there are enough data available to investigate the application of these findings. Evoked god experiences could address existential issues such as finding meaning in life. A review of one's god experiences could provide a means of establishing one's "personal mythology" or "personal theology." Evoked god experiences have already been used with terminal cancer patients (e.g., LSD psychotherapy, see Grof 1980). Finally, it is commendable that science, in this case the neurosciences, is paying attention to mystical experiences. When we compiled our book, *Varieties of Anomalous Experience*, it was not without considerable reflection that my coeditors and I referred to them as "what may be the most influential of all anomalous experiences" (Cardeña et al. 2000, p. 15).

Psi-Related Experiences

If mystical experiences have been the most influential of what Western science considers anomalous experiences, psi-related experiences are probably the least influential. Many Americans claim to have had psi-related experiences (Greeley 1975), yet these experiences generally are seen as curiosities at best, and delusions at worst. They have not spawned major schools of psychotherapy, nor are they given favorable (or even prominent) mention in most psychology texts and courses. As the outrageous comedian Rodney Dangerfield might complain, "Parapsychology don't get no respect!"

While psychology can be defined as the scientific study of behavior and experience, parapsychology (or "psi research") studies those reported anomalies of behavior and experience that appear to stand outside of the currently known explanatory mechanisms that account for organism–environment and organism–organism information and influence flow. Since the end of the nineteenth century, a few valiant investigators have attempted to understand these reports and to determine whether they are worthy of continued attention and investigation (see Rao and Palmer 1987, for a review). The Parapsychological Association (1987), an international body of scientific investigators in the field, has emphasized that "a commitment to the study of psi phenomena does not require assuming the reality of 'non-ordinary' factors or processes." Despite this cautionary statement, parapsychology has been referred to by some critics as a "pseudoscience" (Stanovich 1985) or even a "deviant science" (Ben-Yehuda 1985).

To understand this criticism, it must be recalled that Western science emerged from philosophy and originally proclaimed itself as the search for the understanding of nature. As this quest became more disciplined, greater demands were placed upon scientific undertakings. Today, there is a demand by some critics that psi research produce "replicable" psi experiments and "battle-tested" results before it can be considered a legitimate science. At one level of investigation, there already are "replications" and battle-tested results (see Utts 1991), specifically, the finding that about half of an unselected group will report having had a "psychic experience," supposedly involving those psi phenomena that have been given such labels as "telepathy" (purported anomalous "mind-to-mind" communication), "clairvoyance" (purported anomalous knowledge of a distant, moving, or hidden object or event), "precognition" (purported anomalous knowledge of a future event), and "psychokinesis" (purported anomalous mental influence on a distant, moving, or hidden object). This percentage may vary from one culture, age group, and educational level to the next, but it has been repeated, in one study after another, for the last several decades (see Stokes 1997).

These experiences have been ignored or ridiculed by most behavioral and social scientists. Only occasionally does a representative of mainstream psychology such as Andrew Neher (1990) suggest that this type of experience "is not only potentially significant for our personal lives," but that it also "serves important functions in our society as a whole." Subjective psi experiences (or "psi-related experiences," i.e., those not occurring in controlled laboratory or under "psi task" conditions) appear to interface with heightened sensitivities, creative imagery, self-regulation of body processes, and increased memory (for reviews, see Broughton 1991; Radin 1997; Targ et al. 2000). Therefore, the case for the continuation of psi research is simply that an understanding of these reported experiences is worthy of a disciplined and sustained research effort in order for the social and behavioral sciences to encompass the full dimensions of humanity's potentials.

The Society for Psychical Research, founded in Great Britain in 1882, was the first major organization to attempt to assess psi-related experiences scientifically, beginning with surveys that would later evolve into controlled experiments. The "Report on the Census of Hallucinations," organized by members of the society, analyzed and categorized some 17,000 responses to the question, "Have you ever . . . had a vivid impression of seeing, or being touched . . . , or of hearing a voice; which impression, so far as you could discover, was not due to any external cause?" Affirmative answers were obtained from about 1 in 10 of the respondents, with more visual hallucinations reported than auditory or tactile hallucinations (Sidgwick et al. 1894).

In contrast to questionnaires and surveys, the use of case studies has permitted in-depth examinations of psi-related experiences. The best known of these collections is that of Louisa Rhine (1977); by 1973, her collection of spontaneous cases of alleged psi numbered 12,837. The overall objective of Rhine's studies was to understand what appeared to be the basic psi processes; for example, she identified the main forms in which psi-related experiences were reported as daytime hallucinations, daytime intuitions, realistic dreams, and nonrealistic dreams (i.e., those marked by symbolism and fantasy). The psychiatrist Ian Stevenson (1970) has presented an intensive study of 35 cases of "telepathic impressions," concluding that they shared the same characteristics as did those collected in earlier decades. For example, an interpersonal relationship usually appeared to be necessary for such experiences to occur.

Telepathy, Clairvoyance, and Precognition in Dreams

Most contemporary parapsychologists would agree with their critics that surveys, questionnaires, and case studies are subject to such confounding variables as coincidence, unconscious inference, sensory leakage, embellishment after the fact, falsification of memory, and outright fabrication. However, these anecdotal reports, especially the cases collected by Louisa Rhine, stimulated the psychoanalyst Montague Ullman and me to investigate anomalous dreams in a laboratory setting. Over the years, we obtained results suggesting that some subjects, by means of putative telepathy or clairvoyance, could incorporate pictorial material (unknown to them) that had been randomly selected into their dreams. One meta-analysis of our work yielded data indicating that there was only one chance in several hundred thousand that the results were coincidental (Child 1985). For example, one research participant dreamed about going to Madison Square Garden to buy tickets to a boxing match. On that same night, a psychologist in a distant room had been focusing his attention on a postcard-size reproduction of a boxing match. The postcard was in a sealed envelope and was not randomly selected (from a collection of several postcard-size art prints) until the research participant had retired. His brain waves and eye movements were monitored on an EEG and he was awakened whenever it seemed that he might have been dreaming. The dream reports were tape recorded, transcribed, and evaluated by judges who had not been present during the experimental session.

These experiments were conducted at Maimonides Medical Center in Brooklyn, New York, during the 1960s and 1970s. Our laboratory's basic research procedure was to fasten electrodes to the head of a subject, take him or her to a soundproof room, then randomly select a sealed, opaque envelope containing a postcard-size art print. A psychologist would then take the envelope to a distant room, open it, and study the art print while the sleeping research participant attempted to incorporate material from the art print into his or her dreams. Upon completion of an experimental study, outside judges compared the typed dream reports with the total collection of art prints, attempting to identify the print used on the night of each experiment (e.g., Krippner and Ullman, 1970). Research participants would make their comparisons immediately after awakening, and their results would be recorded by staff members unaware of the actual target picture that had been used. About two out of three times, these experimental sessions produced correct "matches," which, upon further analysis, generally provided statistically significant results.

Such prominent dream researchers as David Foulkes (Belvedere and Foulkes 1971), Gordon Globus (Globus et al. 1968), Calvin Hall (1967), Robert Van de Castle (1971), Keith Hearne (1987), and Kathy Dalton (Dalton et al. 1999) have attempted to repeat these findings. Because the replication rate from these other laboratories was inconsistent, our Maimonides team did not claim to have conclusively demonstrated that communication in dreams is capable of transcending space and time. However, we did open a promising line of investigation (see Sherwood and Roe 2003, for an overview of all the published experiments), and demolished the assumption that psi-related experiences in dreams are an inevitable symptom of mental disorder.

Years later, Michael Persinger and I reviewed the entire body of dream research data from Maimonides Medical Center, investigating possible links between putative psi effects in dreams and geomagnetic activity. Persinger told me that what is called the "geomagnetic field" has several components. The Earth itself creates the main component, as if a huge bar magnet were running through the Earth's core. Regular daily and monthly variations occur. These variations are due to several factors. Weather affects the daily or diurnal variations. Lunar changes affect the monthly variations. Major variations occur due to sunspot activity, as well. Changes in the geomagnetic field can be sudden and unpredictable. The best-known example of charged particles from the sun interacting with the Earth's magnetic field is the aurora borealis, often called the "Northern Lights."

Persinger had conducted an analysis of many spontaneous cases of telepathy and clairvoyance. He found that these psi-related experiences were more likely to occur when the global geomagnetic activity was significantly quieter than the days before or after the reported experience. A day of low amplitude with slow, predictable variations is referred to as a quiet magnetic day. These were the days that were associated with reports of telepathy and clairvoyance at statistically significant levels (Persinger 1987b).

A day of sudden and large amplitude changes is referred to as a magnetically stormy day. Persinger (1989) reported a tendency for reports of poltergeist (so-called noisy ghost) and haunting experiences to occur on these days. One possible component for these experiences is that variety of psi referred to as psychokinesis, which has been studied in the laboratory under psi task conditions. An analysis of some of these experiments has indicated a tendency for them to occur most frequently on magnetically stormy days (see Braud and Dennis 1989).

When I suggested that we analyze data from the Maimonides laboratory, Persinger suggested two hypotheses:

1. Nights on which psi was strong would also be nights that displayed the quietest geomagnetic activity compared to the days before and after.
2. Nights on which psi was weak or absent would not demonstrate this effect.

Persinger and I tested these hypotheses in two ways. First, we examined the initial night that each of the 62 research participants in telepathic and clairvoyance dream experiments spent at our laboratory. For our analysis, we used the results of the matchings made by the research participants themselves. We classified the matches as "High Hits," "Low Hits," "High Misses," and "Low Misses." Geomagnetic measures for the Northern Hemisphere were determined for each night in the study. There were too few "Misses" to yield data adequate for analysis. However, a significant difference was observed between High Hits and Low Hits. High Hits were more likely to occur on quiet magnetic days when there were few electrical storms and sunspots (Persinger and Krippner 1989).

Second, we tested these hypotheses with the matches made by a single research participant named William Erwin, a psychoanalyst who had spent 20 separate, nonconsecutive nights at our laboratory. We assumed that using matches from a single subject would increase the detection of a geomagnetic effect. It would eliminate individual differences, and these were the largest source of variance in these studies.

The typical procedure followed by Erwin was for him to arrive at the laboratory in time to interact with the "transmitter." The transmitter was the person who would spend much of the night looking at the target picture, a postcard of an art print randomly selected after Erwin had gone to bed. The transmitter was isolated from Erwin and spent the night in a distant room. After electrodes were attached to Erwin's head, he parted company with the transmitter and entered a soundproof room.

Two experimenters took turns watching Erwin's brain waves and eye movements on an EEG machine. Near the end of each period of rapid eye movement sleep, Erwin was awakened and asked to describe the dream content that he remembered. His remarks were tape recorded as was a morning interview in which he gave associations to his dream report. Neither Erwin nor the experimenters knew the identity of the target picture.

The tape-recorded remarks were typed and sent to three judges; Erwin also matched his own dreams to the target pictures when the experiment ended. Ten of the nights were High Hits, while the remaining 10 fell outside of this range. One of the High Hits was obtained when the target picture was "School of the Dance" by the French painter Degas. The painting portrays several girls in white ballet costumes in a dance studio. Erwin had had one dream about "being in a class. The instructor was young. She was attractive." A later dream report noted, "There was one little girl who was trying to dance with me" (Ullman et al. 1989).

Using the time period when Erwin was asleep, we found a significant positive correlation between geomagnetic activity and Erwin's matches. The strongest correlations between the score and the geomagnetic activity occurred during the time when most of the dream reports were collected, that is, during the latter part of the night (Krippner and Persinger 1996).

Telepathy and clairvoyance are examples of psi that involve minimal time displacement between the event and the experience. However, precognition is an example of psi that involves significant time displacement between the experience and the event. Alan Vaughan was one of the "sensitives" who obtained many High Hits in our dream studies. Vaughan had been recording his dreams since 1968 when he participated in a parapsychological study focusing on precognition, paying special attention to those dreams that contained what he considered a detailed, literal correspondence to a future event. Most of these dreams, in retrospect, contained what Vaughan considered to be three or more exact details about the event.

Vaughan sent the physicist James Spottiswoode the dates of 61 dreams that he thought were precognitive. Spottiswoode compared the geomagnetic activity on the nights of these dreams with activity 10 days before and 10 days after. There was significantly less geomagnetic activity on the nights of the precognitive dreams than either during the 10 days before or the 10 days after (Krippner et al. 2000). One of these dreams took place when Vaughan was living in Germany. He described the dream to me in a letter, which I received on June 4, 1968. The dream contained many frightening episodes involving the murder of Robert Kennedy. At that time, Kennedy was trying to obtain a nomination for the presidency of the United States. On June 6, Senator Kennedy was assassinated (Ullman et al. 1989, p. 145).

As in several earlier geomagnetic studies of self-reported precognitive dreams, such as those sent in by magazine readers and published with the date of the dream, Vaughan's dreams were not collected under ideal psi task conditions. As a result, they are only suggestive of an association with geomagnetic activity. Also, it should be noted that the procedures used by Persinger and Spottiswoode were somewhat different, and thus, not exactly comparable. However, the association between ostensible psi and geomagnetic field data is strong enough to justify further research under more rigorous conditions.

Persinger (1974) has urged using reported psi phenomena in new and ingenious ways, observing, "Across cultures and throughout history people have been reporting psi-experiences. Let us find out *what they are saying* . . . . It is by looking at the similarities of the verbal behavior that we may find enough consistencies to understand the factors responsible for the *reports*" (p. 13). Persinger (see Schaut and Persinger 1985) has examined several collections of spontaneous cases, including the 35 gathered by Stevenson (1970), reporting that they seemed to occur most frequently when geomagnetic activity was calmer than on the days before or after the experience—and lower than the month's average activity. This approach can be applied to any collection of cases where the date of the alleged experience has been recorded. If repeatable, these effects may help to provide an understanding of the mechanisms underlying psi phenomena, and may even indicate a potentially predictable pattern for such events. In fact, these geomagnetic field perturbations have been reported by other investigators to affect biological systems (e.g., Subrahmanyam et al. 1985).

Psi-Related Experiences and the Brain

A brain-based model of psi experience was proposed in the 1880s by several of the founders of the British Society for Psychical Research (Gurney et al.1886). They proposed that these experiences are constructed by the brain from its own resources (or what might be called "reserve capacities") once the initiating stimulus (e.g., a randomly selected picture in a laboratory test, a real-life crisis situation involving a loved one) sets the constructive mechanisms in motion. This model is surprisingly congruent with some concepts of memory hypothesized a century later. One model holds that memory is "stored" in the brain's cortex, specific memories are evoked by either external stimulation or the lower brain's firing patterns, and the context of the stimuli can access an entire sequence of memories (Teyler and DiScenna 1984).

The psychologist Elizabeth Loftus (1980) adds that incoming information enters short-term memory where it can either be forgotten or maintained by rehearsal and successfully transferred to long-term memory. Retrieving information from long-term memory depends upon cues that enable people to check different parts of their memory bank for the required material. Morton Reiser (1990) uses the term "nodal networks" to describe how memories are stored. These networks of mental representations (e.g., images, words) are organized by "affective links" (e.g., emotions, feelings), and often lead into "associative trains" that enable the retrieval of very early life events. During sleep, the pontine brain stem's periodic firing patterns stimulate the upper brain, primarily the visual–motor cortex, resulting in imagery that congeals into a dream (Hobson 1988). Often, these evoked images contain affect that links them with an important life event from the past through their "associate trains," resulting in a scenario that may be useful in psychotherapy or self-development (Krippner and Dillard 1988).

Very little work has been done regarding possible physiological bases for the manifestation of psi in dreams. Perhaps the initiating stimulus (e.g., the laboratory picture, the real-life event) influences the pontine area's firing pattern, evoking mental representations in the cortex that produces a "match," which engenders conviction that chance factors were not at work. Perhaps the holographic nature of the brain (Pribram 1971) enables the dreamer to "reach out" to incorporate the stimulus that is then "matched" with an appropriate memory. In the Maimonides dream experiments, the "matches" were often direct (as in the case of the Madison Square Garden boxing match dream, and the dancing girl in William Erwin's dream). But sometimes, they were symbolic, as when the randomly selected picture was that of a dead gangster in a coffin and the dream's central image was a dead rat in a cigar box. And, quite frequently, other memories seemed to confound the psi effect in dreams; the associate trains described by Reiser may have led to a waking life experience that matched the initiating stimulus, but the corresponding images were amalgamated with other material that was unrelated to the laboratory picture.

Persinger (1989) has proposed two interpretations of the geomagnetic field effect. The first is that psi is a geomagnetic field correlate; solar disturbances and consequent geomagnetic storms affect this correlate. The second is that the geomagnetic field affects brain receptivity to psi phenomena, which remains constant. In the latter interpretation, psi is always present in space and time, waiting to be accessed by crisis, emotion, or by optimal laboratory stimulus parameters. Geomagnetic activity may affect the detection capacity of the brain for this information, especially the neural pathways that facilitate the consolidation and conscious access to this information. Without this geomagnetic activity, awareness of the psi stimulus might not be as likely and the brain's "latent reserve capacities" would not be utilized.

Taking this argument one step further, Persinger (1989) points out that deep temporal lobe activity exists in equilibrium with the global geomagnetic condition. When there is a sudden decrease in geomagnetic activity, there appears to be an enhancement of processes that facilitate psi reception, especially telepathy and clairvoyance. Increases in geomagnetic activity may suppress pineal melatonin levels and contribute to reductions of cortical seizure thresholds. Indeed, melatonin is correlated with temporal-lobe-related disorders such as depression and seizures. Persinger has postulated that increased geomagnetic activity may contribute to expressive psi, such as spontaneous or laboratory psychokinesis. Some research data (e.g., Braud and Dennis 1989) support this conjecture and Gertrude Schmeidler (1994) has proposed that a psychokinesis subject who is unusually "aroused" (by the geomagnetic activity) would be more effective in the laboratory.

Some parapsychologists specialize in studying purported survival of bodily death, and the apparitions of loved ones reported by the bereaved. Because of melatonin's association with cortical seizures, Persinger hypothesized that bereavement apparitions would be more evident during times of increased geomagnetic activity. Such increases would suppress melatonin levels and contribute to reductions of cortical seizure thresholds, enabling access to cortical memory fragments such as appear in reports of ghosts, phantoms, and other images of dead relatives or loved ones. Persinger's (1988) analysis of over 200 such reports indicated that they tended to occur on days when the geomagnetic activity had increased as compared to the days before and after.

According to Persinger (1989), the stimulation parameters of the human brain's amygdala and hippocampus place them in a unique position to mediate psi experiences. The coherence of endogenous periods of single neurons or groups of neurons within the hippocampus increases the likelihood that they could be "driven" or influenced by external electromagnetic fields of similar pulse or natural frequency. There are also differences in resonance as a function of the sensitivity of the person's brain; the ordinary amygdaloid period is about four cycles per second, while the comparable period of a person with limbic epilepsy is around nine cycles per second.

Of particular relevance to psi is the capacity of the hippocampus to show long-term potentiation—the first step involved in memory storage.

A 400-cycles-per-second electrical stimulation of only 1 second can lead to semipermanent changes in electrical activity and produce observable growth of dendritic spines within 10 minutes. Such quick plasticity indicates that only a few seconds of the appropriate psi-related stimulus could evoke permanent changes in brain microstructures and hence modify memory. Once the memory is consolidated, it could appear as "real" as memory acquired by more traditional pathways.

Psi-Related Experiences and the Temporal Lobes

The two temporal lobes of the brain constitute about 40 percent of the higher functioning area called the cerebrum; thus, there may be a greater potential for dysfunction or anomalous functioning of the temporal lobes than for other lobes. The temporal lobes are well situated for integrating perceptual stimuli of all kinds as well as for integrating various aspects of such cognitive functions as memory, learning, language, sense of self, in addition to emotional, sexual, and aggressive functions. Because of these capacities, psi experiences could also be integrated in the temporal lobes (Neppe 1990).

The deep structures of the temporal lobes are the most electrically unstable portions of the human brain, and temporal lobe lability can be modified by such techniques as meditation. The contribution of temporal lobe processes to psi phenomena have two important implications. First, the phenomenological characteristics of psi experiences, especially spontaneous ones, could be dominated by the functions of the temporal lobes. Such evidence is clearly seen in the propensity for spontaneous psi experiences to occur in dreams, waking imagery, and intense affect that imbues the experience with intense personal meaningfulness (Persinger 1974). Second, the electrical lability of the temporal lobes means that many other stimuli could both compete for neural substrates that facilitate psi experiences as well as simulating experiences resembling psi.

For example, no other brain condition simulates spontaneous psi-related experiences as closely as does limbic temporal lobe epilepsy. If the discharge remains within one lobe and does not propagate to motor regions, no epileptic convulsion will occur. In fact, the observer might not realize the person is experiencing an electrical seizure; nevertheless, there are often clear experiential phenomena that are generated even without afterdischarges (Persinger 1989). However, it would be a mistake to assume that psi experiences are a form of limbic epilepsy; for one thing, psi phenomena are frequently reported during normal nighttime dreaming but epilepsy is amplified and abnormal. Furthermore, the trigger for a psi experience appears to be an external event, while it is an internal event that evokes an epileptic seizure. Nevertheless, their many similarities provide an opportunity for further study.

Persinger (1989) attempted such an investigation by examining two British collections of spontaneous telepathy and clairvoyance reports as well as more recent *Fate* magazine reports, finding that the peak displays of spontaneous psi experiences were reported to have occurred between 2:00 and 4:00 a.m. with a secondary peak around 9:00–11:00 p.m. These were exactly the hours when temporal lobe seizures were most frequently reported before anticonvulsants were introduced into medicine. However, a further peak occurs at 4:00 p.m. (Persinger and Schaut 1988); this peak is not congruent with the data on temporal lobe seizures and its significance remains unknown.

Jan Ehrenwald (1975), a psychoanalyst, was an early explorer of possible linkages between purported psi and deficits in brain functioning that might allow an organism to receive psi-related information. The parapsychologist William Roll (1977; Lowe and Roll 1988) has been another pioneer in the investigation of possible connections between epilepsy and psychokinesis. In one survey of 78 students, he found an increased incidence of olfactory or auditory hallucinatory experiences among the 10 people reporting the highest frequency of purported psi experiences.

In this regard, the neurologist Vernon Neppe (1990) has noted that patients with apparent temporal lobe dysfunctions often claim vivid psi-related experiences. He has developed the Neppe Temporal Lobe Questionnaire, which probes for descriptions of possible temporal lobe symptoms such as auditory hallucinations ("hearing voices"), as well as nonspecific symptoms such as depersonalization. Neppe administered the questionnaire to six people who had reported psi-related experiences and to six others who had not reported these experiences; none of his subjects had major psychiatric histories. The former group had an average of 6.2 temporal lobe symptoms, the latter group had an average of 0.3, and the difference was statistically significant.

The association between temporal lobe symptoms and reported psi-related experiences may suggest a state of physiological continuity of anomalous temporal lobe functioning. These brain processes may lay the groundwork for anomalous experiences and the resulting verbal reports. The independent existence of temporal lobe symptoms implies that the verbal reports of presumed psi are not an inevitable consequence of the brain condition. Nor are these reports limited to people with temporal lobe symptoms; Neppe (1990) interviewed four spiritualist mediums, analyzing their olfactory imagery, as well as similar olfactory imagery reported by the six subjects in his earlier study. A large proportion of these smells were pleasant in nature, and were often described as "perfumed" and "flowery"; pleasant olfactory hallucinations are rarely associated with temporal lobe epilepsy.

The anthropologist Michael Winkelman (1992) studied the ethnographic and phenomenological reports of native shamans' and shamanic healers' altered states of consciousness, finding evidence of unusual temporal lobe discharges. The data resulting from a direct assessment of this hypothesis was sparse but epilepsy has been detected among the Wapogoro, an East African shamanic tribe (Jilek-Aall 1965). Winkelman (1992) also found links between temporal lobe symptoms and reports of "spirit possession" by indigenous mediums. In both groups, there were frequent instances of amnesia, tremors, convulsions, compulsive motor behavior, and other signs that would fall under what Neppe called the "physiological continuity of anomalous temporal lobe functioning." As reported earlier in this essay, there is a provocative literature on epilepsy and its association with mystical experiences (e.g., Taylor 1972).

Psi-Related Experiences and Other Brain Areas

In contrast to the temporal lobes, there is little evidence or logic for claiming that an important role is played by other cerebral cortex areas in psi-related experiences. However, the frontal lobes, as the executive of the cognitive–motor cortex, could logically be associated with putative psychokinesis. The occipital lobes are likely candidates for apparitions, "visions," and perceptions of so-called auras because of their involvement in visual associations. The parietal lobes are involved in visual–spatial distortions such as those that characterize some psi-related reports (Neppe 1983). As a result of an analysis of members of two families with temporal lobe dysfunction, whose members reported psi-related experiences, Neppe suggested that there may be familial predispositions to presumptive psi phenomena (Hurst and Neppe 1981).

Persinger (1989) laments the paucity of literature on brain asymmetry and psi processes. The right hemisphere of the cerebral cortex is more specialized for detection of the spatial relationships between stimulus configurations and their affective associations rather than for the accuracy of detail. Linguistically and analytically, the left hemisphere is more active (Budzynski 1986; Ross and Mesulam 1979), although there are notable exceptions (e.g., many left-handed people; native speakers of Japanese and other languages that incorporate sounds resembling those found in nature). A few parapsychological experiments have utilized target material and conditions geared toward activating the left hemisphere, often obtaining statistically significant results (e.g., Braud and Braud 1977).

Other parapsychological experiments have compared "left hemisphere psi tasks" with "right hemisphere psi tasks," while still others have examined hemispheric dominance during psi tests by recording the eye movements of subjects while they made their responses. When the results of these studies attained statistical significance, they favored the right hemisphere (Schmeidler 1994, pp. 149–150). In one innovative study, the psychologist Robert F. Quider (1984) randomly assigned 40 subjects to four conditions: two with "suggestopedic" (i.e., accelerated learning) instructions for relaxation and visualization (one with music, one without music) and two lacking "suggestopedic" instructions (one with music and one without music). Psi scores did not show an effect of these instructions, but were significantly higher with music than without music when both conditions were combined.

This line of investigation might yield worthwhile results in view of the close connection often reported between mental imagery and presumptive psi (George and Krippner 1984). For example, the psychologists Sheryl Wilson and T.X. Barber (1982) studied 26 individuals who reported vivid (or "eidetic") mental imagery. In comparison with 25 control subjects, these "eidetikers" were not only better hypnotic subjects and engaged in more fantasy both as children and adults, but claimed to have had more clairvoyant, precognitive, and telepathic experiences. In addition, they frequently reported both mystical and OBEs, as well as lucid dreams, and sensing "apparitions" of the dead as well as "auras" of the living.

In the meantime, several parapsychological studies have attempted to identify the brain events that accompany psi, primarily High Hits on telepathy, clairvoyance, and precognition tests. At first, researchers examined brain waves as identified by EEG technology. Three major types of brain waves have been studied: alpha (slow, regular waves typical of relaxed states of casual scanning), beta (rapid, irregular waves found in alert attention), and theta (regular waves, slower than alpha, found in deep relaxation or during some types of mental imagery). Once more sensitive EEGs were developed, it was possible to study the complex form of a single brain response measured in milliseconds; this is referred to as the "evoked response." Positive relationships have been reported between High Hits on psi tasks and alpha activity (Schmeidler 1994, p. 147). When evoked response analysis was done with so-called gifted research participants, the 100-millisecond component of the evoked response sometimes, but not always, showed a significant response to psi stimuli (May et al. 1992; Warren et al. 1992).

Near-Death, Out-of-Body, and "Past-Life" Experiences

Survival of bodily death has been a preoccupation of several parapsychologists since the founding of the Society for Psychical Research. This group's investigation of so-called mediums launched a research focus that now encompasses spontaneous "after-death communication" (ADC) experiences as well as the investigation of messages from deceased loved ones, which mediums purportedly facilitate, or from "angels" and "spirits" that other practitioners "channel" (Drewry 2003; Klimo 1998).

Near-death experiences are profound psychological experiences that occur when experients are close to death or in situations of intense danger. The psychiatrist Bruce Greyson notes that NDEs typically transcend one's personal ego, involve an encounter with a higher or divine principle, and are found cross-culturally (2000, p. 316). Aftereffects of NDEs include an increased interest in spiritual matters, concern for others, and appreciation of life, and a decreased fear of death (p. 319). Stories of tunnels and bright lights abound in NDEs, but may have a basis in the structure and functioning of the brain, the eyes, and other sense organs that operate during these experiences (Woerlee 2004).

Controversy exists over whether NDEs are a reflection of psychological defenses, reactions to medication, the deterioration of one's brain and its functions, or a confirmation of "mind–body" separation. The psychologist Susan Blackmore (2002) has offered an explanatory model of NDEs, which involves links with the dying brain's lack of oxygen, the phenomenology of temporal lobe epilepsy, and the positive emotional effects resulting from endogenous endorphin activity. Greyson, as part of his rebuttal, described a patient whose NDE report correctly described some 20 health care workers in the operating room, most of whom she had never met. Her eyes had been taped shut before the operation began, yet she provided an accurate description of the pneumatic drill that was used to open her skull (2000, p. 340).

The term OBE is an oxymoron for Blackmore, who resonates with the Buddhist view that the ego is illusory. Therefore, neither the OBE experience nor the OBE experient has a "location" (2002, p. 369). Parapsychologist and psychologist Carlos Alvarado, on the other hand, notes that reports of OBEs recur throughout history and across cultures. A variety of contemporary experients have volunteered for psychoneurological testing; their data have differed from baseline waking states, yet the only commonality has been that of relaxation and low arousal (2000, pp. 189–190). In a comprehensive psychoneurological study, 43 NDE experients were found to exhibit more temporal lobe epileptic symptoms as well as more epiletiform EEG activity than non-NDE experients. Contrary to the investigators' expectations, this EEG activity was nearly completely lateralized to the left hemisphere, indicating that NDEs are more closely associated with positive coping styles than with dissociation or posttraumatic stress disorder (Britton and Bootzin 2004). But W.R. Uttal (2001) has written of "the new phrenology," insisting that cognitive constructs and thought processes are too unpredictable to pin down to a specific brain area.

Nonetheless, OBEs have been elicited by electrical stimulation of the temporal cortex (Penfield and Jasper 1954), and Persinger (1995) found a positive relationship with epileptic-like signs. Olaf Blanke, a neurologist, examined six brain-damaged patients at the University Hospital of Geneva, Switzerland, concluding that damage at the function of the temporal and parietal lobes may produce a breakdown of a person's bodily image. The boundary between personal and extrapersonal space may become blurred, and the body is seen occupying positions that do not coincide with the position in which he or she feels it should be placed. Several of the patients suffered from temporal lobe epilepsy; some patients gave the experience a mystical interpretation while others did not (Blanke et al. 2004).

Both NDEs and OBEs may include psi-related experiences, but neither is considered an integral part of parapsychological research. On the other hand, "past-life" experiences (PLEs) are a part of the parapsychological domain, even though anthropologists and ethnologists also investigate these reports. PLEs can be defined as reported experiences or impressions of oneself as a particular person (other than one's current life identify) in a previous time or life (Mills and Lynn 2000, p. 285).

PLEs may occur spontaneously or may be evoked through hypnosis. The psychiatrist Ian Stevenson (1983) has collected hundreds of case reports, cross-culturally, including instances of "xenoglossy" in which hypnotized individuals spoke a language to which they had not apparently been exposed previously. Stevenson (1997) has also identified birthmarks and other biological "markers" that purportedly conform to the experient's previous life. Critics respond that it is difficult to determine whether the "marker" was noticed before the PLE was reported, and that it is possible for people to be exposed to languages (and other unusual learnings) at an early age, and then forget the exposure. This phenomenon, "cryptomnesia" or amnesia for the source of learned information, is one explanation for this rare aspect of PLEs.

NDEs, OBEs, and PLEs all qualify as anomalous experiences. Furthermore, they are of interest to students of religious belief and spiritual practice because they suggest the existence of some aspect of one's psyche that can separate from the physical body, survive physical death, and/or reincarnate as a new personality. Along with mystical experiences, they are the primary anomalies that are of interest to some scholars of religion and spirituality. They are also of increasing interest to health care practitioners due to the accumulating evidence that religious belief and spiritual practice show a positive relationship to such outcomes as longevity, physical health, and rapid recovery from illness (see Krippner 2003).

Implications for Religious Belief and Spiritual Practice

In discussing religious and spiritual issues, I have found it useful to distinguish the two terms. For me, religious pertains to adherence to an organized system of beliefs about the divine (something deemed worthy of veneration and worship) and the observance of rituals, rites, and requirements of that belief system. Spirituality, on the other hand, can be thought of as one's focus on, and/or reverence, openness, and connectedness to, something of significance believed to be beyond one's full understanding and/or individual existence. The relevance of mystical experience is obvious; many mystics are spiritual without being religious, depending on whether their god experience conforms to a religious belief system.

Psi-related experiences are not so easily encompassed by religious belief and spiritual practice. One might consider "miracles" to be the connecting link. However, miracles are defined by theologians as phenomena that do not conform to natural law. If so, they can be described by science's observational methods but can never be fully fathomed by science. As a result, they have little appeal to someone like me who likes to study experiences and events that are not only observable but can be measured, interpreted, explained, and (ultimately) find their place in the organizational schema of the cosmos. At the same time, I take the position that both scientific analysis and disciplined phenomenology can expand the wonder of both human and nonhuman phenomena rather than simply reduce them to equations and formulae. Thus, scientists can study peoples' reactions to alleged miracles and the attitudinal and behavioral changes they might instigate in experients' lives; these investigations can be carried out without buying into the contention that something is going on that is "supernatural" or that "transcends" the natural lawfulness of the universe.

A frequent criticism leveled against parapsychologists is that their work and, consequently, their data are influenced by covert religious agendas. The psychologist and parapsychologist Charles Tart attempted to study this claim by mailing a questionnaire to members of the Parapsychological Association, receiving 77 responses to the 160 questionnaires he mailed out. When queried as to whether spiritual interests played a major role in their entering the field, 49 percent answered negatively, 36 percent positively, and the rest gave responses that fell into "partially" or "unclear" categories. When asked if their current work projects were spiritually motivated, 34 percent answered positively. When asked if they perceived conflicts between mainstream laboratory research and spiritual interests, 1 out of 4 answered positively. Tart confessed that for him, "the spiritual implications of parapsychological data are among the most important aspects of the field," but concluded that this was not an opinion held by a majority of the Parapsychological Association's membership (2003, p. 184).

Tart (1997) has posited the existence of "non-physical worlds" (NPWs) and has proposed technologies for studying them through meditative procedures, OBEs, NDEs, and other altered states of consciousness, bringing back physical evidence of their reality or at least providing reports congruent with other travelers to these NPWs. As a pioneer in the development of transpersonal psychology, Tart (1975) advocates the scientific exploration of psychological phenomena that extend beyond one's ordinary personal and biological identity, as well as their facilitation. Instead of rejecting religion and spirituality as a result of his scientific training, or rejecting science because of his religious and spiritual convictions, Tart has chosen what he calls "a third path," a "leap of faith" based on high-quality scientific evidence and open-minded investigation. This "third path" is frequently encountered among investigators of anomalies from various disciplines.

William Roll (1982) has shared Tart's commitment to rigorous investigative procedures and has assiduously surveyed the literature on life after death. He has reviewed searches for a soul "existing apart from its body" through apparitions of the departed, past-life memories, OBEs, and mediumistic communications, but has concluded that "none have provided evidence for the existence of the departed" after their death (Roll 1997, p. 62). On the other hand, Roll adds that this search assumes that the soul must be separate from the material world, must be distinct from other souls, and must be capable of engaging in conceptual thought. Instead, Roll posits a "longbody" (similar to some Native American beliefs) that includes nature and significant others in its scope, and stretches back in space–time farther than the body's life span. This longbody interacts with a more limited bodily soul, and a "Big Mind" that is much grander than either of them. Roll's model encompasses OBEs (the bodily soul), PLEs (the longbody), and even poltergeist experiences (symptoms of the longbody in distress).

When working with J.B. Rhine, who developed experimental parapsychology in the 1930s, Roll (1997, p. 66) was given a crystal ball as a souvenir. Years later, the physician Raymond Moody, a pioneer investigator of NDEs, saw images in the crystal and soon had created a "psychomanteum" or wall mirror in which grieving survivors claimed to glimpse their departed loved ones (or, in Roll's terms, their loved one's longbodies).

Experiences with the psychomanteum helped shorten the grief period of bereaved individuals (Hastings et al. 2002), thus providing the first practical application of parapsychological data since the U.S. government's controversial "remote viewing" spy forays of the Cold War. Russell Targ, one of the physicists who directed this highly classified program, has written eloquently about how this attempt to apply parapsychology for strategic purposes coincided with his encounter with Dzogchen Buddhism and its goal of timeless awareness and spaciousness, the "truth of the heart" (2004, p. 170). Apparently, Buddhism's tent is large enough to harbor both advocates of psi data such as Russell Targ and such critics of parapsychological data as Susan Blackmore!

Another psychologist and parapsychologist, K.R. Rao, has noted that J.B. Rhine was originally headed for the ministry but ultimately found biology (and eventually psi research) more appealing. Later, he urged the establishment of a new field, the "parapsychology of religion," but it fell on deaf ears because, according to Rao, neither mainstream science nor mainstream religion was comfortable with parapsychology. From time to time, I will encounter a disgruntled scientist who will castigate psi research as a distraction and waste of time, or a religious fundamentalist will label psi phenomena "works of the devil." However, Rao has written at length about the Advaita Hindu tradition and its description of a "transcendental consciousness," which is quite compatible with psi phenomena. When one travels the path to this state, various anomalies may occur; however, these are "guideposts" rather than goals to be attained for themselves (1997, p. 79).

Lest one think that psi research is a harbinger of new religious thought, the philosopher Robert Thouless noted that "the main current of religious thought is still little affected by the progress of parapsychological research." However, he expressed the hope that at least some religiously inclined people could reopen a consideration of "many of the questions that the religious have traditionally regarded as closed" (1977, p. 188). Rhea White (1997) ranks high among the parapsychologists who have given the most thought to psi phenomena and its relevance to spirituality and religion. Her term, "exceptional human experiences" (EHEs) encompasses what have been called mystical, psychic, peak, and flow experiences by other writers. She has described dozens of examples that fall into five major classes: mystical (e.g., conversion and unitive experiences), psychic (e.g., OBEs, precognition), encounters (UFO and angelic experiences), death-related (NDEs and mediumistic experiences), and normal (e.g., empathic and esthetic experiences). One's culture can determine whether or not these experiences are "spiritual," whether they support or challenge the socially approved religious precepts. Indeed, some EHEs are neither spiritual nor religious, yet are "exceptional" in the sense that they have transformative potentials. Without the possibility of changing the experient's life in some way, an OBE, a PLE, or an NDE is simply an "exceptional experience," something to be ignored, dismissed, or briefly discussed with family and friends and then forgotten.

By focusing on the effect of an EHE rather than its veridicality, White has done parapsychology a major service. She has provided a common ground in which scientists and religious scholars, agnostics and the spiritually inclined, as well as parapsychologists and their critics can work together to understand the impact that these anomalies have on people's lives. No immediate judgment as to their etiology is needed; by focusing on the lived experiences themselves, investigators and even lay people can gain a deeper understanding on what it means to be "human," and how one's brief sojourn on this planet can be more meaningful and enjoyable.

I (Krippner 2002) used White's format when I was asked to write my spiritual autobiography for *The International Journal of Transpersonal Studies*. Among the EHEs that were important to me were encounters with indigenous shamans and healers, with plants and animals, with gurus and yogis, and participation in ceremonies that including drinking ayahuasca, ingesting mind-altering mushrooms, and sweating in a Native American *wickiup*. But my spiritual life was also influenced by esthetic EHEs, sexual EHEs, and even one or two psi-related EHEs. I wrote about the insights that emerged from these EHEs: the power of love, the omnipresence of evil, and my conviction that there is a divine "trickster" that keeps us on our toes, that rewards us if we "go with the flow," and that slaps us down if we take ourselves too seriously.

Most mainstream religions see no need to study EHEs or anomalous phenomena; their dogma and their faith are generally sufficient to provide them with the reassurance that they will survive bodily death and that, during their sojourn on earth, their belief system is a valid and useful guide to conduct and morality. However, adherents of many contemporary spiritual practices reject what they consider dogmatic religion, emphasizing personal experience that permits contacts with what they consider the sacred and divine elements of the cosmos. These individuals are more open to supporting and studying research in anomalous phenomena and integrating the results into their spiritual practices and religious beliefs.

Human life upon planet Earth is constantly at risk, perhaps increasingly so as the centuries roll on, as global warming threatens coastal cities, as desertification erodes the land, as pollution corrupts the air and the seas, and as wars and armed conflict waste financial resources, which could be used to avoid ecological disasters. Thoughtful people frequently write about the need for revisioning human attitudes and behaviors. Perhaps these necessary visions will occur in EHEs. Whether they are anomalous or ordinary EHEs makes little difference; they are badly needed and they cannot arrive a day (or night) too soon.

"Postmodern" Science and Psi Research

Parapsychology has pioneered research into several aspects of human behavior and experience that are now a part of mainstream psychology, for example, hypnosis, multiple personalities, anomalous healing. Other topics investigated by the early societies for psychical research (e.g., lucid dreaming, NDEs, OBEs) are beginning to enter the psychological mainstream and the mechanisms for these phenomena are on their way to becoming understood. Perhaps telepathy, clairvoyance, precognition, and psychokinesis will eventually travel the same road. The prefix "para" does not exclude ordinary mechanisms for psi phenomena. Neher (1990) observed that "parapsychology" simply means "alongside of" the mainstream of psychology. After reviewing the two fields, Collins and Pinch (1982) concluded "there is . . . nothing in psychology that definitely makes parapsychology unscientific." They also claimed that "it has not been demonstrated decisively that there are any specific physical principles that conflict with parapsychology." The sociologist Marcelo Truzzi (1985) would consider parapsychology to be a "legitimate scientific enterprise" whether or not psi actually exists, because parapsychologists employ such scientific methods as target randomization, double-blind judging, control groups, and statistical tests. For others (e.g., Leahy and Leahy 1983), methodologically, parapsychology is a science; substantially, the verdict is still out.

This situation illustrates the philosopher Michel Foucault's (1980) observations that power permeates all aspects of science's efforts to obtain knowledge, that scientific legitimacy is inherently political, and that this politicized scientific legitimacy results in definitions, categorizations, and classifications that construct what is considered "reality." Some knowledge comes to be considered "legitimate" in historically specific times and places (Lather 1990) while some knowledge falls by the wayside. The aphorism "knowledge is power" could easily be reversed: power (e.g., political, economic, ideological, religious) determines what is considered to be "knowledge" (and therefore reality) in any given temporal and spatial location. I would apply Foucault's insights to psi research; knowledge accumulated by parapsychologists about anomalous events lacks any type of influential power base; as a result, it fails to be legitimate and to play a role in mainstream scientific discourse.

Postmodern scientists are fond of pointing out that "modern" science holds that only two "stories" about consciousness are credible: "central-state materialism" and the "new epiphenomenalism." Central-state materialism holds that physical science and its laws can explain all mental and physical events. There is an "identity" between mind and brain that will eventually enable all conditions arising in mentation and behavior to be reduced to physics and chemistry. The new epiphenomenalism holds that mental and physical events exist as two distinct domains but most (if not all) mental events exist in a causal relationship with physical events, even "emerging" from conditions that obey the laws of physics and chemistry (Campbell 1984).

Postmodern science, on the other hand, attempts to bypass the mind/body dualism that has characterized modern Western science. For example, the anthropologist Charles Laughlin (Laughlin et al. 1990) has rejected both central-state materialism and the new epiphenomenalism as inappropriate to the phenomena they are trying to address. Rather, he advocates a "structural monism," which holds that mind and body (including experience and behavior) are two imperfect ways of perceiving and knowing the same unknown totality we may call "being." "Spiritual awareness" is one way of knowing the being; "physical" awareness is another way of knowing the being. Neither the so-called spiritual disciplines (theology, psi research, transpersonal psychology, transpersonal anthropology, etc.) nor the more physical disciplines (physics, chemistry, physiology, etc.) can claim to be complete representations or explanations of reality as required by central-state materialism. Furthermore, consciousness does not neatly divide itself into spiritual (noncausal) and mundane (causal) attributes (as required for the new epiphenomenalism).

The structural monist holds that "mind" and "body" (or mind and "brain") are two views of the same reality. Mind is how brain experiences its own functioning, and brain provides the structure of mind. From this perspective, neither the social and psychological sciences nor the neurosciences can be considered as complete accounts of consciousness. Reality is simultaneously composed of many levels, none of which is fundamental. But these levels are born of the analytic mind trying to make sense of an essentially undifferentiated field of systemically related processes. In this manner, Laughlin opens the door to transpersonal considerations, claiming that without them the accounting of consciousness in all its richness would be incomplete.

Chaotic systems analysis is one approach mentioned by Laughlin (1992). Chaos theory is a branch of mathematics specializing in the study of processes that seem so complex that at first they do not appear to be governed by any known laws of principles, but which actually have an underlying order describable by vector calculus and its associated geometry. Examples of chaotic processes include water flowing in a stream or crashing at the bottom of a waterfall, changes in animal populations, and EEG changes in the brain. From the perspective of chaos theory, brain activity during epilepsy is less "chaotic" than normal brain functioning; indeed, the EEG of an epileptic is extremely regular just preceding a seizure (Pool 1989). I (Krippner 1994) have suggested that the bifurcations between various chaotic and non-chaotic attractors might underlie the cyclic sleep changes noted by sleep researchers (e.g., Hobson 1988), with rapid eye movement sleep being more complex chaotically than non-rapid eye movement sleep. The lower dimensionality or greater order of rapid eye movement sleep can be seen from the narrative the dreamer later attributes to the dream. Periodic attractors might represent an attempt to elicit additional images from the dreamer's memory bank, which would facilitate the continuation of a story that is more coherent and more psychologically useful to the dreamer. Because both geomagnetic activity and rapid eye movement sleep are potentially chaotic systems, there would be ample opportunities for psi to enter this confluence of systems, especially if psi itself is chaotic.

Psi research may find greater receptivity on the part of postmodern scientists than it has from modern scientists. In the meantime, the Parapsychological Association has made it clear that labeling an event as a psi phenomenon does not constitute an explanation for an event, but only indicates an event for which a scientific explanation needs to be sought. Furthermore, "regardless of what form the final explanation may take . . . , the study of these phenomena is likely to expand our understanding of the processes often referred to as 'consciousness' and 'mind' and of the nature of disciplined inquiry. . . ." Therefore, the essential touchstones for parapsychologists include the need to stress the speculative nature of the field, to be candid about its controversial status, and not to go beyond what is warranted by the evidence. However, this modesty should be combined with a devotion to scientific procedures and a commitment to the search for discovery and understanding. As the Parapsychological Association concluded, "Parapsychology has a . . . tradition of bringing scientific imagination and rigor to the study of phenomena typically ignored by other investigators. Whatever the eventual outcome of this search may be, it can not help but add to the sum of knowledge about humanity and the human condition" (1987, p. 389).

In his wide-ranging book *The Future of the Body*, the essayist Michael Murphy claims that "we live only part of the life we are given" (1992, p. 3) and catalogs dozens of anecdotal and research reports to demonstrate what can be called the "latent reserve capacities" of the human brain and body. Parapsychological data are placed side by side with evidence from medicine, sports, the martial arts, and the behavioral and social sciences. The examples he gives of voluntary control, self-regulation, transformative practice, and extraordinary human experience indicate not only that modern science has overlooked many human potentials but also that these capacities can provide practical avenues for an acceleration and betterment of human life.

Note

The preparation of this chapter was supported by the Chair for the Study of Consciousness, Saybrook Graduate School and Research Center, San Francisco, California. Gratitude is also expressed to Steven Bauman whose unpublished paper, "Looking for God in All the Right Places," served as an organizational guide for part of this chapter.

References

Albright, C.R. 2000. The "God module" and the complexifying brain. *Zygon*, 35, 735–743.

Alvarado, C.S. 2000. Out-of-body experiences, in E. Cardeña, S.J. Lynn, and S. Krippner (eds.), *Varieties of Anomalous Experience: Examining the Scientific Evidence*. Washington, DC: American Psychological Association, pp. 183–218.

Belvedere, E. and Foulkes, D. 1971. Telepathy and dreams: A failure to replicate. *Perceptual and Motor Skills*, 33, 783–789.

Ben-Yehuda, N. 1985. *Deviance and Moral Boundaries: Witchcraft, the Occult, Science Function, Deviant Sciences and Scientists*. Chicago: University of Chicago Press.

Blackmore, S. 2002. Near-death experiences: In or out of the body? in R. Joseph (ed.), *NeuroTheology: Brain, Science, Spirituality, Religious experience*. San Jose, CA: University Press, California, pp. 361–369.

Blanke, O., Landis, T., Spinelli, L., and Seeck, M. 2004. Out-of-body experience and autoscopy of neurological origin. *Brain*, 127(2), 243–258.

Bratus, B.B. 1990., A. Mikheyev, S. Mikheyev, and Y. Filippov, trans. H. David *Anomalies of Personality: from the Deviant to the Norm* ed.), Orlando, FL: Paul M. Deutsch Press (original work published 1988).

Braud, L.W. and Braud, W.G. 1977. Clairvoyance tests following exposure to a psi conducive tape recording. *Journal of Research in Psi Phenomena*, 1(1), 10–12.

Braud, W.G. and Dennis, S.P. 1989. Geophysical variables and behavior: LVIII. Autonomic activity, hemolysis, and biological psychokinesis: Possible relationships with geomagnetic field activity. *Perceptual and Motor Skills*, 68, 1243–1254.

Britton, W.B. and Bootzin, R.R. 2004. Near-death experiences and the temporal lobe. *Psychological Science*, 15(4), 254–258.

Broughton, R.S. 1991. *Parapsychology: The Controversial Science*. New York: Ballantine.

Budzynski, T.H. 1986. Clinical applications of non-drug-induced states, in B.B. Wolman and M. Ullman (eds.), *Handbook of States of Consciousness*. New York: Von Nostrand Reinhold, pp. 428–460.

Campbell, K. 1984. *Body and Mind*. Notre Dame, IN: University of Notre Dame Press.

Cardeña, E., Lynn, S.J., and Krippner, S. 2000. Introduction: Anomalous experiences in perspective, in E. Cardeña, S.J. Lynn, and S. Krippner (eds.), *Varieties of Anomalous Experience: Examining the Scientific Evidence*. Washington, DC: American Psychological Association, p. 4.

Child, I.L. 1985. Psychology and anomalous observations: The question of ESP in dreams. *American Psychologist*, 40, 1219–1230.

Collins, H.M. and Pinch, T.J. 1982. *Frames of Meaning: The Social Construction of Extraordinary Science*. London: Routledge and Kegan Paul.

Dalton, K., Steinkamp, F., and Sherwood, S.J. 1999. A dream GESP experiment using dynamic targets and consensus vote. *Journal of the American Society for Psychical Research*, 93, 145–166.

d'Aquili, E.G. 1993. Apologia pro sciptura sua, or maybe we got it right after all. *Zygon*, 28, 177–200.

———. 2002. The neuropsychology of aesthetic, spiritual and mystical states in R. Joseph (ed.), *NeuroTheology: Brain, Science, Spirituality, Religious Experience*. San Jose, CA: University Press, California, pp. 233–242.

Derr, J.S. and Persinger, M.A. 1989. Geophysical variables and behavior: LIV. Zeitoun (Egypt) apparitions and the Virgin Mary as tectonic strain-induced luminosities. *Perceptual and Motor Skills*, 68, 123–128.

Doblin, R. 1991. Pahnke's "Good Friday experiment": A long-term follow-up and methodological critique. *Journal of Transpersonal Psychology*, 23, 1–28.

Drewry, M.D.J. 2003. Purported after-death communication and its role in the recovery of bereaved individuals: A phenomenological study. *Proceedings of the Academic of Religion and Psychical Research 28th Annual Conference*. Bloomfield, CT: Academy of Religion and Psychical Research, pp. 74–87.

Ehrenwald, J. 1975. Cerebral localization and the psi syndrome. *Journal of Nervous and Mental Diseases*, 161, 393–398.

Flournoy, T. 1994. *From India to the Planet Mars: A Case Study in Multiple Personality with Imaginary Languages*. Princeton, NJ: Princeton University Press (original work published 1900).

Foucault, M. 1980. *Power/Knowledge: Selected Interviews and Other Writings, 1972–1977* (C. Gordon, ed. and trans.). New York: Pantheon.

George, L. and Krippner, S. 1984. Mental imagery and psi phenomena: A review, in S. Krippner (ed.), *Advances in Parapsychological Research*, vol. 4. Jefferson, NC: McFarland, pp. 64–82.

Geschwind, N. 1983. Interictal behavioral changes in epilepsy. *Epilepsia*, 24 (suppl.), S23–S30.

Globus, G.G., Knapp, P., and Skinner, J. (1968). An appraisal of telepathic communication in dreams. *Psychophysiology*, 4, 365.

Greeley, A.M. 1975. *The Sociology of the Paranormal: A Reconnaissance*. Beverly Hills, CA: Sage.

Greyson, B. 2000. Near-death experiences, in E. Cardeña, S.J. Lynn, and S. Krippner (eds.), *Varieties of Anomalous Experience: Examining the Scientific Record*. Washington, DC: American Psychological Association, pp. 315–352.

Grof, S. 1980. *LSD Psychotherapy*. Pomona, CA: Hunter House.

———. 1998. *The Cosmic Game: Explorations of the Frontiers of Human Consciousness*. Albany: State University of New York Press.

Gurney, E., Myers, F.W.H., and Podmore, F. 1886. *Phantasms of the Living*. London: Trubner.

Hall, C.S. 1967. Experiments with telepathically influenced dreams. *Zeitschrift fur Parapsychologie und Grenzgebite der Psychologie*, 10, 18–47.

Hastings, A., Hutton, M., Braud, W., Bennett, C., Berk, I., Boynton, T., Dawn, C., Ferguson, E., Goldman, A., Greene, E., Hewett, M., Lind, V., McLellan, K., and Steinbach-Humphrey, S. 2002. Psychomanteum research: Experiences and effects on bereavement. *Omega: Journal of Death and Dying*, 43(3), 211–228.

Hearne, K. 1987. A dream-telepathy study using a home "dream machine." *Journal of the Society for Psychical Research*, 54, 139–142.

Helminiak, D.A. 1984. Neurology, psychology, and extraordinary religious experiences. *Journal of Religion and Health*, 23, 33–46.

Hitt, J. 1999. This is your brain on God. *Wired*, 7, 11.

Hobson, J.A. 1988. *The Dreaming Brain*. New York: Basic Books.

Hurst, L.A. and Neppe, V.M. 1981. A familial study of subjective paranormal experience in temporal lobe dysfunction subjects. *Parapsychological Journal of South Africa*, 2(2), 56–64.

James, W. 1958. *The Varieties of Religious Experience: A Study in Human Nature*. New York: New American Library (original work published 1902).

Jilek-Aall, L. 1965. Epilepsy in the Wapogoro tribe of Tanganyika. *Acta Psychiatrica Scandanavia*, 41, 57–86.

Joseph, R. 2002. The limbic system and the soul: Evolution and the neuroanatomy of religious experience. *Zygon*, 36, 105–136.

Jung, C.G. 1970. On the psychology and pathology of so-called occult phenomena in, *The Collected Works of C.G. Jung*, vol. 1. Princeton, NJ: Princeton University Press (original work published 1902), pp. 6–91.

Klimo, J. 1998. *Channeling: Investigations on Receiving Information from Paranormal Sources*, 2nd ed. Berkeley: North Atlantic.

Krippner, S. 1994. Humanistic psychology and chaos theory. *Journal of Humanistic Psychology*, 34, 48–61.

———. 2002. Dancing with the trickster: Notes for a transpersonal autobiography. *International Journal of Transpersonal Studies*, 21, 1–18.

———. 2003. Spirituality and healing, in D. Moss, A. McGrady, T.C. Davies, and I. Wickramasekera (eds.), *Handbook of Mind-Body Medicine for Primary Care*. Thousand Oaks, CA: Sage, pp. 191–201.

Krippner, S. and Dillard, J. 1988. *Dreamworking*. Buffalo, NY: Bearly.

Krippner, S. and Persinger, M. 1996. Evidence for enhanced congruence between dreams and distant target material during periods of decreased geomagnetic activity. *Journal of Scientific Exploration*, 10, 487–493.

Krippner, S. and Ullman, M. 1970. Telepathy and dreams: A controlled experiment with electro-encephalogram-electro-oculogram monitoring. *Journal of Nervous and Mental Disease*, 151, 394–403.

Krippner, S., Vaughan, A., and Spottiswoode, S.J.P. 2000. Geomagnetic factors in subjective precognitive experiences. *Journal of the Society for Psychical Research*, 64, 109–118.

Landis, B.O., Spinelli, L., and Seeck, M. 2004. Out-of-body experience and autoscopy of neurological origin. *Brain*, 127, 242–243.

Lather, P. 1990. Postmodernism and the human sciences. *The Humanistic Psychologist*, 18, 64–84.

Laughlin, C.D. 1992. *Scientific Explanation and the Life Field: A Biogenetic Structural Theory of Meaning and Causation*. Sausalito, CA: Institute of Noetic Sciences.

Laughlin, C.D., McManus, J., and d'Aquili, E.G. 1990. *Brain, Symbol & Experience: Toward a Neurophenomenology of Human Consciousness*. Boston: Shambhala.

Leahy, T.H. and Leahy, G.E. 1983. *Psychology's Occult Doubles: Psychology and the Problem of Pseudoscience*. Chicago: Nelson-Hall.

Loftus, E. 1980. *Memory*. Reading, MA: Addison-Wesley.

Lowe, L.M. and Roll, W.G. 1988. Psi survey research project findings. *Proceedings, International Conference on Psychic Research, December 2–5, 1988*. Carrollton, GA: West Georgia College.

MacDonald, W.L. 1994. The popularity of paranormal experiences in the United States. *Journal of American Culture*, 1, 35–42.

Makarec, K. and Persinger, M.A. 1985. Temporal lobe signs: Electroencephalographic validity and enhanced scores in special populations. *Perceptual and Motor Skills*, 50, 831–842.

Mandel, A. 1980. Toward a psychobiology of transcendence: God in the brain, in J. Davidson and R. Davidson (eds.), *The Psychobiology of Consciousness*. New York: Plenum, pp. 379–463.

Masters, R.E.L. and Houston, J. 1973. Subjective realities, in B. Schwartz (ed.), *Human Connections and the New Media*. Englewood Cliffs, NJ: Prentice Hall, pp. 88–106.

May, E.C., Luke, W.W., Trask, V.V., and Frivold, T.J. 1992. Phenomenological Research and Analysis in L.A. Kenkel and G.R. Schmeidler (eds), *Research in Parapsychology 1990*. Metuchen, NJ: Scarecrow Press, pp. 69–70.

McClenon, J. 1994a. Surveys of anomalous experiences: A cross-cultural analysis. *Journal of the American Society for Psychical Research*, 88, 117–135.

———. 1994b. *Wondrous Events: Foundations of Religious Beliefs*. Philadelphia: University of Pennsylvania Press.

Mills, A. and Lynn, S.J. 2000. Past-life experiences, Phenomenological Research and Analysis, in E. Cardeña, S.J. Lynn, and S. Krippner (eds.), *Varieties of Anomalous Experience: Examining the Scientific Evidence*. Washington, DC: American Psychological Association, pp. 283–313.

Murphy, M. 1992. *The Future of the Body: Explorations into the Future Evolution of Human Nature*. Los Angeles: Jeremy P. Tarcher.

Murphy, M. and Donovan, W. 1996. *The Physical and Psychological Effects of Meditation: A Review of Contemporary Research with a Comprehensive Bibliography, 1991–1996*, 2nd ed. Sausalito, CA: Institute of Noetic Sciences.

Neher, A. 1990. *The Psychology of Transcendence*. New York: Dover.

Neppe, V.M. 1983. Temporal lobe symptomatology in subjective paranormal experients. *Journal of the American Society for Psychical Research*, 77, 1–30.

———. 1990. Anomalistic experience and the cerebral cortex, in S. Krippner (ed.), *Advances in Parapsychological Research*, vol. 6. Jefferson, NC: McFarland, pp. 168–183.

Newberg, A. and d'Aquili, E. 2001. *Why God Won't Go Away*. New York: Ballantine.

Pahnke, W.N. 1966. Drugs and mysticism. *International Journal of Parapsychology*, 8, 295–314.

Parapsychological Association. 1987. Terms and methods in parapsychological research. *Journal of Humanistic Psychology*, 29, 394–399.

Penfield, W. and Jasper, H. 1954. *Epilepsy and the Functional Anatomy of the Human Brain*. Boston: Little, Brown.

Persinger, M.A. 1974. *The Paranormal. Part I. Patterns*. New York: MSS Information Corporation.

———. 1987a. *Neurophysiological Bases of God Beliefs*. New York: Praeger.

———. 1987b. Spontaneous telepathic experiences from phantasms of the living and low global geomagnetic activity. *Journal of the American Society for Psychical Research*, 81, 23.

———. 1988. Increased geomagnetic activity and the occurrence of bereavement hallucinations: Evidence for melatonin-mediated microseizuring in the temporal lobe? *Neuroscience Letters*, 88, 271–274.

———. 1989. Psi phenomena and temporal lobe activity: The geomagnetic factor, in L.A. Henkel and R.E. Berger (eds.), *Research in Parapsychology 1988*. Metuchen, NJ: Scarecrow Press, pp. 121–156.

———. 1995. Out-of-body-like experiences are more probable in people with elevated complex partial epileptic-like signs during periods of enhanced geomagnetic activity: A nonlinear effect. *Perceptual and Motor Skills*, 80, 563–569.

Persinger, M.A. and Krippner, S. 1989. Dream ESP experiments and geomagnetic activity. *Journal of the American Society of Psychical Research*, 83, 101–106.

Persinger, M.A. and Schaut, G.B. 1988. Geomagnetic factors in subjective telepathic, precognitive, and postmortem experiences. *Journal of the American Society for Psychical Research*, 82, 217–235.

Pool, R. 1989. Is it healthy to be chaotic? *Science*, 243, 604–607.

Pribram, K.H. 1971. *Languages of the Brain*. Englewood Cliffs, NJ: Prentice-Hall.

Quider, R.F. 1984. The effect of relaxation/suggestion and music on forced-choice ESP scoring. *Journal of the American Society for Psychical Research*, 78, 241–262.

Radin, D. 1997. *The Conscious Universe: The Scientific Truth of Psychic Phenomena*. San Francisco: HarperEdge.

Ramachandran, V.S. and Blakeslee, S. 1998. *Phantoms in the Brain*. New York: Quill.

Rao, K.R. 1997. Reflection on religion and anomalies of consciousness, in C.T. Tart (ed.), *Body, Mind, Spirit: Exploring the Parapsychology of Spirituality*. Charlottesville, VA: Hampton Roads, pp. 68–82.

Rao, K.R. and Palmer, J. 1987. The anomaly called psi: Recent research and criticism. *Behavioral and Brain Sciences*, 10, 539–551.

Reiser, M.F. 1990. *Memory in Mind and Brain: What Dream Imagery Reveals*. New York: Basic Books.

Rhine, L.E. 1977. Research methods with spontaneous cases, in B.B. Wolman (ed.), *Handbook of Parapsychology*. New York: Van Nostrand Reinhold, pp. 59–80.

Roll, W.G. 1977. Poltergeists, in B.B. Wolman (ed.), *Handbook of Parapsychology*. New York: Van Nostrand Reinhold, pp. 382–413.

———. 1982. The changing perspective on life after death, in S. Krippner (ed.), *Advances in Parapsychological Research*, vol. 3. New York: Plenum, pp. 147–291.

———. 1997. My search for the soul, in C.T. Tart (ed.), *Body, Mind, Spirit: Exploring the Parapsychology of Spirituality*. Charlottesville, VA: Hampton Roads, pp. 50–67.

Ross, E.D. and Mesulam, M.M. 1979. Dominant language functions of the right hemisphere. *Archives of Neurology*, 36, 144–148.

Schaut, G.B. and Persinger, M.A. 1985. Subjective telepathic experiences, geomagnetic activity and the ELF hypothesis. Part I. Data analysis. *Psi Research*, 4(1), 4–20.

Schmeidler, G.R. 1994. ESP experiments 1978–1992: The glass is half full, in S. Krippner (ed.), *Advances in Parapsychological Research*, vol. 7. Jefferson, NC: McFarland, pp. 104–197.

Sherwood, S.J. and Roe, C.A. 2003. Dream ESP studies since the Maimonides programme. *Journal of Consciousness Studies*, 10, 85–109.

Sidgwick, H., Sidgwick, E., and Johnson, A. 1894. Report on the Census of Hallucinations. *Proceedings of the Society for Psychical Research*, 10, 25–422.

Spanos, N.P., Cross, P.A., Dickson, K., and Dubreuil, S.C. 1993. Close encounters: An examination of UFO experiences. *Journal of Abnormal Psychology*, 102, 624–632.

Stace, W.T. 1960. *Mysticism and Philosophy*. Philadelphia: Lippincott.

Stanovich, K. 1985. *How to Think Straight about Psychology*. Glenview, IL: Scott Foresman.

Stevenson, I. 1970. *Telepathic Impressions*. Charlottesville: University Press of Virginia.

———. 1983. *Unlearned Language: New Studies in Xenoglossy*. Charlottesville: University Press of Virginia.

———. 1997. *Biology and Reincarnation: A Contribution to the Etiology of Birthmarks and Birth Defects*. Westport, CT: Praeger.

Stokes, D. 1997. Spontaneous psi phenomena, in S. Krippner (ed.), *Advances in Parapsychological Research*, vol. 8. Jefferson, NC: McFarland, pp. 6–87.

Subrahmanyam, S., Sanker Narayan, P.V., and Srinivasan, T.M. 1985. Effect of magnetic micropulsations on the biological systems—a bioenvironmental study. *International Journal of Biometereology*, 29, 293–305.

Targ, E., Schlitz, M., and Irwin, H.J. 2000. Psi-related experiences, in E. Cardeña, S.J. Lynn, and S. Krippner (eds.), *Varieties of Anomalous Experience: Examining the Scientific Evidence*. Washington, DC: American Psychological Association, pp. 219–252.

Targ, R. 2004. *Limitless Mind: A Guide to Remote Viewing and Transformation of Consciousness*. Novato, CA: New World Library.

Tart, C.T. (ed.). 1975. *Transpersonal Psychologies*. New York: Harper and Row.

Tart, C.T. 1997. On the scientific study of nonphysical worlds, in C.T. Tart (ed.), *Body, Mind, Spirit: Exploring the Parapsychology of Spirituality*. Charlottesville, VA: Hampton Roads, pp. 214–219.

Tart, C.T. 2003. Spiritual motivations of parapsychologists? Empirical data. *Journal of Parapsychology*, 67, 181–184.
Taylor, D.C. 1972. Mental state and temporal lobe epilepsy. *Epilepsia*, 13, 727–765.
Teyler, T.J. and DiScenna, P. 1984. The topological anatomy of the hippocampus: A clue to its function. *Brain Research Bulletin*, 12, 711–719.
Thouless, R.H. 1977. Implications for religious studies, in S. Krippner (ed.), *Advances in Parapsychological Research*, vol. 1. New York: Plenum, pp. 175–190.
Truzzi, M. 1971. Definition and dimensions of the occult: Toward a sociological perspective. *Journal of Popular Culture*, 5, 635–646.
———. 1985. A skeptical look at Paul Kurtz' analysis of the scientific status of parapsychology. *Journal of Parapsychology*, 44, 35–55.
Ullman, M., Krippner, S., and Vaughan, A. 1989. *Dream Telepathy*, 2nd ed. Jefferson, NC: McFarland.
Uttal, W.R. 2001. *The New Phrenology: On the Localization of Cognitive Processes*. Cambridge, MA: MIT Press.
Utts, J. 1991. Replication and meta-analysis in parapsychology. *Statistical Science*, 6, 363–386.
Van de Castle, R.L. 1971. The study of GESP in a group setting by means of dreams. *Journal of Parapsychology*, 35, 312.
Wade, J. 2004. *Transcendent Sex: When Lovemaking Opens the Veil*. New York: Paraview Pocket Books.
Warren, C.A., McDonough, B.E., and Don, N.S. 1992. Partial replication of single-subject event-related potential effects in a psi task. *Proceedings of Presented Papers: 35th Annual Parapsychological Association Convention*. Durham, NC: Parapsychological Association, pp. 315–334.
White, R.A. 1995. Exceptional human experiences and the experiential paradigm. *ReVision*, 18, 18–25.
———. 1997. Exceptional human experiences, in C.T. Tart (ed.), *Body, Mind, Spirit: Exploring the Parapsychology of Spirituality*. Charlottesville, VA: Hampton Roads, pp. 83–100.
Wilson, S.C. and Barber, T.X. 1982. The fantasy-prone personality: Implications for understanding imagery, hypnosis, and parapsychological phenomena. *Psi Research*, 1(3), 944–1116.
Winkelman, M. 1986. Trance states: A theoretical model and cross-cultural analysis. *Ethos*, 14, 174–203.
Winkelman, M.J. 1992. *Shamans, Priests and Witches: A Cross-Cultural Study of Magico-Religious Practitioners*. Tempe: Arizona State University.
Woerlee, G.M. 2004, May–June. Darkness, tunnels, and light. *Skeptical Inquirer*, pp. 28–32.
Wulff, D.M. 2000. Mystical experiences, in E. Cardeña, S.J. Lynn, and S. Krippner (eds.), *Varieties of Anomalous Experience: Examining the Scientific Evidence*. Washington, DC: American Psychological Association, pp. 397–440.

CHAPTER FOUR

Sacred Emotions

ROBERT A. EMMONS

Introduction

Given that religion is a human universal, evolutionary and cultural perspectives on the emotions are incomplete without a comprehensive understanding of the role of religion. Religion provides context and direction for emotion and the influence of religious systems on emotional experience and expression is considerable. For example, religions prescribe certain emotions and proscribe others. Religion also modulates the expression of emotion. This chapter will examine emotions and emotional processes that normally occur in the context of religion. Recent scientific research on religion and emotion will be highlighted.

There has been rapid growth in the psychology of religion (Emmons and Paloutzian 2003) and the psychology of emotion (Lewis and Haviland-Jones 2000) in recent years, and so one would expect to see considerable scholarship directed toward the interface of these two fields. While a literature search using PsycINFO database for the period 1988–2002 returned 2,875 citations for the term religion and 5,116 for the term emotion, a scant 5 citations include both terms! The range of emotional phenomena is vast, and we cannot attempt to do justice to this vastness within a single chapter. Because of the recent emergence of the scientific study of positive emotions, we will emphasize the role of religion in the generation and regulation of emotional experience, focusing primarily on positive emotional experience. The study of positive emotions is a major trend in contemporary affective science (Fredrickson 2001), and we wish to highlight the many ways in which the psychology of religion can contribute to a growing understanding of positive emotions and the functions of positive emotions in people's lives.

This chapter has several purposes: to present a brief historical overview on the study of emotion and religion; to review recent research on emotions typically considered to be religious; to document the various ways in which religion might modulate emotional experience; to consider various functions that religious emotions might serve. Our overriding concern is to sketch the newest emerging lines of research that show promise of contributing significantly to the psychology of religion and of emotion during the next several years. We begin first by describing what we mean by emotion.

What is Emotion? Levels of Emotional Experience

Any discussion of religion and emotion presupposes an understanding of what emotion is. The field of affective science has been moving toward standardized terminology that provides researchers and clinicians with a common frame of reference. Thus, before beginning our presentation of the literature on religion and emotion, it might be helpful to familiarize the reader with what is meant by the concept of emotion, and how an emotion differs from other related affective phenomena. In doing so, we will borrow from the recent conceptual analysis of Rosenberg (1998). She proposed that the common forms of affective experience could be structured into three hierarchical levels of analysis: affective traits, moods, and emotions.

Rosenberg (1998) placed affective traits at the top of the hierarchy of affective phenomena. She defined affective traits as stable predispositions toward certain types of emotional responding that set the threshold for the occurrence of particular emotional states. For example, hostility is thought to lower one's threshold for experiencing anger, or happiness could be thought of as lowering one's threshold for experiencing pleasant affect. Affective traits are relatively stable components of personality, which are consistently expressed over time and across situations. Some of the research that we review in this chapter will be focused on this level of the affective hierarchy.

In contrast to affective traits, emotions are "acute, intense, and typically brief psychophysiological changes that result from a response to a meaningful situation in one's environment" (Rosenberg 1988, p. 250). Emotions are a subset of a larger class of affective phenomena (Fredrickson 2001). They are discrete states that involve the appraisal of the personal meaning of a circumstance in a person's environment. Both the type of emotion experienced, and its intensity, depend on cognitive interpretation or appraisal of the situation. Such appraisal involves not only assessing the nature of the external situation or event that might cause the emotional response, but also the responses of other people exposed to that same situation or event. Emotions typically motivate a particular course of action; each discrete emotion triggers a particular action tendency (Fredrickson 2001). A major division between types of emotions are affect program theories and propositional attitude theories (Griffiths 1997; Roberts 2003). Affect programs pertain to the basic, universal emotions such as anger, disgust, joy, sadness, and fear while the latter category contain a wider range of cognitively complex emotions including guilt, shame, pride, and gratitude. Basic emotions are universal and innate. There exists for each a recognizable facial expression and distinct physiological patterning. The higher, cognitively complex emotions depend heavily on cognitive appraisals and are assumed to exhibit greater cultural variation. Religion, at least when it comes to the generation of emotion, appears to have more do with the latter than with the former.

Rosenberg considered moods, which wax and wane, fluctuating throughout or across days, as subordinate to affective traits, but as superordinate to discrete emotion episodes. Moods are subtle and less accessible to conscious awareness than are emotions (i.e., one is less likely to be aware of anger as a mood than as an emotion). Despite their subtlety relative to emotions, however, moods are important because they are expected to have broad, pervasive effects on consciousness that emotions simply cannot because of their relatively short duration (Rosenberg 1998). Because the majority of research on religion and affect has been at the level of affective traits of discrete emotions, we will have comparatively little to say about religion and mood.

Conceptualizing Links between Religion and Emotion

The connection between religion and emotion is a long and intimate one. For one, religion has always been a source of profound emotional experience. Commenting on this historical association, Pruyser writes, "there is something about emotion that has always had a great appeal to the religionist" (1967, p. 142). Religion likely influences both the generation of emotion and the regulation of emotional responses. We will discuss religion and the generation of specific, discrete emotions later. Links between religion and emotion can also be seen in religious attitudes toward emotional experience and expression. Watts (1996) distinguishes between two main notions about the role of emotions in religious life. The charismatic movement stresses the cultivation of intense positive emotions and their importance in religious experience and collective religious rituals (see also McCauley 2001), whereas the contemplative tradition stresses a calming of the passions and the development of emotional quietude. In addition to these two approaches to regulating emotions, there is the ascetic view (Allen 1997) that links religion with greater awareness of emotion (possible emotional intelligence, to use a contemporary term) and the creative expression of emotion.

Silberman (2003) suggests three ways in which religious and spiritual meaning systems influence emotion. First, religion prescribes appropriate and inappropriate emotions and their levels of intensity. For example, within Judaism, people are encouraged to love God with all of their hearts (Deut. 6:5) and to serve God with joy (Deut. 28: 47). Second, beliefs about the nature and attributes of God may give rise to specific emotions as well as influence overall emotional well-being. For example, a belief about a loving personal God may have a positive effect on emotional well-being, while a belief about a punitive vengeful God may have the opposite effect. Third, religion offers the opportunity to experience a uniquely powerful emotional experience of closeness to the sacred (Otto 1928).

Concerning the intensity issue, Ben Ze'ev (2002) hypothesizes that religion influences the intensity of emotion in three ways. First, religious belief systems influence the meaningfulness attached to events. To the degree to which people perceive a divine influence on daily events, these events will be perceived as more meaningful and hence capable of generating stronger emotions than ordinary events. Second, according to Ben Ze'ev, religious and nonreligious persons differ in their perceptions deservingness for life events. Because of the belief that events signify God's intention and will, religious individuals are more likely to be accepting of life events than nonreligious individuals, and deservingness is typically associated with less intense emotional reactions. Third is the issue of controllability. Religious persons, according to Ben Ze'ev, typically believe that God directs and controls everyday events. Personal controllability is positively associated with emotional intensity; thus, all things being equal, the emotional intensity of religious individuals would be lower than that of nonreligious individuals. These are intriguing hypotheses that need to be empirically tested.

Religion and the Generation of Emotion

The role of emotion in religion was found to be central in several prominent accounts of religious experience. Jonathan Edwards described the function of religious emotions in his theological classic *A Treatise Concerning Religious Affections* (1746/1959). Edwards was so struck by the evidentiary force of emotion that he made it a cornerstone of his theology, as exemplified in this quote: "The Holy Scriptures do everywhere place religion very much in the affections; such as fear, hope, love, hatred, desire, joy, sorrow, gratitude, compassion, and zeal" (p. 96). These affections were divided into two groups according to whether they were characterized by approval (gratitude, love, joy) or disapproval (hatred, fear, sorrow). Thus, an important appraisal dimension for Edwards was approval/liking versus disapproval/rejection (Pruyser 1967). Rather than belief, which was seen as intellectual and heart-less by Edwards, these affections were to be taken as the signs of genuine spiritual experience. A review of his contributions (Hutch 1978) suggests that considerable insights can be gained from a reading of his insights into the nature of religious emotions.

Schleiermacher's (1799) notable treatise on religion also placed emotion at the center of conscious religious experience. Feeling was central. Reverence, humbleness, gratefulness, compassion, remorse, and zeal were described as essential elements of religious experience by Schleiermacher. In agreement with Edwards, Schleiermacher viewed intellectual beliefs as overly rational and lacking in spontaneity; the heart of religion was seen as the heart, not the head (Pruyser 1967, p. 140).

Arnold was quite possibly the first psychology of emotion theorist to write extensively about positive human emotions in her book *Emotion and Personality*. In the chapter on positive emotions, she included a section on religious emotions in which she noted that in addition to the prototypical religious emotions of reverence and awe, which Otto (1917/1958) and others had identified, several other emotions can be experienced toward God (which was her criteria for a religious emotion). In particular, love, joy, and happiness are "reactions to overwhelming abundance, an infinity, of the good and the beautiful" (1960, p. 328) and contain "a hint of eternity" (p. 160). Clearly, these emotions are imbued with a spiritual significance for Arnold. They serve the function of motivating people toward states of perfection and total fulfillment. Her phenomenological analysis of happiness as a religious feeling and its differentiation from joy, serenity, and contentment was an early, important contribution to understanding differences between discrete positive emotions.

What Makes Emotions Sacred?

What does it mean to say that certain emotions or emotional experiences are sacred? We can identify several characteristics of sacred emotions. First, sacred emotions are those emotions that are more likely to occur in religious (e.g., churches, synagogues, mosques) *settings* than in nonreligious settings. However, this does not mean that sacred emotions cannot be experienced in nonreligious settings. Second, sacred emotions are those that are more likely to be elicited through spiritual or religious activities or *practices* (e.g., worship, prayer, meditation) than by nonreligious activities. However, this does not mean they cannot be activated through nonreligious channels as well. Third, sacred emotions are more likely to be experienced by *people* who self-identify as religious or spiritual (or both) than by people who do not think of themselves as either spiritual or religious. However, sacred emotions can be felt (on occasion) by people who do not think of themselves as religious or spiritual. Fourth, sacred emotions are those emotions that religious and spiritual *systems* around the world have traditionally sought to cultivate in their adherents. Fifth and last, sacred emotions are those emotions experienced when individuals imbue seemingly secular aspects of their lives (e.g., family, career, events) with a spiritual significance (Mahoney and Pargament 1999; Pargament 2002).

The search for the sacred is the defining feature of religion (Hill et al. 2000). The term "sacred" refers to a divine being, a divine object, an ultimate reality, or an Ultimate Truth as perceived by the individual (Hood et al., p. 68). Pargament (1999) has argued that conceiving of spirituality in terms of an ability to imbue everyday experiences, goals, roles, and responsibilities with sacredness opens new avenues for empirical exploration. Furthermore, perceiving aspects of life as sacred is likely to elicit spiritual emotions.

Spiritual emotions such as gratitude, awe and reverence, love and hope are likely to be generated when people perceive sacredness in various aspects of their lives. Mahoney et al. (1999) found that when marital partners viewed their relationship as imbued with divine qualities, they reported greater levels of marital satisfaction, more constructive problem solving behaviors, decreased marital conflict, and greater commitment to the relationship, compared to couples who did not see their marriage in a sacred light. Similarly, Tarakeshwar et al. (2001) found that a strong belief that nature is sacred was associated with greater pro-environment beliefs and a greater willingness to protect the environment. A plausible hypothesis to be tested in future research is whether sanctification of the environment leads to the experiencing of more frequent and intense sacred emotions such as awe and wonder in nature.

Specific Sacred Emotions

Gratitude

Gratitude has been defined as "the willingness to recognize the unearned increments of value in one's experience" (Bertocci and Millard 1963, p. 389), and "an estimate of gain coupled with the judgment that someone else is responsible for that gain" (Solomon 1977, p. 316). At its core, gratitude is an emotional response to a gift. It is the appreciation felt after one has been the beneficiary of an altruistic act. Some of the most profound reported experiences of gratitude can be religiously based or associated with reverent wonder toward an acknowledgment of the universe (Goodenough 1998), including the perception that life itself is a gift. In the great monotheistic religions of the world, the concept of gratitude permeates texts, prayers, and teachings. Worship with gratitude to God for the many gifts and mercies are common themes, and believers are urged to develop this quality. A religious framework thus provides the backdrop for experiences and expressions of gratitude.

McCullough and colleagues (2001) recently reviewed the classical moral writings on gratitude and synthesized them with contemporary empirical findings. They suggested that the positive emotion of gratitude has three moral functions: it serves as a moral barometer (an affective readout that is sensitive to a particular type of change in one's social relationships, the provision of a benefit by another moral agent that enhances one's wellbeing), a moral motivator (prompting grateful people to behave prosocially themselves), and a moral reinforcer (that increases the likelihood of future benevolent actions). McCullough et al. (2002) found that measures of gratitude as a disposition were positively correlated with nearly all of the measures of spirituality and religiousness, including spiritual transcendence, self-transcendence, and the single-item religious variables. The grateful disposition was also related to measures of spiritual and religious tendencies. Although these correlations were not large (i.e., few of them exceeded $r = .30$), they suggest that spiritually or religiously inclined people have a stronger disposition to experience gratitude than do their less spiritual/religious counterparts. Thus, spiritual and religious inclinations may facilitate gratitude, but it is also conceivable that gratitude facilitates the development of religious and spiritual interests (Allport et al. 1948) or that the association of gratitude and spirituality/religiousness is caused by extraneous variables yet to be identified. The fact that the correlations of gratitude with these affective, prosocial, and spiritual variables were obtained using both self-reports and peer-reports of the grateful disposition suggests that these associations are substantive and not simply the product of mono-method biases in measurement. This study may be also be useful for explaining why religiously involved people are at a lower risk for depressive symptoms or other mental health difficulties.

McCullough et al. (2002) found that people who reported high levels of spirituality reported more gratitude in their daily moods, as did people higher in religious interest, general religiousness, and intrinsic religious orientation. Interestingly, however, the extrinsic, utilitarian religious orientation and quest seeking religious orientation were not significantly correlated with the amount of gratitude in daily mood. These findings suggest that people high in conventional forms of religiousness, especially people for whom religion is a fundamental organizing principle (i.e., people high in intrinsic religiousness) and people who report high levels of spiritual transcendence, experience more gratitude in their daily moods than do their less religious/spiritual counterparts. Watkins et al. found that the trait gratitude correlated positively with intrinsic religiousness and negatively with extrinsic religiousness. The authors suggest that the presence of gratitude may be a *positive* affective hallmark of religiously and spiritually engaged people, just as an absence of depressive symptoms is a *negative* affective hallmark of spiritually and religiously engaged people. They likely see benefits as gifts from God, "as the first cause of all benefits" (Watkins et al. 2003, p. 437).

Awe and Reverence

Few would disagree that the emotions of awe and reverence are central to religious experience. Awe was the cornerstone of Otto's (1917/1958) classic analysis of religious experience. The essence of religious worship, for Otto, was the overpowering feeling of majesty and mystery in the presence of the holy, which is at the same time fascinating and dreadful. This juxtaposition of fear and fascination is a hallmark of religious awe (Wettstein 1997).

Several philosophers of emotion have offered conceptual analyses of awe where they define awe and distinguish it from reverence and related states. Roberts describes awe as a *sensitivity to greatness*, accompanied by a sense of being overwhelmed by the object of greatness, and reverence as "an acknowledging subjective response to something excellent in a personal (moral or spiritual) way, but qualitatively above oneself" (2003, p. 268). The major distinction between awe and reverence, for Roberts, is that awe could equally be experienced in response to something perceived as vastly evil as to something vastly good, but reverence is typically reserved for those things or persons esteemed worthy of it, in a positive or moral sense. Similarly, Woodruff states, "Reverence is the well-developed capacity to have the feelings off awe, respect and shame when these are the right feelings to have" (2001, p. 8). Solomon (2002) argues that awe is passive whereas reverence is active: to be awestruck implies paralysis, while reverence leads to active engagement and responsibility toward that which a person reveres.

In contrast to these substantial theological and philosophical writings, little research in the psychology of religion has focused on either awe or reverence as a religious emotion. Many psychologists mention awe in their studies of religious experiences, but few have attempted to study it systematically. Maslow included the experience of awe under the broad umbrella of "peak experiences" (1964, p. 65), an umbrella that included "practically everything that, for example, Rudolf Otto defines as characteristic of religious experience" (p. 54). Several other studies have included awe under the slightly less broad category of mystical experiences, but since awe is not the purpose of these studies, their research and conclusions are difficult to utilize with respect to awe. For example, though Hardy (1979) lists awe, reverence, wonder as a category of religious experience recorded in his database, his examples merely include a description of the "it" of a particular mystical experience or mention awe as an aftereffect of the experience. Interestingly, Hardy found that awe was not a particularly frequently reported experience: awe, reverence, and wonder occurred in 7 percent of reported religious experiences that he collected, compared to 21 percent for joy and happiness and 25 percent for peace and security. Likewise, when Hood (1975) included awe as an item on his mysticism scale, he was not interested in the experience of awe per se, but in the mystical experience that might (or might not) produce awe.

Keltner and Haidt have recently offered a prototypical approach to awe, which represents an important new contribution. According to their definition, an awe experience includes both a *perceived vastness* (whether of power or magnitude) and a *need for accommodation*, which is an "inability to assimilate an experience into current mental structures" (2003, p. 304). Variation in the valence of an awe experience depends on whether the stimulus is appraised in terms of beauty, exceptional ability, virtue, perceived threat, or supernatural origin. In contrast, those experiences that do not include both perceived vastness and need for accommodation are not occurrences of awe, but are simply members of the awe family. For example, surprise involves accommodation without vastness. Feelings of deference involve vastness without accommodation. Unfortunately, there is very little empirical research on awe, and until this changes anything we say about awe as a religious emotion must be restricted to what can be gleaned from sacred writings.

As the study of awe is still in its early stages, future research should begin with the prototype approach to awe offered by Keltner and Haidt (2003) and the definition of reverence offered by philosophers and theologians

(Roberts 2003; Woodruff 2001), and go on to develop tests to measure individual differences in these experiences. Once a reliable measure of awe and reverence exists, individual differences in these experiences can be explored, as well as their relation to religion and spirituality, their developmental antecedents, and their association with emotional and physical well-being.

Wonder

Wonder is another emotion that has received scant empirical attention by psychologists but has a significant spiritual thrust. Bulkeley defined wonder as "the emotion excited by an encounter with something novel and unexpected, something that strikes a person as intensely powerful, real, true, and/or beautiful" (2002, p. 6). Brand (2001) provided a phenomenological account of wonder–joy: profound and deeply moving experiences of positive emotions where there is a co-occurrence of feelings of wonder, joy, gratitude, awe, yearning, poignancy, intensity, love, and compassion. They are an opening up of the heart to the persons or profound circumstances being witnessed and are triggered by a variety of circumstances. Experiences of wonder are a significant feature of many of the world's religious, spiritual, and philosophical traditions (Bulkeley 2002). Bulkeley proposes that the experience of wonder involves a two-fold process of (1) a sudden decentering of the self when faced with something novel and unexpectedly powerful, followed by (2) and ultimate recentering of the self in response to new knowledge and understanding. It is evident that the wonder that Bulkeley describes and the sense of awe described by Haidt and Keltner have much in common; it will be up to future research to establish the unique properties of these overlapping states.

Hope

Hope is a theological virtue, one of the Big 3, along with faith and charity. In Christian theology, hope is looking forward to the eternal world where the kingdom of God will be ushered in: "Let us hold unswervingly to the hope we profess, for he who promised is faithful" (Heb. 10:23, *NIV*). In its religious context, hope provides respite during trials, brings perseverance during challenges, and provides assurance of eternal joy.

Hope research has burgeoned over the past decade, with studies indicating numerous positive effects on mental and physical health (see Snyder et al. 2002, for a review). In this light, whenever religion fosters or hinders hope, one would expect significant positive effects on the whole person. In current research, the construct of hope is often couched in terms of goals, with hope requiring the thought of a goal, perceived pathways to those goals (pathway thoughts), and motivation (agency thoughts) to follow through to the goal. Snyder et al. (2002) use this understanding of hope to explain the link previously found between religion or religious involvement and health or well-being: religions provide adherents with goals, paths to those goals and incentives to reach those goals, either for good or for ill.

Sethi and Seligman (1993) found that among nine Jewish, Christian, and Muslim groups, the more fundamentalist the group was, the more hopeful and optimistic were the sermons, the liturgy, and the average participant's outlook. This finding of greater hope in persons of fundamentalist faiths is an intriguing one, given that fundamentalism is often associated with a more constricted and less spontaneous approach to life. Could it be that persons in conservative faith systems tend to present overly positive images of themselves and thus deny negative emotions? Bullard and Park (2001) tested the hypothesis that fundamentalism (measured in terms of adherence to Protestant orthodoxy) is related to the overt expression of emotions. They used a frequently employed measure of emotional expressiveness that classifies respondents into high anxious, low anxious, repressor, or defensively high-anxious categories. Fundamentalism was associated with anxiety such that the low fundamentalism group was more likely to be highly anxious; no significant patterns were found between the other three expressive styles and fundamentalism. Thus, the occurrence of greater positive emotions found in fundamentalist faiths is not due to the non-expression or repression of negative affect. This study is the only one that has examined whether adherence to religious doctrine is associated with styles of emotional expression.

Religion and the Regulation of Emotion

Emotion regulation refers to the processes by which individuals influence which emotions they have, the intensity of these emotions, and how they are expressed (Gross 1999). The regulatory process may be conscious or unconscious, intentional or unintentional. Emotions, both positive and negative, can be transformed or regulated by intentionally engaging in spiritual practices. Religious teachings and texts contain information concerning how emotions should be handled. The importance of emotion regulation in everyday life provides a legitimate rationale for examining the role of religion in this process. Emotional regulation techniques that have their rationales in religious traditions can modulate everyday emotional experience (Schimmel 1997; Watts, 1996) providing spiritual rationales and methods for handling problematic emotions such as anger, guilt, and depression. Watts and Williams (1988, chapter 6) draw parallels between religious and clinical approaches to emotional control and cite meditational training as an activity with origins in both Western and Eastern contemplative religions. Positive emotional benefits have been reported in Zen meditation (Gillani and Smith 2001) and the cultivation of transpersonal states long associated with spiritual and religious traditions (McCraty et al. 1998). Baer (2003) reviewed the literature on mindfulness-meditation interventions and found that these interventions appear to alleviate a variety of negative emotional states (primarily anxiety and depression) and may be efficacious in cultivating positive states such as compassion.

Thayer et al. (1994) examined the success of several behavioral and cognitive strategies for regulating unpleasant moods and raising energy levels. One category of strategies were labeled as religious/spiritual, though there was no information provided as to what these specific religious and spiritual strategies actually were. As mood management techniques, these were found to be more common in older participants than in younger ones and were particularly effective for reducing nervousness, tension, and anxiety. Although spiritual and religious activity was not among the most common behaviors used to reduce tension and anxiety, it was rated as the most successful. In a factor analysis, religious and spiritual techniques loaded on a pleasant distraction factor; this factor was found to be the most effective strategy for mood change. Interestingly, although low in absolute frequency (study participants were doctoral level psychotherapists), religious practices were rated as the single best method of regulating unpleasant moods.

Forgiveness

Forgiveness is a religiously based technique that has been shown to be powerful in regulating negative emotions. Pargament (1997) suggests that forgiveness is religious in that (1) religion lends a spiritual significance to the act of forgiving, and (2) religion offers role models and concrete methods to facilitate forgiveness. Forgiveness as a contemporary psychological or social science construct has generated popular and clinical interest as well as empirical investigation (for reviews, see McCullough et al. 2000; Witvliet et al. 2001, 2002). The scientific literature on forgiveness is growing rapidly across a number of areas of psychology, including the social–clinical interface (McCullough 2001), although clinical applications of forgiveness probably still bear little connection to empirical research.

There have been a handful of studies that have been explicitly designed to examine the impact of forgiveness on the remediation of negative emotions. Witvliet and her colleagues (2001, 2002) examined subjective emotions and emotional physiology during forgiving and unforgiving imagery. In their initial study, Witvliet et al. (2001) found that when participants visualized forgiving responses toward people who had offended them, they experienced significantly less anger, sadness, and overall negative arousal compared with when they rehearsed the offense or maintained a grudge. Paralleling the self-reports were greater SNS arousal (skin conductance and blood pressure increases) and facial tension during unforgiving imagery. A follow-up study examined the emotions of transgressors (Witvliet et al. 2002). When transgressors imagined seeking forgiveness from their victims, they reported lower levels of sadness, anger, and guilt and higher levels of hope and gratitude *if* they imagined the victims genuinely forgiving them. Imagining reconciliation rather than forgiveness led to a similar reduction in negative emotions (anger, sadness, guilt) and increase in positive ones (gratitude, hope, empathy).

Forgiveness interventions have also been shown to be successful in alleviating depression, anxiety, and grief in male partners of post-abortion women (Coyle and Enright 1997) and depression and anxiety in incest survivors (Freedman and Enright 1996). In the latter study, the intervention group also showed significant gains in overall levels of hopefulness, suggesting, as did the work of Witvliet and colleagues, that forgiveness is involved in facilitating positive emotions as well as reducing negative emotions. The ability of forgiveness interventions to increase certain positive emotions is one of the more surprising findings in the research literature on forgiveness to date.

In one of the few cultural studies on forgiveness, Huang and Enright (2000) examined forgiveness and anger in Taiwanese sample. Adults recalled an incident of deep, interpersonal hurt and their affective state was recorded both during and after recall. The researchers found that when participants granted forgiveness unconditionally out of a sense of compassion, self-reported levels of anger were lower than when they forgave out of a sense of duty or obligation. Thus, the effectiveness *of* forgiveness to reduce negative emotions is contingent upon the motivation *for* forgiveness.

Mindfulness

A number of philosophical, psychological, and spiritual traditions, both East and West, highlight the importance of mindfulness, but are there really adaptational and mental health benefits to being more conscious of what's happening in the here and now? Mindfulness—an enhanced attention to and awareness of the present—is currently the subject of innumerable books, seminars, and workshops designed to facilitate this quality of consciousness as a means to helping people live more authentic, happier lives. But very little research has examined its direct role in psychological health and well-being.

Brown and Ryan (2003) first developed a self-report instrument, called the Mindful Attention Awareness Scale (MAAS), to measure mindfulness, and administered it to subjects ranging from college students to working adults to Zen meditators to cancer patients. In mindfulness, which Brown and Ryan showed is a unique quality of consciousness, two experiences work in tandem: attending to present, ongoing events, and experiences while allowing new events and experiences to come into awareness. In their research, Brown and colleagues have found that more mindful individuals, as measured by the MAAS, have a greater self-regulatory capacity and higher levels of well-being.

Regarding self-regulation, Brown and Ryan (study 3) showed that those who are more mindful are more attuned to their emotions, as reflected in a higher concordance between their explicit, or self-attributed emotional states and implicit, or nonconscious emotions. Because implicit measures are not susceptible to conscious control and manipulation, this suggests that more mindful individuals are more attuned to their implicit emotions and reflect that knowledge in their explicit, affective self-descriptions. This is consistent with theory positing that present-centered awareness and attention facilitates self-knowledge—a crucial element of integrated functioning.

A number of studies have shown that mindfulness has direct relations to well-being outcomes, as well. For example, Brown and Ryan (study 1) report that similar to other personal qualities, mindfulness can be cultivated and enhanced, or neglected and allowed to diminish. Brown and Ryan (study 2) also showed that people who actively cultivated a heightened attention to and awareness of what's taking place in the present moment through meditative practices had higher levels of mindfulness. And in a clinical study with early-stage cancer patients who received training in mindfulness as the central element of an eight-week stress reduction program (study 5), those individuals who showed greater increases in mindfulness, as assessed by the MAAS, showed greater declines in mood disturbance and stress.

Religion and Emotion: Remaining Issues

Functions of Religious Emotions

Current models of emotions typically aim to describe the form and function of emotions in general. Despite this aim, many models are formulated around prototypic and negative emotions like fear and anger. For instance, key to many theorists' models of emotions is the idea that emotions are, by definition, associated with specific action tendencies. What functions do religious emotions serve? Noting that traditional models based on specific action tendencies did not do justice to positive emotions, Fredrickson (2001) developed an alternative model for the positive emotions that better captures their unique effects. She called this the broaden-and-build theory of positive emotions because positive emotions appear to *broaden* people's momentary thought–action repertoires and *build* their enduring personal resources. Whereas the narrowed mindsets of negative emotions carry direct and immediate adaptive benefits in situations that threaten survival, the broadened mindsets of positive emotions, which occur when people feel safe and satiated, are beneficial in other ways. Specifically, these broadened mindsets carry indirect and long-term adaptive benefits because broadening *builds* enduring personal resources (Fredrickson 1998).

Fredrickson (2001) analyzed the functions of several distinct positive emotions. Joy, for instance, creates the urge to play, push the limits, and be creative—urges evident not only in social and physical behavior, but also in intellectual and artistic behavior. Interest, a phenomenologically distinct positive emotion, creates the urge to explore, take in new information and experiences, and expand the self in the process. Contentment, a third distinct positive emotion, creates the urge to savor current life circumstances, and integrate these circumstances into new views of self and of the world. And gratitude, a fourth distinct positive emotion, creates the urge to creatively repay kindness. These various thought–action tendencies, to play, explore, savor and integrate, and repay kindness, each represent ways that positive emotions broaden habitual modes of thinking or acting. In general terms, then, positive emotions appear to enlarge the cognitive context (Isen 1987), an effect recently linked to increases in brain dopamine levels (Ashby et al. 1999).

Finding positive meaning is perhaps the most reliable path to cultivating positive emotions (Ryff and Singer 1998). To the extent that religions offer their believers worldviews that help them to find positive meaning in both ordinary daily events (e.g., appreciating nature) and major life challenges (e.g., finding benefit in a cancer diagnosis), they also cultivate positive emotions such as joy, serenity, awe, gratitude, and hope. According to the broaden-and-build theory, these positive emotions should in turn broaden people's mindsets, making them more creative and integrative in their thinking, and build and replenish critical personal and social resources, such as resilience, optimism, and social support. These resources, a wide range of studies have shown, enhance health and well-being.

In future research, it will be important to conceptually and empirically distinguish secular positive emotions (i.e., positive emotions felt outside religious or sacred contexts) from one or more categories of religious or sacred positive emotions, which might include positive emotions felt in religious services toward God or a higher power, toward other believers, or otherwise connected to that which believers imbue with a sense of the sacred. Religious practices may be distinctly human ways of initiating upward spirals that enhance spiritual growth as well as health and well-being.

Are Religious Emotions Unique?

A perennial issue in the psychology of religion pertains to the uniqueness of emotions that are labeled as religious. Are these a separate class of emotions or simply ordinary emotions felt in religious contexts or elicited through religious rituals such as prayer and worship? Consider these statements from William James:

> In the psychologies and in the philosophies of religion, we find the authors attempting to specify just what entity it is. One man allies it to the feeling of dependence; one makes it a derivative from fear; others connect it with the sexual life; others still identify it with the feeling of the infinite; and so on. Such different ways of conceiving it ought of themselves to arouse doubt as to whether it possibly can be one specific thing; and the moment we are willing to treat the term "religious sentiment" as a collective name for the many sentiments which religious objects may arouse in alternation, we see that it probably contains nothing whatever of a psychologically specific nature. There is religious fear, religious love, religious awe, religious joy, and so forth. But religious love is only man's natural emotion of love directed to a

religious object; religious fear is only the ordinary fear of commerce, so to speak, the common quaking of the human breast, in so far as the notion of divine retribution may arouse it; religious awe is the same organic thrill which we feel in a forest at twilight, or in a mountain gorge; only this time it comes over us at the thought of our supernatural relations; and similarly of all the various sentiments which may be called into play in the lives of religious persons. As concrete states of mind, made up of a feeling plus a specific sort of object, religious emotions of course are psychic entities distinguishable from other concrete emotions; but there is no ground for assuming a simple abstract "religious emotion" to exist as a distinct elementary mental affection by itself, present in every religious experience without exception. As there thus seems to be no one elementary religious emotion, but only a common storehouse of emotions upon which religious objects may draw, so there might conceivably also prove to be no one specific and essential kind of religious object, and no one specific and essential kind of religious act. (1902/1958, pp. 39–40)

For James, what makes religious emotion *religious* are ordinary emotions felt under circumstances that make it apparent to the person that God or a higher power is involved.

Emotion and Spiritual Transformation

A major research area in the psychology of religion has always been conversion or transformation (Paloutzian et al. 1999). Many theories of the processes underlying spiritual transformations have been offered, but virtually all converge on the importance of the affective basis of spiritual transformation (Oatley and Djikic 2002). In these perspectives, emotions are seen as agents of transformation in the spiritual self. While the emphasis is generally placed on the role of negative emotions in triggering spiritual changes, positive emotions may play an important role as well. For example, Allport et al. (1948) found that gratitude was the fourth most cited reason among youth for turning to religion, and the role of gratitude and goodness (as well as awe and wonder) in G.K. Chesterton's (1936) adult conversion to Catholicism is legendary.

The study by Ullman (1989) is frequently cited as supporting the hypothesis that conversion is more based on emotion than it is on an intellectual search process. In examining conversion among four different faith groups, Ullman found that the converts reported a greater degree of emotional distress in childhood than did the nonconverts, and were more likely to say that emotional stress was a more important factor in their conversion than was a cognitive quest.

The research that exists is suggestive of links between emotion and transformation, but much more needs to be done. There is a great need for longitudinal studies of both emotion as a motivator and changes in affective traits as a consequence of transformation. Future research should also focus more on positive emotions, both as motivators of change as well as potential consequences of change. The measurement of positive emotions has improved considerably in recent years and researchers have established and well-validated measures to draw upon and incorporate into their research designs.

Religious Emotions in Emotion History: Two Illustrations

A vastly different approach to emotion and religion can be found in the field of *emotions history* (Stearns and Lewis 1998). Emotions history examines the experience and expression of emotions amongst American subcultures during specific historical contexts, and seeks to discern the prevailing affective climate in these groups during these periods. The formal study of history of emotions is a relatively new discipline, and two recent studies warrant mention here as illustrations of how religious emotions are influenced by historical context.

Working under the assumption that Judaism requires emotional involvement and transactions with God, Mayer (1994) engaged in a lexigraphic study of emotion trends in biblical texts. He classified nine emotion terms (happiness, anger, fear, sadness, love, hate, contempt, guilt, and envy) in the books of the Hebrew Bible and examined changes in the frequency of their occurrence over the period of the twelve centuries during which, according to general scholarly agreement, the books were written. The primary purpose of the study was to see whether emotion changed over time. As the centuries progressed, Mayer found a systematic increase in references to happiness; no other emotion was shown to systematically increase or decrease over this time period. Although he considers a number of alternative hypotheses and is cognizant of the perils and limitations of a psychohistorical analysis, Mayer suggests that this finding can be taken as evidence of the positive psychological benefits of a highly religious culture, and advocates a historical analysis of emotion and religion for understanding factors that influence emotion in the present.

A second study sought to describe the predominant emotions experienced and expressed in American Pentecostal women in the first half of the twentieth century (Griffith 1998). This qualitative study records the expressions of pious emotions in the lives of southern, rural, and poor female members of the Pentecostal church. One of the primary hallmarks of the Pentecostal faith is the natural and authentic expression of emotion, and the Pentecostal movement has traditionally sought to provoke and sustain strong emotions in believers. Narratives of conversion, reports of healing experiences, and responses to prayers revealed a high occurrence of emotions pertaining to praise, gratitude, love, joy, and exuberant happiness. Griffith hypothesized a dual role for these emotions: (1) they defined an ethic of separation setting apart believers from nonbelievers and from other Christian sects, thus enhancing commitment to the in-group, and (2) were essential elements in constructing a testimony for communicating one's faith to others and for providing assurance and certainty of one's own faith. This study, along with Mayer's (1994), are examples of how historical and theological contexts shape emotion and provide important clues about the function of religious emotions in everyday life.

Future Directions and Conclusion

There are two trends that are likely to have a significant impact on emotion research within the psychology of religion in the near future.

First, further progress in religion and emotion is likely to be spurred on by the current vigorous activity in the field of religion and health (Koenig et al. 2001). Researchers are examining mechanisms that explain the effects of religious practices on health. Borrowing from the broaden-and-build theory (Fredrickson 2001), sacred positive emotions can serve as resources that a person can draw upon in times of need, including coping with stress and dealing with and recovering from physical illness. It is also plausible, for example, that the biology of emotions and related states activated during religious worship (praise, reverence, awe, gratitude, love, hope) could have neuroendocrine or immunological consequences, thus potentially accounting for the salubrious effects of religious practices on health outcomes. Any examination of the neurobiology of these states will have to rely upon the phenomenological properties of worship as well, thus producing new insights at this level of analysis.

Second, the growing cognitive science of religion field (Andresen 2001; Pyysiäinen and Anttonen 2002) is likely to open new vistas for understanding the functions of emotion in religious contexts and in religious cognition. The role of emotions in the adoption and transmission of religious beliefs currently plays a prominent role in several cognitive theories of religion (Andresen 2001), particularly in accounting for the provocativeness of religious rituals (McCauley 2001). Much of this work focuses on religion as counterintuitiveness (Boyer 2001) and emotional responses to counterintuitive representations. Research by Boyer has shown that counterintuitive representations are more effectively recalled than ordinary or even unusual representations, and this may be due to their ability to arouse strong emotions. Emotion is also assumed to play a pivotal role in resolving doubts concerning religious representations (beliefs) and in enhancing commitment to the object of those representations. Franks (2003) cites the example of positive emotions in response to perceived answers to prayers as serving to reduce doubts about the benevolence of God. Connecting the act of prayer to the experience of positive emotion provides at least a temporary resolution in the mind of the believer who may have doubted God's benevolence. Given the pervasiveness of religious doubts

(Clark 1958; Hunsberger et al. 2002), an incorporation of the role of emotion might contribute to the understanding of both the development and the resolution of questions and doubts concerning religious doctrines.

In each of these two cases, it is clear that progress will require collaboration between psychologists who specialize in religion and experts in evolutionary biology, neuroscience, philosophy, anthropology, and cognitive science, so that developments in the psychology of religion take into account and build upon advances in these related scientific disciplines. It will also be necessary to take an approach of downward causation, in which individual beliefs and socioreligious contexts regulate biological systems of the body. Successful researchers who contribute to the next generation of knowledge at the interface of religion and emotion are thus likely to be schooled not just in the sciences but in theology as well.

References

Allen, D. 1997. Ascetic theology and psychology, in R.C. Roberts and M.R. Talbot (eds.), *Limning the Psyche: Explorations in Christian Psychology*. Grand Rapids, MI: Eerdmans, pp. 297–316.

Allport, G.W., Gillespie, J.M., and Young, J. 1948. The religion of the post-war college student. *Journal of Psychology*, 25, 3–33.

Andresen, J. (ed.) 2001. *Religion in Mind: Cognitive Perspectives on Religious Belief, Ritual, and Experience*. Cambridge, UK: Cambridge University Press.

Arnold, M.B. 1960. *Emotion and Personality*. New York: Columbia University Press.

Ashby, F.G., Isen, A.M., and Turken, U. 1999. A neuropsychological theory of positive affect and its influence on cognition. *Psychological Review*, 106, 529–550.

Baer, R.A. 2003. Mindfulness training as a clinical intervention: A conceptual and empirical review. *Clinical Psychology: Science and Practice*, 10, 125–143.

Ben Ze'ev, A. 2002. *The Subtlety of the Emotions*. New York: Cambridge University Press.

Bertocci, P.A. and Millard, R.M. 1963. *Personality and the Good*. Philadelphia, PA: David Mckay.

Boyer, P. 2001. *Religion Explained*. New York: Basic Books.

Brand, W. 2001. Experiencing tears of wonder-joy: Seeing with the heart's eye. *Journal of Transpersonal Psychology*, 33, 99–111.

Brown, K.W. and Ryan, R.M. 2003. The benefits of being present: Mindfulness and its role in psychological well-being. *Journal of Personality and Social Psychology*, 84, 822–848.

Bulkeley, K. November 2002. *The Evolution of Wonder: Religious and Neuroscientific Perspectives*. Paper presented at the Annual Meeting of the American Academy of Religion, Toronto, Canada.

Bullard, P.L. and Park, C.L. 2001. Emotional expressive style as a mediator between religion and health. *International Journal of Rehabilitation and Health*, 4, 201–214.

Chesterton, G.K. 1936. *The Autobiography of G.K. Chesterton*. New York: Sheed and Ward.

Clark, W.H. 1958. *The Psychology of Religion*. New York: Harper and Row.

Coyle, C.T. and Enright, R.D. 1997. Forgiveness intervention with post-abortion men. *Journal of Consulting and Clinical Psychology*, 65, 1042–1046.

Edwards, J. 1746/1959. *Religious Affections*. J.E. Smith (ed.). Series title: His Works, vol. 2. New Haven: Yale University Press.

Emmons, R.A. and Paloutzian, R.F. 2003. The psychology of religion. *Annual Review of Psychology*, 54, 377–402.

Franks, B. 2003. The nature of unnaturalness in religious representations: Negation and concept combination. *Journal of Cognition and Culture*, 3, 41–68.

Fredrickson, B.L. 1998. What good are positive emotions? *Review of General Psychology*, 2, 300–319.

Fredrickson, B.L. 2001. The role of positive emotions in positive psychology: The broaden-and-build-theory of positive emotions. *American Psychologist*, 56, 218–226.

Freedman, S.R. and Enright, R.D. 1996. Forgiveness as an intervention goal with incest survivors. *Journal of Consulting and Clinical Psychology*, 64, 983–992.

Gillani, N.B. and Smith, J.C. 2001. Zen meditation and ABC relaxation theory: An exploration of relaxation states, beliefs, dispositions, and motivations. *Journal of Clinical Psychology*, 57, 839–846.

Goodenough, U. 1998. *The Sacred Depths of Nature*. New York: Oxford University Press.

Griffith, R.M. 1998. "Joy unspeakable and full of glory": The vocabulary of pious emotion in the narratives of American Pentecostal women, 1910–1945, in P.N. Stearns and J. Lewis (eds.), *An Emotional History of the United States*. New York: New York University Press, pp. 218–240.

Griffiths, P.E. 1997. *What Emotions Really Are*. Chicago, IL: University of Chicago Press.

Gross, J.J. 1999. Emotion regulation: Past, present, future. *Cognition and Emotion*, 13, 551–573.

Hardy, A. 1979. *The Spiritual Nature of Man: A Study of Contemporary Religious Experience*. Oxford, UK: Clarendon Press.

Hill, P.C., Pargament, K.I., Wood, R.W., Jr., McCullough, M.E., Swyers, J.P., Larson, D.B., and Zinnbauer, B.J. 2000. Conceptualizing religion and spirituality: Points of commonality, points of departure. *Journal for the Theory of Social Behavior*, 30, 51–77.

Hood, R.W. 1975. The construction and preliminary validation of a measure of reported mystical experience. *Journal for the Scientific Study of Religion*, 14, 29–41.

Huang, S.T. and Enright, R.D. 2000. Forgiveness and anger-related emotions in Taiwan: Implications for therapy. *Psychotherapy*, 37, 71–79.

Hunsberger, B., Pratt, M., and Pancer, S.M. 2002. A longitudinal study of religious doubts in high school and beyond: Relationships, stability, and searching for answers. *Journal for the Scientific Study of Religion*, 41, 255–266.

Hutch, R.A. 1978. Jonathan Edwards' analysis of religious experience. *Journal of Psychology and Theology*, 6, 123–131.

Isen, A.M. 1987. Positive affect, cognitive processes, and social behaviour, in L. Berkowitz (ed.), *Advances in Experimental Social Psychology, Vol. 20* (pp. 203–253). San Diego, CA: Academic Press.

James, W. 1902/1958. *The Varieties of Religious Experience*. New York: Longmans.

Keltner, D. and Haidt, J. 2003. Approaching awe: A moral, spiritual, and aesthetic emotion. *Cognition and Emotion*, 17, 297–314.

Koenig, H.K., Larson, D., and McCullough, M.E. 2001. *Handbook of Religion and Health*. New York: Oxford University Press.

Lewis, M. and Haviland-Jones, J.M. (eds.) 2000. *Handbook of Emotions*, 2nd ed. New York: Guilford.

Mahoney, A., Pargament, K.I., Jewell, T., Swank, A.B., Scott, E., Emery, E., and Rye, M. 1999. Marriage and the spiritual realm: The role of proximal and distal religious constructs in marital functioning. *Journal of Family Psychology*, 13, 321–338.

Maslow, A. 1964. *Religions, Values, and Peak Experiences*. New York: Penguin.

Mayer, J.D. 1994. Emotion over time within a religious culture: A lexical analysis of the Old Testament. *The Journal of Psychohistory*, 22, 235–248.

McCauley, R.N. 2001. Ritual, memory, and emotion: Comparing two cognitive hypotheses, in J. Andresen (ed.), *Religion in Mind: Cognitive Perspectives on Religious Belief, Ritual, and Experience*. New York: Cambridge University Press, pp. 115–140.

McCraty, R., Barrios-Choplin, B., Rozman, D., and Atkinson, M. 1998. The impact of a new emotional self-management program on stress, emotions, heart rate variability, DHEA and cortisol. *Integrative Physiological and Behavioral Science*, 33, 151–170.

McCullough, M.E., Emmons, R.A., and Tsang, J. 2002. The grateful disposition: A conceptual and empirical topography. *Journal of Personality and Social Psychology*, 82, 112–127.

McCullough, M.E., Kilpatrick, S.D., Emmons, R.A., and Larson, D.B. 2001. Is gratitude a moral affect? *Psychological Bulletin*, 127, 249–266.

McCullough, M.E., Pargament, K.I., and Thoresen, C.E. (eds.) 2000. *Forgiveness: Theory, Practice and Research*. New York: Guilford Press.

Oatley, K. and Djikic, M. 2002. Emotions and transformation: Varieties of experience of identity. *Journal of Consciousness Studies*, 9, 97–116.

Otto, R. 1958. *The Idea of the Holy*. J.W. Harvey (trans.). London: Oxford University Press (original work published 1917).
Paloutzian, R.F., Richardson, J.T., and Rambo, L.R. 1999. Religious conversion and personality change. *Journal of Personality*, 67, 1047–1080.
Pargament, K.I. 1997. *The Psychology of Religion and Coping*. New York: Guilford Press.
Pargament, K.I. 1999. The psychology of religion *and* spirituality? Yes and no. *The International Journal for the Psychology of Religion*, 9, 3–16.
Pargament, K.I., and Mahoney, A. (in press). Sanctification as a vital topic for the psychology of religion. *International Journal for the Psychology of Religion*.
Pruyser, P.W. 1967. *A Dynamic Psychology of Religion*. New York: Harper and Row.
Pyysiäinen, I. and Anttonen, V. (eds.) 2002. *Current Approaches in the Cognitive Science of Religion*. New York: Continuum.
Roberts, R.C. 2003. *Emotions: An Essay in Aid of Moral Psychology*. New York: Cambridge University Press.
Rosenberg, E.L. 1998. Levels of analysis and the organization of affect. *Review of General Psychology*, 2, 247–270.
Ryff, C. and Singer, B. 1998. The role of purpose in life and personal growth in positive human health, in P.T.P. Wong and P. Fry (eds.), *The Human Quest for Meaning: A Handbook of Psychological Research and Clinical Applications*. Mahwah, NJ: Erlbaum, pp. 213–235.
Schimmel, S. 1997. *The Seven Deadly Sins: Jewish, Christian, and Classical Reflections on Human Psychology*. New York: Oxford University Press.
Schleiermacher, F. 1799. *On Religion: Speeches to Its Cultured Despisers* J. Oman, (trans.) London: Kegan Paul, Trench, Tubner.
Sethi, S. and Seligman, M.E.P. 1993. Optimism and fundamentalism. *Psychological Science*, 4, 256–259.
Silberman, I. 2003. Spiritual role modeling: The teaching of meaning systems. *The International Journal for the Psychology of Religion*, 13, 175–195.
Snyder, C.R., Sigmon, D.R., and Feldman D.B. 2002. Hope for the sacred and vice versa: Positive goal-directed thinking and religion. *Psychological Inquiry*, 13(3), 234–238.
Solomon, R.C. 1977. *The Passions*. Garden City, NY: Anchor Books.
Solomon, R.C. 2002. *Spirituality for the Skeptic: The Thoughtful Love of Life*. New York: Oxford University Press.
Stearns, P.N. and Lewis, J. (eds.) 1998. *An Emotional History of the United States*. New York: New York University Press.
Tarakeshwar, N., Swank, A.B., Pargament, K.I., and Mahoney, A. 2001. Theological conservatism and the sanctification of nature: A study of opposing religious correlates of environmentalism. *Review of Religious Research*, 42, 387–404.
Thayer, R.E., Newman, J.R., and McClain, T.M. 1994. Self-regulation of mood: Strategies for changing a bad mood, raising energy, and reducing tension. *Journal of Personality and Social Psychology*, 67, 910–925.
Ullman, C. 1989. *The Transformed Self: The Psychology of Religious Conversion*. New York: Plenum Press.
Watkins, P.C. 2003. Gratitude and happiness: Development of a measure of gratitude and relationships with subjective well-being. *Social Behavior and Personality*, 31, 431–452.
Watts, F. and Williams, M. 1988. *The Psychology of Religious Knowing*. Cambridge, UK: Cambridge University Press.
Watts, F.N. 1996. Psychological and religious perspectives on emotion. *International Journal for the Psychology of Religion*, 6, 71–87.
Wettstein, H. 1997. Awe and the religious life. *Judaism: A Quarterly Journal of Jewish Life and Thought*, 46, 387–407.
Witvliet, C.V., Ludwig, T.E., and Bauer, D.J. 2002. Please forgive me: Transgressors' emotions and physiology during imagery of seeking forgiveness and victim responses. *Journal of Psychology and Christianity*, 21, 219–233.
Witvliet, C.V., Ludwig, T.E., and Vander Laan, K.L. 2001. Granting forgiveness or harboring grudges: Implications for emotion, physiology, and health. *Psychological Science*, 12, 117–123.
Woodruff, P. 2001. *Reverence: Renewing a Forgotten Virtue*. New York: Oxford University Press.

CHAPTER FIVE

Where Neurocognition Meets the Master: Attention and Metacognition in Zen

TRACEY L. KAHAN AND PATRICIA M. SIMONE

> To follow the lead of Nietzsche, James, and Jung, if the higher spiritual traditions of humanity might actually refer to something important . . . then some sort of account of how they could occur as an expression of the structure of the human mind will be necessary.
> —H.T. Hunt, *On the Nature of Consciousness*

A Zen Story

In a classic Zen story, retold by Kapleau, a man approached Ikkyu, a Zen master, and asked for the highest wisdom.

> Ikkyu immediately took his brush and wrote the word "Attention."
> "Is that all?" asked the man. "Will you not add something more?"
> Ikkyu then wrote twice running: "Attention. Attention."
> "Well, remarked the man rather irritably, "I really don't see much depth or subtlety in what you have just written."
> Then Ikkyu wrote the same word three times running: "Attention. Attention. Attention." (1965, pp. 10–11)

As this story suggests, in Zen Buddhism, the systematic training of attention is itself spiritual practice and a path to wisdom (see, e.g., Aitken 1982; Austin 1998; Cleary 1995; Kapleau 1965; Novak 1990; Rani and Rao 1986; Walsh and Vaughn 1993). Zen teaches that through the diligent practice of zazen(Zen sitting meditation), practitioners: settle the body and the mind (Aitken 1982); cultivate present-centered awareness, or "mindfulness" (Hanh 1976); observe the "habits" of mind and, ultimately, gain insight into the "true nature of being" (Aitken 1982; Hanh 1998; Okumura 1985;

Suzuki 1970). Although numerous religious and meditative traditions consider the training of attention an important aspect of spiritual practice, nowhere is attention more central than in Zen Buddhism (see, e.g., Austin 1998; Rani and Rao 1998; Zaleski and Kaufman 1997). Dogen Kigen Zenji, one of the pioneers of Zen in Japan, goes so far as to say "Zazen is itself enlightenment" (cited by Aitken 1982, p. 14).

Within Buddhism, there are two perspectives on *how* attention is "trained" through meditative practice. In one, the development of attention is treated as "the training of good habits." It is improved in the same way a muscle is strengthened in weight training so that it can then perform harder and longer work without tiring (Varela et al. 1991, p. 24; Wallace 1991). The other perspective is that the ability to sustain attention is part of the basic nature of the human mind, but that this "true mind" has been obscured by unnecessary mental activity and prior conditioning (e.g., Aitken 1982). Thus, rather than developing new attention skills, zazen gradually eliminates these unnecessary obscuring habits of the mind. These perspectives are not necessarily mutually exclusive. Rather, they may reveal a developmental process in zazen; just as one must rehabilitate a muscle that has atrophied from underuse, so may the disciplined "practice" of sustained attention in zazen yield progressively deeper insights into the nature of mind and the human condition (see, especially, Cleary 1995; Varela et al. 1991).

Chapter Overview

In this chapter, we explore the "training" of attention in the Zen Buddhist practice of sitting meditation (zazen) from the perspectives of cognitive psychology and cognitive neuroscience. In doing so, we integrate three propositions:

1. *zazen* cultivates present-centered awareness or *sustained selective attention* to one's moment-by-moment experience;
2. the practice of sustained selective attention in *zazen* interacts with executive processes in order to *regulate* attention;
3. as the ability to *sustain* present-centered awareness increases, the need for attentional regulation decreases, and *more cognitive resources* are available for "simply noticing" whatever occurs in one's moment-by-moment experience.

The essence of our argument is as follows. Attention serves to augment activity within specific areas of the brain that are required to carry out particular cognitive functions. This increased activity is what allows us to focus on a target and to ignore competing distractions (Edelman 1989; Posner and Rothbart 1991; LaBerge 1995). Attention is a limited commodity, meaning that if our attentional resources are consumed, performance will suffer. What consumes attentional resources? Selecting *in* what to attend to, selecting *out* what to ignore, maintaining arousal and interest over time, and monitoring attention to ensure that we have not been distracted and that the selected target is still being attended to.

Through the regular practice of zazen, as the mind settles and becomes less distractible, one becomes better able to sustain attention in the present moment (e.g., on the breath).[1] As fewer attentional resources are required to "regulate" attention—to continually return the attention to the breath when it has been distracted—more attentional resources are available to attend to experiences arising in the present moment, including, ironically, the same thoughts, feelings, memories, sensations that would have distracted us previously. The difference is that one simply observes these events dispassionately as they arise and fall away or, as Cheri Huber (2003) says, one "notices," "accepts," "embraces," and "lets go." With a reduced need to regulate attention, one's attentional capacity is devoted to present experience. One can "notice" more aspects of experience as they occur without being distracted by reactions, judgments, or internal commentaries on that experience (i.e., without being drawn away from the present moment). This experience, known in the Buddhist literature as "sarana" or freedom from conditioning (Aitken 1982), has also been termed "bare attention" (Suzuki 1962) or the "beginner's mind" (Suzuki 1970).

We begin with a background discussion of attention from the perspective of cognitive psychology and cognitive neuroscience. We then endeavor to consider zazen within this theoretical and empirical framework. Finally, we consider some of the implications of our approach, including implications for cognitive neuroscience and Western approaches to Zen.

What is Attention?

> Everyone knows what attention is. It is the taking possession by the mind, in clear and vivid form, of one out of what seem several simultaneously possible objects or trains of thought. Focalization, concentration of consciousness are of its essence. It implies withdrawal from some things in order to deal effectively with others.
> —William James, *The Principles of Psychology*

Muriel Brown noted as early as 1930, and it still true today: William James likely did not have scientists in mind when he said, "Everyone knows what attention is." Most people do have a sense of what attention is; for example, when we tell someone to "pay attention," or "focus" on what they are doing, or notice what is going on. Also, we are quite aware of attention when there is a failure to attend and we miss important information such as someone's name or the proper exit off the highway. However, attention has been difficult to scientifically define and study. Defining attention is a challenge because although attention seems to be one unitary function, it is more likely a collection of complex and interrelated processes (see Parasuraman, 1998; Posner and Boies 1971; van Zomeren and Brouwer 1994).

Attention as a Selective Mechanism

We are under constant bombardment from external sensory stimulation and internal processing of events such as thoughts, memories, and emotions. Because of computational limitations in the nervous system (see Neibur and Koch 1998), there is a limit to how much information we can process at any given time. Because we cannot process everything in parallel, we need a cognitive mechanism that allows us to select relevant information and ignore irrelevant information. Thus, attention provides a "selective" mechanism to help us handle this information "bottleneck" (Broadbent 1958). For example, if your intention (goal) is to listen to an important lecture and the people behind you are talking, you must make a "selection." What you select will be influenced by your intentions (top–down or conceptually driven processing) and by characteristics of the stimulus event (bottom–up or data-driven processing). Thus, even if your intention is to follow the lecture (top–down process), you may be compelled to attend to the conversation if it is especially loud (bottom–up process) (Solso 2001).

Attention may be viewed as a mental resource that is both limited in amount ("capacity") and flexible (see, especially, Kahneman 1973; Lavie and Fox 2000). In the lecture example, if the couple behind you is discussing something of interest to you, you may allocate some attention to listening in on their conversation while also continuing to listen to the speaker. Psychologists refer to this as "dividing" attention (see Hirst 1986; Spelke et al. 1976). Attending to multiple sources of information in this way can be accomplished without compromising comprehension, as long as one's entire attentional capacity has not been exceeded. On the other hand, if the lecture is especially complex and requires a great deal of concentration, more attentional capacity is consumed (Kahneman 1973). You would likely not have sufficient capacity to also follow the conversation. In fact, you might have to allocate some attention to actively ignoring the conversation so as not to be distracted from the lecture. If the combination of listening to the lecture and ignoring the conversation requires more attentional capacity than is available, your understanding of the lecture will be compromised. In short, once attentional capacity has been exceeded, performance suffers.

Sustaining Attention (Vigilance)

> As interest is sustained, so will attention be maintained.
> —William James, paraphrased in *Attentional Processing*

Even without distraction from a nearby conversation, you may not be able to sustain your attention to the lecture, especially if it is long or your motivation, interest, or arousal wanes. Your ability to maintain attention over time is related to your level of arousal, which is influenced by many factors, including interest (James 1890; LaBerge 1995), motivation (LaBerge 1995), stress-induced mental fatigue (Kaplan 1987), even circadian rhythms and sleep debt (Dement 1999; Monk 1991). In particular, it appears that low levels of arousal impact vigilance performance by limiting the availability of attentional resources (Humphreys and Revelle 1984; Matthews, et al. 1990). In other words, if you are not sufficiently motivated to pay attention to the lecture, distraction is likely and more attentional resources will be needed to select (and re-select) the lecture. With fewer attentional resources to sustain attention on the lecture, comprehension will likely suffer (Duffy 1962; Parasuraman et al. 1998, p. 233).

In our coming discussion of attention and meditation, we build on the assumption that attention is a mental resource with a limited capacity and focus on two aspects of attention that consume this capacity. As mentioned previously, due to the limits of our information processing resources, *selectivity* of attention is essential (Desimone and Duncan 1995). Once a target is selected, it is *sustained* attention (vigilance) that enables us to maintain goal-directed behavior over time (see Robbins 1998).

Cognitive–Behavioral Research on Selective Attention

In *selective attention* tasks, participants are asked to select and focus on what is relevant (a target) and ignore what is irrelevant (distractors). In a classic selective attention study by Treisman and Gelade (1980), subjects were presented with one blue X (target) among few or many distractors (red Os and blue or red Xs). Subjects were slower to locate the target in the presence of many distractors. Also, subjects were less accurate in selecting a target in the presence of distractors and performance declined further with the addition of more distracting information (Treisman and Gelade 1980). These findings indicate that irrelevant information is processed and consumes attentional resources that would otherwise be directed to processing the target.

What is distracting about irrelevant information? Irrelevant information (distractors) may pull our attention away from what we were intending to focus on, as discussed in the lecture example. However, distractors compete for our attention even when we are successful in selecting the appropriate target. Steven Tipper(1985) was one of the first to systematically study the effects of distractors on behavior. In his classic study, he presented participants with a pair of trials. In the first trial, the prime trial, subjects were presented with outline drawings of two objects and asked to name the target. The target object was outlined in green and the distractor in red. For example, for a green dog and a red cup, the subject should ignore the distractor (cup) and name as the target "dog." The second trial, called the probe, presented either two new stimuli (control condition) or re-presented the distractor (cup) as a target. In this case, the former distractor (cup outlined in red) was now presented as a target (cup outlined in green). A new stimulus, umbrella, was presented as a distractor (outlined in red). Tipper found that subjects were slower to respond to the cup in the probe trial compared to the control condition in which the target was not previously presented as a distractor. He concluded that although selection of the dog in the prime trial was accurate, processing of the distractor (cup) also occurred and had a carry-over effect on the next trial when the cup was re-presented as the target.

More recently, Nilli Lavie and her colleagues (Lavie and Cox 1997; Lavie and Fox 2000) have demonstrated that the processing of irrelevant information depends upon how much attentional capacity is required to select the target. Subjects were presented with prime and probe displays. For the prime, subjects had to decide which of three possible target letters was presented in one of six possible locations in the center row of the display while ignoring the peripheral distractor. Figure 5.1 illustrates this procedure.

Perceptual load was manipulated in the prime trial by adding nontargets in the other center row locations. This should increase the amount of

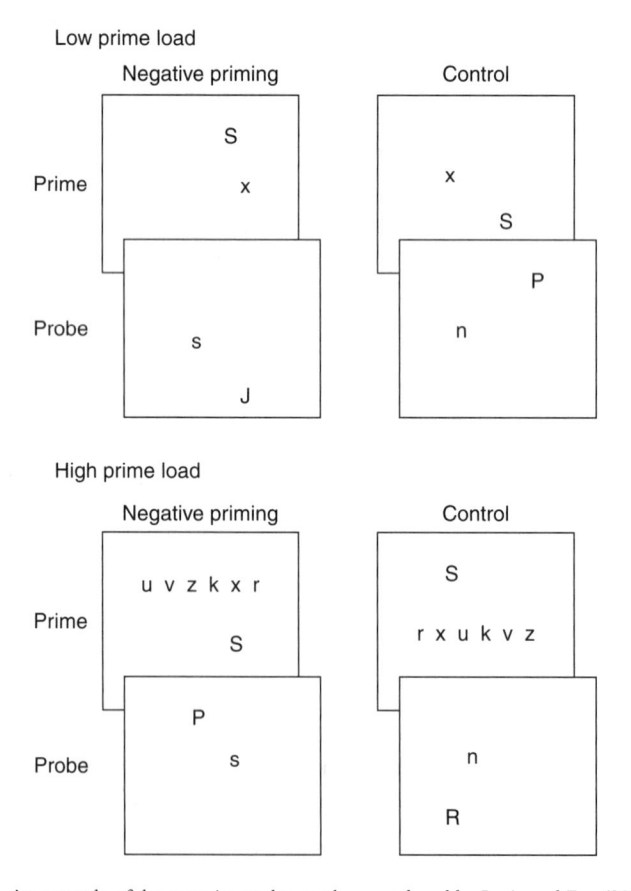

Figure 5.1 An example of the experimental procedure employed by Lavie and Fox (2000).

attention required to select the target because more items must be selected out; thus, more of the available capacity is consumed. On a relevant probe display, the distractor was re-presented as the target. They found that when the perceptual load was low (few nontargets), responses to the distractor as a target were slowed, meaning that the distractor was processed in the prime trial. When perceptual load was high (with five nontargets in the prime display), reaction times were *not* slowed when the distractor became the target, meaning that the distractor was *not* processed in the prime trial. These results suggest that when one is presented with one target and one distractor, little attentional capacity is needed to process the target and the remaining attentional capacity allows processing of the distractor. However, when there are many targets, attentional resources are devoted to processing the targets and little capacity remains to process the distractor (see, especially, Lavie and Fox 2000).

Taken together, the studies described above demonstrate that nontarget and distracting information compete with target information for attentional resources (Lavie and Fox 2000; Tipper 1985; Treisman and Gelade 1980). Of particular importance to our later discussion of attention and meditation, this research suggests that when a target task is simple, attentional capacity demands are low and it is available to process irrelevant information, leading to interference and perhaps distraction.

"Divided attention" tasks offer another way to investigate capacity limitations of attention. These tasks require participants to attend to two (or more) target tasks at the same time. Typically, one task is defined as the primary task and the other as the secondary task. The question is whether performance on the primary task is affected by the introduction of the secondary task.

Johnston and Heinz's (1975, 1978) research shows that the process of *selecting* a target "consumes" attentional capacity, and more capacity is consumed when the process of selection requires more initial processing. Similarly, if one or more of the tasks is novel or complex, more attentional capacity is consumed. In either case, if attentional capacity is exceeded, performance will be compromised (also see Lavie and Fox 2000; Spelke et al. 1976). However, with practice, participants are often able to "automatize" a task (or tasks), thereby "freeing capacity" and thus improving performance (also see Hirst et al. 1980; Schneider and Shiffrin 1977; Shiffrin and Schneider 1977). Automatizing a task can be beneficial, for example, when it provides a mental shortcut. LaBerge (1975) provides a helpful description of this process:

> For example, imagine learning the name of a completely unfamiliar letter. This is much like learning the name that goes with the face of a person recently met. When presented again with the visual stimulus, one recalls a time-and-place episode which subsequently produces the appropriate response. With further practice, the name emerges almost at the same time as the episode. This "short-circuiting" is represented by the formation of a direct line between the visual and name codes.

The process still requires attention... As more and more practice accumulates, the direct link becomes automatic. (Mandler 1954, p. 52; cited by Solso et al. 2005, pp. 96–97)

Such mental shortcuts are especially helpful for procedural tasks like driving a car, playing a musical instrument, or engaging in an athletic pursuit. That is, unless one has automatized the task or skill inaccurately. Anyone who has attempted to remove the "uhs" and "ums" from his or her professional presentations or to change an ineffective golf swing knows how difficult it is to "deautomatize" a complex skill (Deikman 1982). Deautomatizing these skills would require a shift from divided attention to "focused attention" (Treisman and Gelade 1980) or from "automatic processing" to "controlled processing" (Schneider and Shiffrin 1977; Shiffrin and Schneider 1977). In any case, returning one's awareness and cognitive control to a task that has become habitual or automatic requires and consumes considerable attentional resources. This process is readily seen in the classic Stroop Effect, first demonstrated by J.R. Stroop in 1935. In a Stroop task, subjects are asked to name the color of the ink that stimuli are printed in. The Stroop Effect is the persistent finding that when stimuli are simple patches of color, subjects are much faster to name the color of the ink of each patch than when stimuli are color words printed in a different color (e.g., the word "red" printed in blue). The Stroop Effect presumably occurs because of a response conflict between naming the color of the ink, which requires selective attention to a particular feature of the stimuli, and the automatic process of reading (also see MacLeod 1997).

Cognitive–Behavioral Research on Sustained Attention

Selection is an obvious and important aspect of attention. What happens when we must not only select, but also *maintain* that selection for a period of time? The prolonged attention to a task is called sustained attention, or vigilance (see LaBerge 1995). Comparatively little research has examined sustained attention.

Mackworth's Clock Test (Mackworth 1950) has been used to study sustained attention. In this task, participants watch a clock for two hours and monitor the jumps of the second hand. The participant's task is to respond when the second hand jumps two rather than only one second, a variation which occurs only 3–5 percent of the time and in an unpredictable manner. Typical findings are that the major decrement in performance is within the first 15 minutes, with a more gradual decline thereafter (Teichner 1974) (see Parasuraman, et al. 1998, for a review of related research).

David LaBerge (1995) has a somewhat different conceptualization of sustained attention. He uses "preparatory" attention to describe goal-directed attention sustained over a period of seconds or longer. In other words, preparatory attention allows us to maintain attention with an associated expectation in mind. For example, while stopping at a red light, you may sustain your attention on the light in preparation for it to turn green. In the lecture/conversation example described earlier, you could use "preparatory attention" to sustain your focus on the lecture because you expect to hear something new or useful to you. LaBerge points out that "maintenance" attention is very different, and has hardly been studied. Maintenance attention is used when observing an object with no immediate goal in mind, such as when watching waves crashing in the ocean, observing a bee pollinating a flower, or listening to music. The difference is that whereas preparatory attention is sustained attention to "what will be," maintenance attention is sustained attention to "what is"; it is attending to an object for the sake of the moment, without consideration of upcoming events. This conceptualization is similar to "receptive awareness" or "presence-openness" in the meditation literature (see Hunt 1995, chapter 11).

The Interplay of Attention and Metacognition in the Regulation of Attention

The varieties of attention discussed above are important to the selection and maintenance of goal-directed behavior (Parasuraman 1998). However, attention also interacts with intentionality, self-monitoring, and self-regulation—higher-order central executive and metacognitive processes which also consume cognitive resources (see, especially, Baars 1988, 1997; Baddeley 1987; Flavell 1979; Nelson and Narens 1990). One must set a goal or intention: understanding a written passage, for example. One must choose an attentional strategy to help achieve the goal, such as choosing to return one's attention to the beginning of a sentence in order to better understand the meaning of the sentence. Also, one must monitor his or her progress or outcomes to ensure the intention is fulfilled.

We suggest that in zazen, especially in early meditative practice, there is an interplay between attention and three important executive or metacognitive processes: *intentionality* (setting a goal or intention); *self-monitoring* (monitoring one's behavior in relation to that intention); and *self-regulation* (choosing a response that moves toward fulfilling one's intention) (see, especially, Metcalfe and Shimamura 1994; Nelson and Narens 1990; Umitla 1988; Underwood 1997).

Neural Mechanisms of Attention and Executive Functions

What is the evidence that the attentional and executive systems are interdependent, yet dissociable? Studies involving brain damage, neuroimaging, such as positron emission tomography (PET) in humans, as well as single-unit recordings in animals have helped determine which brain regions are more or less involved in various aspects of attentional processing (Corbetta 1998; Motter 1998; Posner and DiGirolamo 1998, 2000; Rafal 1996).

In neurocognitive and meditative research, researchers tend to agree on the existence of two main attention centers in the brain. One center, located in the posterior region of the brain, is responsible for bringing attention to stimuli and taking attention away. The second, located more anteriorly, is responsible for regulatory functions of attention (Austin 1998; d'Aquili and Newberg 1999; Posner and DiGirolamo 2000; Posner and Peterson, 1990; Posner and Rothbart 1991). These two systems are highly interconnected and they influence processing throughout the brain, yet they can also work independently.

According to Posner and colleagues (Posner and DiGirolamo 2000; Posner and Peterson 1990; Posner and Rothbart 1991), the Posterior Attention Network includes connections among the pulvinar nucleus of the thalamus, the collicular nuclei of the midbrain, and the posterior parietal lobe. Damage to any one of these structures affects the orienting of attention. Orienting of attention involves engaging attention to a target, disengaging attention from a target, and moving attention. Therefore, the Posterior Network is important for *selection*: bringing attention to a target and switching attention to a new target.

In meditation research, Austin (1998) highlights the importance of connections between the subcortical pulvinar nucleus of the thalamus and the posterior parietal lobe in attention. In particular, the pulvinar nucleus is responsible for engaging attention and the posterior parietal lobe for disengaging attention (p. 370). Additionally, d'Aquili and Newberg suggest that the Orientation Association Area, which is located in the posterior parietal lobe, is responsible for generating a sense of space and allowing orientation to incoming stimuli (1999, p. 112). This Orientation Association Area is closely related, if not identical, to the Posterior Attention Network proposed by Posner and colleagues (see Posner and Peterson 1990).

Multiple lines of evidence suggest that the Anterior Attention Network proposed by Posner and colleagues, comprised of prefrontal cortex (PFC) and anterior cingulate, is responsible for *control* over our attention (see Posner and DiGirolamo 1998, 2000; Posner and Rothbart 1991). Anatomically, the PFC is situated and connected with other brain regions so as to impart top–down influence throughout the brain (see Miller 2000). It is connected with the temporal lobes for executive control of voluntary recall of memories (Tomita et al. 1999), and it is intricately connected to the posterior parietal lobe, involved in the disengaging of attention (see Desimone and Duncan 1995). Humans with damage to PFC lose the ability to control where their attention is directed (e.g., WCST, Stroop test) (see Braver et al. 2001), and these patients become highly distractable. Similarly, "patients who have sustained damage to their right frontal lobe cannot pay attention during a monotonous, repetitive task" (Austin 1998, p. 274, citing research by Willkins, et al. 1987; Woods and Knight 1986).

Single-cell recording studies (Fuster 1989; Goldman-Rakic 1987) found that cells in dorsolateral PFC were active during the delay in delayed response tasks. The dorsolateral PFC has also been implicated in context processing and maintaining task-relevant information in working memory (Braver et al. 2001). Activity of the dorsolateral PFC is mediated by the neurotransmitter dopamine; drugs that block the activity of dopamine (such as haloperidol, an antipsychotic) impair cognitive control, while drugs that enhance dopamine (such as amphetamine) can improve cognitive control (see Braver et al. 2001, p. 749). The dorsolateral PFC is also connected with the area of the brain involved in pleasure, the ventral tegmental area and nucleus accumbens, which rely on dopaminergic input. Therefore, cognitive control and the dopamine reward pathway are related (Miller 2000). In addition to the dorsolateral PFC, the anterior cingulate cortex is also involved in attentional processing and is activated during times of target or error detection (Posner and Rothbart 1991). Anterior cingulate cortex and dorsolateral PFC are also active during dual task situations (Posner and DiGirolamo 1998, 2000). Interestingly, once a task is well-practiced, executive control of attention is no longer required and dorsolateral PFC and anterior cingulate areas are no longer active (Abdullaev and Posner 1997). As reported by Posner and DiGirolamo: "[the anterior cingulate cortex] is especially active during tasks that require some thought and is reduced or disappears as tasks become routine" (1998, p. 411).

In their neurobiological model of meditation, d'Aquili and Newberg (1999) include an Attention Association Area. This system, located in the PFC, is akin to the Anterior Attention Network proposed by Posner and colleagues (see Posner and Peterson, 1990). D'Aquili and Newberg emphasize the connections between PFC and the limbic system, which is important for the modulation of emotion, as well as other connections described earlier.

Posner and Peterson (1990) also proposed a third attentional network—the Vigilance Network—which allows one to sustain attention over a period of time. This sustained attention is achieved by activating the cortex from the locus coeruleus, a nucleus in the midbrain, which is part of the ascending activating systems (Kolb and Whishaw 2003). These cortical projections include target regions of the Anterior (dorsolateral PFC and anterior cingulate cortex) and Posterior (posterior parietal lobe) Attention Networks. For example, a PET scan study has shown that a vigilance task, in which participants maintained attention in order to detect frequent auditory targets, activated the right lateral midfrontal lobe and decreased activation in the anterior cingulate (Cohen et al. 1988). The Vigilance Network is intricately connected with both the Posterior Attention Network (Morrison and Foote 1986) and the Anterior Attention Network (Posner and Rothbart 1986).

Integrating Neurocognition and Phenomenology: An Example

So, how do attention and metacognition work together in every day experience? In our daily lives, a quiet environment with few distractions is highly unusual. Most often, we find ourselves trying to manage information "overload." As Baars (1997) points out:

> In an overload situation (e.g., when working memory is already completely loaded), "metacognition" will be impaired or impossible, and all the activities that require metacognition—self-monitoring, skepticism, deciding what to pay attention to next—may be lost. (1997, p. 102)

We rely on the three networks of attention when put in an overload situation. The Vigilance Network from the locus coeruleus provides excitatory input to the cortex. This network connects both with the Posterior and Anterior Attention Networks to help maintain arousal and focus. The Posterior Network selects the focus of attention. The cingulate cortex in the Anterior Network is activated when a target is detected. The Posterior Network is also responsible for making shifts in attention driven by "bottom–up processes," such as when an "orienting" of attention is guided by particular stimulus features. Or the Anterior Network may direct the Posterior Network to shift attention based on intention or other top–down processes (see Posner and Rothbart 1991).

The Anterior Attention Network is also responsible for our capacity to ignore distracting information, to monitor and also regulate where attention is. All of these attentional processes (arousal, selecting, monitoring) consume attentional resources. As Posner and DiGirolamo (1998, 2000) have demonstrated, as one becomes more familiar and practiced with particular tasks, more of the processing become automatic, there is less need to regulate attention, and, hence, less reliance on the Anterior Attention Network.

Pause for a moment and focus on the words "North Dakota." Think of *nothing but* the words North Dakota for 30 seconds. Close your eyes for this brief exercise, then return to our discussion. This task requires sustained, selective attention; *sustained* because we asked you to hold this thought for a period of time, and *selective* because you were to focus on one thing and ignore all other irrelevant information. This task also requires metacognitive skills: you have to set the "intention" to think about North Dakota; you need to "monitor" whether you are, in fact, thinking about North Dakota; and you must return your attention to North Dakota if you get distracted (regulate).

So, what was your experience with the North Dakota exercise? Were you distracted by unrelated thoughts, feelings, or sensations? Did your mind move to things associated with North Dakota? Did you think of its capital

(Bismarck), its location in the United States relative to your location, or that Bismarck is a type of donut? Did you think of something completely unrelated? Did you even notice that your attention wandered from the words North Dakota? If you are like most of us, you found it difficult to sustain just the one thought without your attention being automatically drawn to some other thought, sensation, emotion, or memory. Perhaps the last time you had a donut you became sick and you found yourself remembering this incident. Now you have moved from a state in North America to a memory of an illness, with no clear path from one point to the other.

The intention may be to direct our attention (e.g., to North Dakota), but attention is easily, even automatically, distracted by associated thoughts, feelings, sensations, or memories, as illustrated in figure 5.2. This example shows how we can move swiftly from one related or unrelated topic to the next without being consciously aware of any one thought, or of the chain of

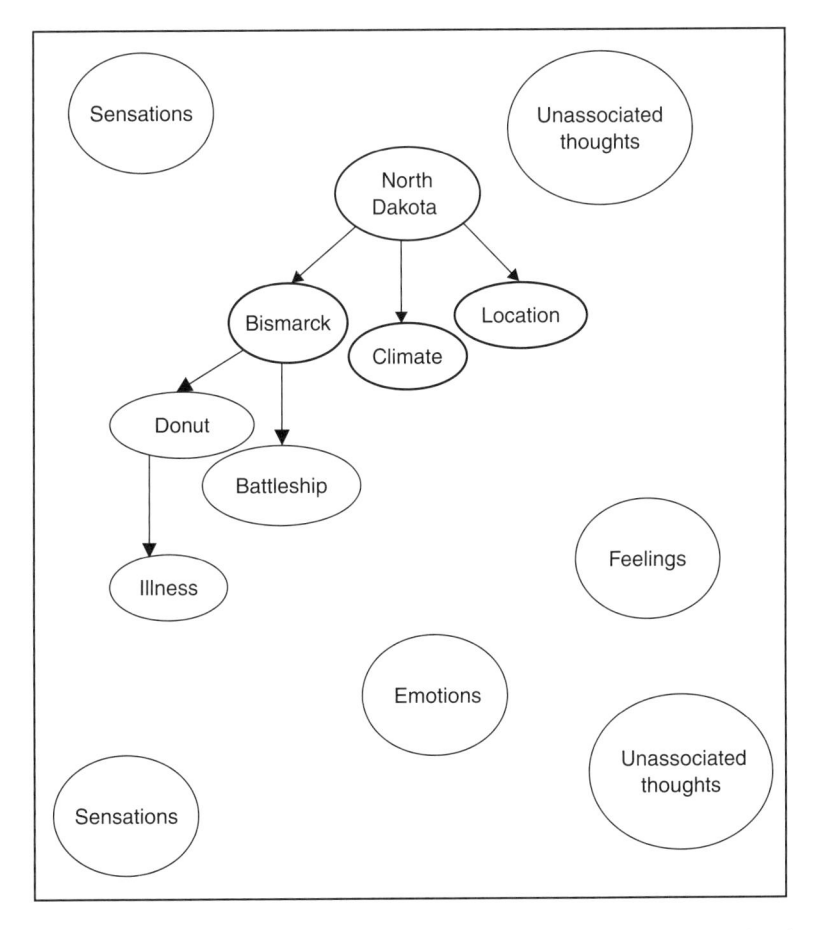

Figure 5.2 Illustration of representative associations automatically activated in the "North Dakota" selective, sustained attention task.

associations (see LaBerge 1995). Ellen Langer (1989, 2000) terms this process "mindlessness." Normally, we would not question our ability to sustain our attention on the two words North Dakota, nor would we be particularly interested in the outcome of such a simple thought exercise. Consider, however, that unless your attention remained focused solely on the words North Dakota for those 30 seconds, you were distracted and did not successfully complete this brief task!

Why is attention so easily distracted in the North Dakota exercise? We suggest that it is largely because of how attentional resources are allocated. As mentioned previously, Lavie and colleagues demonstrated that distracting information is more likely to be processed when the target does not utilize all of our capacity or cognitive resources for attention. Since the target thought, North Dakota, does not require much of your attentional capacity, distractors will compete for your attention. With the remaining capacity, you might process and be distracted by an external event, such as the telephone ringing in the next room. Also, persistent distraction arises from the associated thoughts, feelings, memories, and sensations that are automatically activated when, in this case, you hear North Dakota. Attentional focus may be drawn away from North Dakota in order to more fully process these associated thoughts or to follow thoughts associated with the associates. The notion of a *train* of thought is certainly apt here! (Also see the semantic activation model of memory developed by Collins and Loftus 1975.) Alternatively, we may find that we cannot sustain our attention to North Dakota, and our attention simply wanders to an unrelated thought, feeling, sensation, or external event. In our view, all of the attentional and regulatory processes mentioned here consume cognitive resources: selecting in a target (selection); selecting out a distractor (ignoring); maintaining focus on a target (sustaining attention); and checking or intentionally changing where one's attention is directed (regulation). So even if one is successful in avoiding distraction from North Dakota, because so many attentional resources are devoted to monitoring, regulating, maintaining interest, etc., minimal attentional resources remain to concentrate on North Dakota.

Attention and Metacognition in Zazen

> To master the vast process of thought, to erect a temple of intellectual understanding from the top of which we "see" as never before, we must first clear the site.
> —Christmas Humphreys, *A Western Approach to Zen*

In zazen, an individual maintains a specific posture for a period of 30–40 minutes usually in a setting where external environmental stimuli and potential distractions are minimized (see, especially, Okumura 1985 for a description of the formal instructions for zazen). In the words of Dogen Zenji:

> In our zazen, it is of primary importance to sit in the correct posture. Then, regulate the breathing and calm down. In Hinayana, there are two elementary ways (of beginner's practice): one is to count the breaths, and the other is to contemplate the impurity of the body. (Cited by Okumura 1985, p. 29)

Numerous accounts of the phenomenology of zazen describe the inherent distractibility of mind, or the "monkey mind" (see, especially, Aitken 1982; Humphreys 1971; Suzuki 1970). Thus, even though a novice meditator may engage in zazen with an *intention* to focus on the breath, he or she soon discovers that the mind is restless and reactive and not easily given to the seemingly simple task of "following the breath."

Our earlier discussion of attention foreshadowed three reasons why distraction should be highly likely during zazen. First, selecting the breath as the focus of attention seems to be a simple, undemanding task. As such, it should not consume much attentional capacity. With considerable capacity available to process irrelevant information, we would expect distraction to occur readily, even when external distractors have been minimized (see Lavie and Cox 1997; Lavie and Fox 2000). Second, distraction during zazen is likely even for those highly motivated to sustain their attention on the breath. As we saw in the research on sustained attention, a large decrement in sustained attention occurs within the first 15 minutes and performance continues to decline thereafter (Parasuraman et al. 1998). Finally, because of the inherent potential for distraction during zazen, the executive system must be actively engaged in order to regulate attention. Thus, in addition to attentional processing (selection and sustaining), zazen also entails the metacognitive processes of *intentionality, self-monitoring,* and *self-regulation* (see, especially, Metcalfe and Shimamura 1994; Nelson 1992; Nelson and Narens 1990). Each of these processes consume resources, leaving minimal resources available for the moment.

With continued practice, the novice begins to notice what distracts her and even to label the type of distraction ("thinking," "emotion," "judgment") without being drawn into the extended associative process typically triggered by the content of a distracting thought, sensation, or feeling. she develops the ability to return her attention to the breath (to her ongoing, moment-by-moment experience) without judgment or commentary (see Cleary 1995; Hanh 1976; Kapleau 1965). Because attending to the breath demands few attentional resources, considerable cognitive resources are available for metacognitive processes. These metacognitive processes include monitoring and regulating, for example, *what* the mind is attending to, such as the breath or the rise and fall of thoughts, feelings, sensations, and *how* the mind is attending/selecting objects of attention, as well as *how* attention is returned to the breath when distractions occur. Although these cognitive and metacognitive processes themselves consume cognitive resources (see, especially, Baars 1997, 1988; Lavie and Fox 2000), sufficient capacity is presumably available with a task as simple as "counting the breath."

Thus, we see that zazen involves the practice of *sustained, selective attention* (to one's moment-by-moment experience) and the practice of sustained, selective attention requires metacognitive skills, especially for novice practitioners.

Before one begins the practice of meditation, one typically exerts little executive control over cognition due to automaticity of cognition (see Bargh and Chartrand 1999; Langer 1989, 2000). Shortcuts and automatic associations gained through a lifetime of experience move us away from experiencing the moment. As Blackstone and Josipovic note, "In Zen practice, we are trying to become beginners, to experience life without the interference of our whole accumulation of opinions and ideas" (1986, p. 12).

In early meditative practice, the novice attempts to develop executive control over attention and cognition (Wallace 1991). These supervisory and regulatory processes consume resources, limiting the attentional capacity available for attending to the moment. With practice, however, one may become more "mindful." In contrast to our usual experience of mental distractibility, when one is mindful, one experiences "an enhanced attention to and awareness of" what is happening in the moment (Brown and Ryan 2003, p. 822; also see Hanh 1976; Langer 1989, 2000). Mindfulness is akin to what LaBerge (1995) described as maintenance attention. Mindfulness (present-centered awareness) makes it possible to observe the mind dispassionately and to discover for oneself fundamental insights into human nature. As Robert Aitken explains: "We are concerned with realizing the nature of being, and zazen has proved empirically to be the practical way to settle down to the place where realization is possible" (1982, p. 14).

The practice of zazen, then, is considered the doorway to an enlightened understanding of our mental and emotional conditioning and how this conditioning gives rise to dissatisfaction ("suffering") and belief in a separate, unchanging self (Cleary 1995; Hanh 1998; Okumura 1985).

Empirical Evidence of the Interplay of Attention and Metacognition in Zazen

> The thought-machine must be brought under control, in order that it may be rightly used to raise consciousness to its limits and beyond. The operative word is "control". Look once more at the ideal, "to let the mind abide nowhere," to use it—as a bird flying free, as a car which is never stuck to the road it uses.
>
> —Humphreys, *A Western Approach to Zen*

In addition to phenomenological reports, cognitive–behavioral studies of attention in novice and practiced meditators offer evidence that the interplay between attention and executive functions differs for novice and practiced meditators.

A study conducted by Valentine and Sweet on the effects of concentrative and receptive meditation on sustained attention tested the following hypotheses:

1. Since both concentrative and receptive meditation involve the training of attention, it was predicted that people practicing either type of meditation would show superior performance on an attentional task when compared with controls.
2. Since increased practice of meditation should train attention further, it was predicted that long-term meditators would show superior performance, again when compared with controls (1999, p. 63).

The first hypothesis was examined by comparing the performance of concentrative and mindfulness meditators with a control group of non-meditators on Wilkins's Counting Test sets 1 and 2. Wilkins's Counting Test is a vigilance task and requires sustained focused attention. In the test, a series of trials made up of binaural auditory tones, randomly varying in length from 2 to 11 bleeps, are delivered at different rates. The task is to count the bleeps and report the number presented at the end of each series. The sets consist of auditory stimuli presented at a relatively slow rate (0.25 Hertz) during an 18-minute session. To test the second hypothesis described above, the performance of long-term and short-term meditators was compared. The 19 meditators were members of a Buddhist center, 8 males and 11 females, with a mean age of 33. Participants were classified as concentrative meditators if they agreed with the statement: "I focus my attention as far as possible to a single point—a mental image, a perceptual object, breath, sound, or thought" (p. 63). Long-term meditators were those who had practiced for 25 months or more and short-term meditators were those who had practiced for 24 months or less. The control group consisted of 24 second-year college students, 14 females and 10 males, with a mean age of 22. They were considered to be comparable on an intellectual level with the meditators.

The results of Valentine and Sweet's study confirmed their hypotheses. First, meditators' performance was better than that of controls on sets 1 and 2 of Wilkins's Counting Test. Mean total estimates were significantly higher for meditators (M score of 197.48) than for controls (M score of 169.87) (p. 65). Also, long-term meditators showed further increments in attention in comparison with shortterm meditators with mean scores on the Wilkins's test of 202.6 vs. 190.77, respectively. The authors concluded that the practice of zazen appears to lead to improvements in sustained selective attention. It must be noted, however, that results from cross-sectional studies (those that compare the characteristics of non-randomly assigned groups) must be interpreted with caution due to the likelihood of important confounds. For example, Valentine and Sweet's samples of long-term meditators were not comparable to the short-term meditators on important variables such as age, distribution of male/females, and selection process. The observed group differences in attentional skills may well be associated with meditative practice, but these differences may be more strongly associated with other variables. As well, *changes* in attentional skills cannot be discovered in a cross-sectional study. Change over time can only be reliably revealed in experimental or quasi-experimental studies that involve repeated measures on the same individuals and where careful attention is paid to insuring that the experimental and comparison group(s) are comparable. These qualifications do not invalidate Valentine and Sweet's conclusions; they simply underscore the need for additional, controlled studies and the importance of replicating important findings. Other research has been conducted that is consistent with Valentine and Sweet's conclusions. For example, several studies found that meditators perform better on standard psychological attention tests (e.g., Davidson et al. 2003; Novak 1990; Rani and Rao 1986).

The above cognitive–behavioral research provides evidence consistent with the hypothesis that the meditative practice of zazen improves one's ability to sustain selective attention (also see Wallace 1991). We are also interested in the neurobiological correlates of zazen and in whether neural changes occur as a person becomes more experienced at Zen meditation. We saw earlier that the attentional capacities of practiced meditators differ from those of non-meditators. Meditation has been shown also to have carryover effects to psychological well-being, behavior, and physiological functions (see Brown and Ryan 2003; Davidson et al. 2003; Epstein 1995; Rosenberg 2004; Shapiro et al. 1998). If meditation affects behavior, and behavior relies on the activity of the brain, then it follows that meditation is associated with lasting changes in brain activity. Although Austin published "Zen and the Brain" in 1988, research into the neurobiological basis of meditation is still in its infancy. Much of the empirical work examining the biology of meditation has focused on EEG recordings (see Blackmore 2004, for a summary of this work). Other than Austin's (1988) intensive case study and preliminary studies reported by d'Aquili and Newberg (1999), little research on attention and zazen has been conducted using newer and more powerful techniques.

Further Speculations on Zen and the Brain

> Buddhism is, above all, a method of inquiry into oneself. That inquiry supposedly reveals the emptiness and impermanence of all phenomena, the illusory nature of self, and the origins and ending of suffering.
> —S. Blackmore, *Consciousness*

In our previous discussion, we noted the importance of two independent, yet interrelated, attention networks in attentional functioning. One system, located in the parietal lobes, is responsible for orienting, engaging, and disengaging attention. This attentional system was called the Posterior Network by Posner and Peterson (1990) and the Orientation Association Area by d'Aquili and Newberg (1999). The second system, located in the front of the brain in the frontal lobes, was called the Anterior Attention Network by Posner and Peterson (1990) and the Attention Association Area by d'Aquili and Newberg (1999) and is responsible for executive control of attention.

Recall that Posner and DiGirolamo (1998, 2000) reported that once a task was well-practiced, activity in the dorsolateral PFC and anterior cingulate decreased, indicating a decrease in executive functions. Surely, the dorsolateral PFC and anterior cingulate cortex also make appropriated adjustments as attentional skills are developed with meditative practice. For example, whereas these areas would be involved in regulating and controlling attention in novice practitioners, the need for cognitive control would be reduced, perhaps even eliminated, with long-term practice. Alan Wallace's description of the nine stages of attentional training ("samatha") in Tibetan Buddhist practice is consistent with such a perspective: "With the attainment of the ninth state [of attentional training] called *balanced placement*... only an initial impulse of will and effort is needed at the beginning of each meditation session; for after that, uninterrupted, sustained attention occurs effortlessly" (1999, p. 182).

In terms of neural activity, even in advanced stages of attentional training, we would expect activity in the PFC. In fact, an increase in frontal lobe activity in meditators has been reported. Using PET scan techniques, Austin (1998) noted that during rest, the brain activity of an experienced meditator (himself) was different from that of a non-meditator who was also resting. In particular, he noted increased activity in his frontal lobes compared to non-meditators. This increased frontal lobe activity has also been reported during meditation (d'Aquili and Newberg 1999) in meditators compared with non-meditators.

What might this increase in activity mean? Since executive functions are no longer consuming the cognitive resources that were required to regulate and monitor attention, these resources become available to "fully engage the moment"; hence, the corresponding increase in frontal lobe activity such as reported by Austin (1998) and d'Aquili and Newberg (1999). In other words, the long-term practice of zazen does not involve the automatization of executive functions, but rather a change in how these resources are deployed; that is, there is less attentional regulation and more noticing of the varied qualities of present-moment experience.

The Vigilance Network (locus coeruleus input to cortex) also influences the activity of the Anterior Attention Network, as demonstrated by Cohen et al. (1988). In particular, they found that a vigilance task, in which participants maintained attention in order to detect infrequent auditory targets, caused a decrease in activity of the anterior cingulate. Recall that anterior cingulate activity is related to target detection. Posner and Rothbart (1991) speculate the decrease in anterior cingulate activity during sustained attention tasks results in the feeling of being "empty-headed" in that one may suspend other cognitive activity in order to avoid stray thoughts, which may otherwise interfere with target detection. This situation exemplifies what

LaBerge (1995) called "preparatory" attention. Preparatory attention involves maintaining attention with an associated expectation in mind, such as detecting infrequent targets as in the vigilance task mentioned earlier. Attention is sustained because an upcoming event (the target) is of interest.

Maintenance attention, according to LaBerge (1995) involves sustaining attention over a period of time with no goal or expectation in mind. Attention, in this case, is sustained, not with the expectation that something interesting *will* happen, but rather because what is happening *now* is of interest. LaBerge suggests some examples of events that may invoke maintenance attention, such as "observing ocean waves, flames leaping in a fireplace, a bird in flight, a series of pictures in an art museum" (p. 92). LaBerge suggests that attention is maintained under these conditions because of internal cognitive processes and the characteristics of the external stimulus. Does one feel empty-headed during tasks involving maintenance attention, as they might during vigilance tasks requiring preparatory attention? Likely no, because the person is fully aware and engaged with what is happening in the moment. In other words, in a vigilance task, one can likely engage preparatory attention rather "mindlessly." On the other hand, when one's maintenance attention is engaged, attention is fully devoted to present experience and one is mindful. To our Western sensibilities, mindfulness may seem like a small accomplishment. However, in Buddhist practice, mindfulness is central. As Thich Nhat Hanh explains: "Mindfulness is the energy that sheds light on all things and all activities, producing the power of concentration, bringing forth deep insight and awakening. Mindfulness is at the base of all Buddhist practice" (1974, p. 26).

In contrast to Western studies of attention, where there is little or no discussion of the self who is attending, in Zen, the nature of mind and self are themselves explored and questioned: "Who" is doing the attending? What is this self composed of? Where is this self? Is the self really as stable over time as we tend to believe? (See, especially, Aitken 1982; Blackmore 2004; Varela et al. 1991.) The aim of Zen, according to Humphreys, is "to reach the end of thought and then, by the power of the thought-machine itself, to break out of its limitations into Prajna-awareness, a direct vision of 'things as they are' " (1971, p. 132).

Directions for Future Research

In the final chapters of her recent comprehensive text on consciousness, Susan Blackmore discusses the interests shown by many psychologists and neuroscientists in the Buddhist, and especially Zen, methods of investigating the nature of the mind:

> Buddhism is, above all, a method of inquiry into oneself. That inquiry supposedly reveals the emptiness and impermanence of all phenomena, the illusory nature of self, and the origins and ending of suffering ... Within Buddhism, psychologists have found both methods

and theories that touch on the deepest mysteries in the psychology of consciousness. (2004, p. 402)

Several other contemporary cognitive scientists and neuroscientists have also highlighted the potential of meditative practice, both for one's personal development and as a means to deepen scientific understanding of cognition and consciousness (see, especially, Austin 1998; d'Aquili & Newberg 1999; Hunt 1995; Varela et al. 1991).

For example, consider the number of recent models of consciousness and cognition that discuss executive functions at length (e.g., Austin 1998; Baars 1988, 1997; John 2003; Newman and Baars 1993; Rees et al. 2002; Solms and Turnbull 2002; Umitla 2000). These theories vary in the proposed specific relations among attention, awareness, and consciousness. Couple this fruitful research with the burgeoning interest in the neuroscience of subjective experience and evermore sophisticated brain imaging technologies, and the time is ripe to extend the boundaries and contributions of cognitive neuroscience.

We hope that we have successfully made the case for the value of a neurocognitive consideration of the training of attention in zazen. Such an approach reveals "attention" in zazen as considerably more nuanced than was suggested by the opening Zen story: a variety of attentional skills is intentionally practiced and developed (selective attention, sustained attention, maintenance attention); executive and metacognitive skills are also required in order to monitor and regulate attention; and intentionality (volition) guides and sustains the difficult practice of present-centered awareness. Ideally, this chapter will make a constructive contribution to future investigations of the complex relationships among the component cognitive processes associated with meditation and consciousness, investigations that of necessity will converge phenomenological, cognitive, behavioral, and neural approaches.

Notes

The authors are grateful to Matthew Freeland and Shabana Palla for research assistance during earlier stages of this work. Some of the ideas discussed in the section on metacognition and mindfulness practice were first presented at the 2000 Tuscon Conference "Towards a Science of Consciousness."

1. As is often noted by those writing about the practice of zazen, we acknowledge the inevitable difficulty in trying to even peripherally describe the often ineffable experiences that arise during meditative practice. In our own way and in relation to our own depth of experience with zazen, we are attempting to characterize the seemingly simple activity of bringing one's attention to the present moment. Inevitably, one must undertake his or her own direct investigation of the nature of mind.

References

Abdullaev, Y.G. and Posner, M.I. 1997. Time course of activating brain areas in generating verbal associations. *Psychological Science*, 8, 56–59.
Aitken, R. 1982. *Taking the Path of Zen*. New York: North Point Press.

Austin, J.H. 1998. *Zen and the Brain.* Cambridge, MA: MIT Press.
Baars, B. 1988. *A Cognitive Theory of Consciousness.* Cambridge, MA: MIT Press.
———. 1997. *In the Theater of Consciousness: The Workspace of the Mind.* New York, NY: Oxford University Press.
Baddeley, A. 1987. *Human Memory: Theory and Practice.* London: Psychology Press.
Bargh, J. and Chartrand, T. 1999. The unbearable automaticity of being. *American Psychologist,* 54, 462–479.
Blackmore, S. 2004. *Consciousness: An Introduction.* New York: Oxford University Press.
Blackstone, J. and Josipovic, Z. 1986. *Zen for Beginners.* New York: Writers' and Readers' Publishing.
Braver, T.S, Barch, D.M., Keys, B.A., Carter, C.S. Cohen, J. D., Kaye, J. A., 2001. Context processing in older adults: Evidence for a theory relating cognitive control to neurobiology in healthy aging. *Journal of Experimental Psychology: General,* 130 (4), 746–763.
Broadbent, D. 1958. *Perception and Communication.* London and New York: Pergamon Press.
Brown, K.W. and Ryan, R.M. 2003. The benefits of being present: Mindfulness and its role in psychological well-being. *Journal of Personality and Social Psychology,* 84 (4), 822–848.
Brown, M.W. 1930. Continuous reaction as a measure of attention. *Child Development,* 1, 255–291.
Cleary, T. (trans.) 1995. *Minding Mind: A Course in Basic Meditation* (especially Chapter 4: Zen Master Dogen: A generally recommended mode of sitting meditation, pp. 22–26. Boston, MA: Shambhala.
Cohen, R.M., Semple, W.E., Gross, M., Holcomb, H.J., Dowling, S.M., and Nordahl, T.E. 1988. Functional localization of sustained attention. *Neuropsychiatry, Neuropsychology and Behavioral Neurology,* 1, 3–20.
Collins, A.M. & Loftus, E.F. 1975. A spreading activation theory of semantic processing. *Psychological Review,* 82, 407–428.
Corbetta, M. 1998. Fronto-parietal cortical networks for directing attention and the eye to visual locations: Identical, independent, or overlapping neural systems? *Proceedings of the National Academy of Science, USA,* 95, 831–838.
D'Aquili, E. & Newberg, A.B. 1999. *The Mystical Mind: Probing the Biology of Religious Experience.* Minneapolis, MN: Fortress Press.
Davidson, R.J., Kabat-Zirm, J., Schumacher, J., Rosenkranz, M., Muller, D., Santorelli, S. F., et al. 2003. Alterations in brain and immune function produced by mindfulness meditation. *Psychosomatic Medicine,* 65, 564–570.
Deikman, A.J. 1982. *The Observing Self.* Boston: Beacon Press.
Dement, W. 1999. *The Promise of Sleep.* New York: Dell Publishing.
Desimone, R. and Duncan, J. 1995. Neural mechanisms of selective visual attention. *Annual Review of Neuroscience,* 18, 193–222.
Duffy, E. 1962. *Activation and Behavior.* New York: Wiley.
Edelman, G.M. 1989. *The Remembered Present.* New York: Basic Books.
Epstein, M. 1995. *Thoughts Without a Thinker: Psychotherapy from a Buddhist perspective.* New York: Basic Books.
Flavell, J.H. 1979. Metacognition and cognitive monitoring. *American Psychologist,* 34, 906–911.
Fuster, J.M. 1989. *The Prefrontal Cortex: Anatomy, Physiology and Neuropsychology of the Frontal Lobe.* New York: Raven Press.
Goldman-Rakic, P.S. 1987. Circuitry of primate prefrontal cortex and regulation of behavior by representational memory, in F. Plum and V. Mountcastle (eds.), *Handbook of Physiology—The Nervous System,* vol. 5. Behesda, MD: American Physiological Society, pp. 373–417.
Hanh, T.N 1974. *Zen Keys.* New York, NY: Double Day & Co.
———. 1976. *The Miracle of Mindfulness.* Boston: Beacon Press.
———. 1998. *The Heart of the Buddha's Teachings.* Berkeley, CA: Parallax Press.
Hirst, W. 1986. The psychology of attention, in J.E. LeDoux and W. Hirst (eds.), *Mind and Brain.* Cambridge, England: Cambridge University Press, pp. 105–141.
Hirst, W., Spelke, E.S., Reaves, C.C., Caharack, G., and Neisser, U. 1980. Dividing attention without alternation or automaticity. *Journal of Experimental Psychology: General,* 109, 98–117.
Huber, C. 2003. *When You're Falling, Dive: Acceptance, Freedom and Possibility.* Murphys, California: Keep it Simple Books.
Humphreys, C. 1971. *A Western Approach to Zen.* Wheaton, IL: The Theosophical Publishing House.

Humphreys, M.S. and Revelle, W. 1984. Personality, motivation, and performance: A theory of the relationship between individual difference and information processing. *Psychological Review*, 91, 153–184.

Hunt, H.T. 1995. *On the Nature of Consciousness: Cognitive, Phenomenological, and Transpersonal Perspectives.* New Haven: Yale University Press.

James, W. 1890/1950. *The Principles of Psychology*, vol. 1. New York: Dover.

John, E.R. 2003. A theory of consciousness. *Current Directions in Psychological Science*, 12(6), 244–250.

Johnston, W.A. and Heinz, S.P. 1975. Depth of non-target processing in an attention task. *Journal of Experimental Psychology*, 5, 168–175.

———. 1978. Flexibility and capacity demands of attention. *Journal of Experimental Psychology: General*, 107, 420–435.

Kahneman, D. 1973. *Attention and Effort*. Englewood Cliffs, NJ: Prentice-Hall.

Kaplan S. 1987. Cited in: *The Experience of Nature: A Psychological Perspective*. New York: Cambridge University Press.

Kapleau, P. (ed.) 1965. *The Three Pillars of Zen: Teaching, Practice, and Enlightenment*. Boston: Beacon Press.

Kolb, B. and Whishaw, I.Q. 2003. *Fundamentals of Human Neuropsychology* (especially pp. 384–385). New York, NY: Worth Publishers.

LaBerge, D. 1975. Acquisition of automatic processing in perceptual and associative learning, in P.M.A. Rabbit and S. Dornic (eds.), *Attention and Performance V*. London: Academic Press, pp. 50–64.

———. 1995. *Attentional Processing: The Brain's Art of Mindfulness*. Cambridge, MA: Harvard University Press.

Langer, E.J. 1989. *Mindfulness*. Reading, MA: Addison Wesley.

———. 2000. Mindful learning. *Current Directions in Psychological Science*, 9 (6), 220–223.

Lavie, N. and Cox, S. 1997. On the efficiency of visual selective attention: Efficient visual search leads to inefficient distractor rejection. *Psychological Science*, 8(5), 395–398.

Lavie, N. and Fox, E. 2000. The role of perceptual load in negative priming. *Journal of Experimental Psychology: Human Perception and Performance*, 26(3), 1038–1052.

Mackworth, N.H. 1950. Researches on the measurement of human performance. *Medical Research Council Special Report Series 268*. London: His Majesty's Stationery Office.

MacLeod, C. (March/April, 1997). Is your attention under your control? The diabolic Stroop effect. *Psychological Science Agenda*, 6–7.

Matthews, G., Davies, D.R., and Less, J.L. 1990. Arousal, extraversion, and individual differences in resources. *Journal of Personality and Social Psychology*, 59, 150–168.

Metcalfe, J. and Shimamura, A.P. (eds.) 1994. *Metacognition: Knowing about Knowing*. Cambridge, MA: MIT Press.

Miller, E.K. 2000. The prefrontal cortex and cognitive control. *Nature Reviews Neuroscience*, 1, 59–65.

Monk, T.H. 1991. *Sleep, Sleepiness and Performance*. New York, NY: John Wiley.

Morrison, J.H. and Foote, S.L. 1986 Noradrenergic and serotonergic innervation or cortical, thalamic and tectal structures in old and new world monkeys. *Journal of Comparative Neurology*, 143, 117–118.

Motter, B.C. 1998. Neurophysiology of visual attention, in R. Parasuraman (ed.), *The Attentive Brain* Cambridge, MA: MIT Press, pp. 51–70.

Neibur, E. and Koch, C. 1998. Computational architectures for attention, in R. Parasuraman (ed.), *The Attentive Brain*. Cambridge, MA: MIT Press, pp. 163–186.

Nelson, T.O. 1992. *Metacognition: Core Readings*. Boston, MA: Allyn and Bacon.

Nelson, T.O. and Narens, L. 1990. Metamemory: A theoretical framework and new findings, in G.H. Bower (ed.), *The Psychology of Learning and Motivation*, vol. 26 New York: Academic Press, pp. 125–173.

Newman, J. and Baars, B. 1993. A neural attentional model for access to consciousness: A global workspace perspective. *Concepts in Neuroscience*, 4 (2), 255–290.

Novak, P. (Summer 1990). The practice of attention. *Parabola*, 5–12.

Okumura, S. (trans.) 1985. *Shikantaza: An Introduction to Zazen*. Tokyo, Japan: Sotoshu Shomuchu.

Parasuraman, R. 1998. The attentive brain: Issues and prospects, in R. Parasuraman (ed.), *The Attentive Brain*. Cambridge, MA: MIT Press, pp. 1–15.

Parasuraman, R., Wark, J.S., and See, J.E. 1998. Brain systems of vigilance, in R. Parasuraman (ed.), *The Attentive Brain*, Chapter 11. Cambridge, MA: MIT Press, pp. 221–256.

Posner, M.I. and Boies, S.J. 1971. Components of attention. *Psychological Review*, 78, 391–408.

Posner, M.I. and DiGirolamo, G.J. 1998. Executive attention: Conflict, target detection, and cognitive control, in R. Parasuraman (ed.), *The Attentive Brain*. Cambridge, MA: MIT Press, pp. 401–423.

———. 2000. Cognitive neuroscience: Origins and promise. *Psychological Bulletin*, 126 (6), 873–889.

Posner, M.I. and Peterson, S.E. 1990. The attention system in the brain. *Annual Review of Neuroscience*, 13, 25–42.

Posner, M.I. and Rothbart, M.K. 1991. Attentional mechanisms and conscious experience, in Millner, A.O. and Rugg, M.O. (eds.), *The Neuropsychology of Consciousness*, Chapter 5. New York: Academic Press, pp. 92–111.

Rafal, R.D. 1996. Visual attention: Converging operations from neurology and psychology, in A.F Kramer, M.G.H. Coles and G.D. Logal (eds.), *Converging Operations in the Study of Visual Selective Attention*. Washington: American Psychological Association, pp. 139–192.

Rani, N.J. and Rao, P.V.K. 1986. Meditation and attention regulation. *Journal of Indian Psychology*, 14 (1,2), 26–30.

Rees, G., Kreiman, G., and Koch, C. 2002. Neural correlates of consciousness in humans. *Neuroscience*, 3, 261–270.

Robbins, T. 1998. Arousal and attention: Psychopharmacological and neuropsychological studies in experimental animals, in R. Parasuraman (ed.), *The Attentive Brain*. Cambridge, MA: MIT Press, pp. 198–220.

Rosch, E. 1997. Mindfulness meditation and the private (?) self. In U. Neisser and D. Jopling (eds.), *Culture, Experience, and the Conceptual Self*. Cambridge, England: Cambridge University Press, pp. 185–202.

———1999. Is wisdom in the brain? *Psychological Science*, 10 (3), 222–224.

Rosenberg, E. 2004. Mindfulness and consumerism, in T. Kasser and A.D. Kanner (eds.), *Psychology and Consumer Culture*. Washington, DC: APA Books, pp. 107–125.

Schneider, W. and Shiffrin, R.M. 1977. Controlled and automatic human information processing: Detection, search and attention. *Psychological Review*, 84, 1–66.

Shapiro, S.L., Schwartz, G.E., and Bonner, G. 1998. Effects of mindfulness-based stress reduction on medical and premedical students. *Journal of Behavioral Medicine*, 21, 581–599.

Shiffrin, R.M. and Schneider, W. 1977. Controlled and automatic human information processing: II, Perceptual learning, automatic attending, and a general theory. *Psychological Review* 84, 127–190.

Solms, M. and Turnbull, O. 2002. *The Brain and the Inner World: An Introduction to the Neuroscience of Subjective Experience*. New York, NY: Other Press.

Solso, R.L. 2001. *Cognitive Psychology*, 6th ed. Boston, MA: Allyn and Bacon.

Solso, R.L., Maclin, M. K., and Maclin, O. H. 2005. *Cognitive Psychology*, 7th ed. Boston, MA: Allyn & Bacon.

Spelke, E., Hirst, W., and Neisser, U. 1976. Skills of divided attention. *Cognition*, 4, 215–230.

Stroop, J.R. 1935. Studies of interference in serial verbal reactions. *Journal of Experimental Psychology*, 18, 643–662.

Suzuki, S. 1962. *The Essentials of Zen Buddhism*. New York: E. P. Dutton.

———. 1970. *Zen Mind, Beginner's Mind*. New York: Weatherhill.

Teichner, W.H. 1974. The detection of a simple visual signal as a function of time on watch. *Human Factors*, 16, 339–353.

Tipper, S.P. 1985. The negative priming effect: Inhibitory priming by ignored objects. *Quarterly Journal of Experimental Psychology*, 37A, 571–590.

Tomita, H., Ohbayashi, M., Nakahara, K., Hasegawa, I., and Miyashita, Y. 1999. Top–down signal from prefrontal cortex in executive control of memory retrieval. *Nature*, 401, 699–703.

Treisman, A. and Gelade, G. 1980. A feature integration theory of attention. *Cognitive Psychology*, 12, 97–136.

Umitla, C. 1988. The control operations of consciousness, in A.J. Marcel and E. Bisiach (eds.), *Consciousness and Contemporary Science*. New York: Oxford University Press, pp. 334–355.

———. 2000. Conscious experience depends on multiple brain systems. *European Psychologist*, 5 (1), 3–11.

Underwood, T. 1997. On knowing what you know: Metacognition and the act of reading. *The Clearing House*, 71(2), 77–80.

Valentine, E.R. and Sweet, P.L. 1999. Meditation and attention: A comparison of the effects of concentrative and mindfulness meditation on sustained attention. *Mental Health, Religion, & Culture*, 2, 59–70.

Van Zomeren, A.H. and Brouwer, W.H. 1994. *Clinical Neuropsychology of Attention*. New York: Oxford University Press.

Varela, F.J., Thompson, E., and Rosch, E. 1991. *The Embodied Mind: Cognitive Science and Human Experience*. Cambridge, MA: MIT Press.

Verfaellie, M., Bowers, D., and Heilman, K. 1988. Hemispheric asymmetries in mediating intention, but not selective attention. *Neuropsychologia*, 26, 521–531. [Cited by Austin 1998, re: point that attention is different from intention and where intention is defined as: "(normal), deliberate, anticipatory preparation for action," p. 275.]

Wallace, A. 1991. The Buddhist tradition of Samatha: Methods for refining and examining consciousness. *Journal of Consciousness Studies*, 6(2–3), 175–187.

Walsh, R. and Vaughn, F. (eds.) 1993. *Paths Beyond Ego: The Transpersonal Vision* (especially pp. 47–55). New York: Tarcher-Putnam.

Wilkins, A., Shallice, T., and McCarthy, R. 1987. Frontal lesions and sustained attention. *Neuropsychologia*, 25, 359–365. [Cited by Austin 1998, re: claim that points with frontal lobe damage have difficulty focusing in a boring, monotonous task.]

Woods, D. and Knight, R. 1986. Electrophysiologic evidence of increased distractibility after dorsolateral prefontral lesions. *Neurology*, 36, 212–216. [Cited by Austin 1998.]

Zaleski, P. and Kaufman, P. 1997. *Gifts of the Spirit: Living the Wisdom of the Great Religious Traditions* (especially pp. 1–11). New York: HarperCollins.

CHAPTER SIX

From Chaos to Self-Organization: The Brain, Dreaming, and Religious Experience

DAVID KAHN

Introduction

It is well accepted in the scientific community that mind arises out of brain activity. By mind we mean our thinking, our awareness of our surroundings, how we perceive the world, how we act in the world, our self-awareness, our spiritual and religious experiences, in short, everything that makes us human.

By brain activity we include the electrical activation of brain neurons, the release of neurotransmitters by neurons, the change in brain chemistry, the electrical and chemical communication between individual neurons and between populations of neurons, and the communication between different regions of the brain.

How is mind related to brain activity? Normally, brain activity is present. Is mind a result of lots of brain activity, or maybe, only certain kinds of brain activity?

When we are dreaming, brain activity is high but different than when we are awake. Some say we are not "in our right mind" because the brain isn't operating as it does when we are awake. A person with a brain injury may be rendered "mindless" because of a physical disruption in brain functioning. Chemical substances can put us "out of our mind" by changing how the brain functions. Meditation practice is thought to be "mindful" because of the kind of brain activity that results. How is our mind affected by our brain activity? And how is our brain activity affected by our mind?

In this chapter, we discuss how our mind is related to our brain activity. The discussion will provide examples that show the relationship between brain activity and mind, examples that illustrate that a changed brain activity results in a changed mind, and that meditation and other mindful practices result in a changed brain activity. We also attempt to answer the question of *how* mind emerges from brain activity. We will suggest (as have others, see, e.g., Freeman 1999; Haken 1996, 2002; Kelso 1995) that the brain is a self-organizing system, and that mind emerges from brain activity through the self-organizing properties of brain activity. Through the process of self-organization, a population of interacting neurons may bring into existence complex patterns, conceivably, even mind. To support this claim, we shall cite evidence that intelligent behavior can emerge without intelligent agents.

Neurons and Conditions for Self-Organization

Neurons are the communication cells of the brain. They are capable of transmitting electrical pulses and currents, and are like other cells in the body except for one important difference. The neuron sends out thread-like extensions of itself, called dendrites, that carry electrical current. The dendrites generate and transmit this electrical current to other neurons when they receive electrical pulses from the axon of another neuron.

There are a lot of neurons in the brain; the estimate is a hundred billion. And there are many connections between them, at least ten thousand connections for each neuron. This interconnectivity provides an enormous potential for electrical communication in the brain. Yet the neurons in the brain are not completely interconnected. And they are not completely independent of each other either. This balance between a fully interconnected system and a completely independent one is necessary for self-organization to occur. If the system were fully interconnected, additional neural communication would be severely limited. An example of a fully interconnected system is a crystal—there is little room for additional communication between the elements of a crystal; it is complete; and will not self-organize into a different structure. Another example of a highly interconnected system from which structure is not likely to emerge is deep delta sleep. In this stage of sleep, there is global brain-wave synchrony, which deprives the brain of a sufficient number of independent parts to permit adequate communication. Little if any coordinated mental activity emerges in deep delta sleep. This is contrasted with rapid eye movement (REM) sleep where there is brain-wave activity with small-scale coherency but no overall synchrony (Kahn and Hobson 1993; Kahn et al. 2000a, 2002a). In REM, the mind is actively producing coordinated mental activity, which is manifested as dreaming.

If the system, on the other hand, consisted of interacting but independent elements, neural communication would again be severely limited. An example of interacting but independent elements is a gas consisting of independently moving molecules. Such a system will not self-organize, this time for the opposite reason, namely, the lack of relationship between the independently moving gas molecules. There has to be some integration or functional connectivity and communication between the molecules for them to participate in their own self-organization. In short, there needs to be a balance between integration and differentiation (Bressler and Kelso 2001; Tononi and Edelman 1998; Tononi et al. 1998).

Bressler and Kelso (2001) call this regime, where there is a balance between integration and differentiation, metastable. The dynamics of a metastable regime makes use of the balance between integrating and segregating influences, between interdependence and independence. The advantage of metastable regimes is that they allow for transitions between states to occur rapidly. Metastability (being on the edge of stability) can be essential for a system that needs to respond quickly and flexibly to input, as does the brain, so long as the system remains in a stable state when it needs to. The normal brain has evolved to be able to do both—it can remain in a stable state when it is to the benefit of the organism to do so, but can also rapidly self-organize into a new state when that is to the benefit of the organism.

Some Basics of Self-Organization

Self-organization is one of those well-documented, if still not fully understood, phenomena of nature. Perhaps the two most famous early pioneers in this field are Ilya Prigogine, who won the 1977 Nobel Prize for his discovery and interpretation of self-organized structures (Nicolis and Prigogine 1977, 1989), and Hermann Haken (1981), who explored the theoretical basis for self-organization. Basically, these researchers showed that unless a system is in thermodynamic equilibrium, it could become unstable when its environment changes. But only isolated systems that are not in contact with the environment can truly ever be in thermodynamic equilibrium. Such systems can only be created in the laboratory where good approximations to thermodynamic equilibrium can actually be achieved. For example, a gas of uniform temperature inside an enclosure isolated from the environment will maintain its initial homogeneous temperature, or very quickly equilibrate to a homogeneous temperature, if not initially so. It is then said to be in thermodynamic equilibrium. But outside the laboratory, systems are open, that is, they are in contact with the environment, and hence are not in thermodynamic equilibrium. All living systems are open systems, are not in thermodynamic equilibrium, and are, therefore, candidates for self-organization if they become unstable because of changes in the environment. And living systems are prone to becoming unstable because of their nonlinear nature. In a nonlinear system, a small input can have large and unexpected consequences, that is, even the smallest perturbation can cause a nonlinear system to become unstable and to, therefore, potentially self-organize into a new state. Examples of nonlinear systems include the self-organizing Zhabotinsky chemical reaction (Nicolis and Prigogine 1977), tornadoes, the all-or-nothing property of neuronal firing, laser action, shock waves, and traffic jams (Scott 2004). By contrast, linear systems are generally stable to changes in the environment. If a linear system receives an input of a given magnitude, it will respond in a predictable way. Slowly pressing down on the accelerator of your car is an example of a system designed to be linear—the further the pedal is depressed, the proportionately faster the car moves.

While it is true that a nonlinear system can be very sensitive to changes in its environment and can, therefore, abruptly become unstable when environmental changes push it beyond its threshold for stability, it is also true that even a nonlinear system once established may become stable to changes in its environment. Such a system is said to be in a basin of attraction or in a potential well; the system has achieved relative stability to outside changes. Obvious examples in the world of psychology include learned coping mechanisms that are difficult to change despite changed environments.

Examples of Self-Organization

Bénard Instability

An example that illustrates self-organization is the so-called Bénard instability. In this example, water is placed in a pan and heated from below causing a temperature gradient to exist between the bottom and top surfaces of the water (the top surface is cooler than the bottom surface). At first, the water molecules move faster and with greater energy as the temperature gradient increases. However, when the temperature gradient becomes great enough, the nonlinearity of this system permits this increased temperature difference to completely destabilize the system. As the threshold for stability is crossed, the freely moving water molecules self-organize into a new state, which is characterized not by randomly moving water molecules, but by large hexagonal structures each made up of billions of individual water molecules. If the walls are also heated, further structure ensues, changing from a hexagonal to a spiral pattern (Haken 1981, 1996; Nicolis and Prigogine 1977). What is remarkable is that billions of randomly moving water molecules have cooperated and self-organized to create a more complex highly structured system (the hexagonal or spiral pattern). All this without being "told" ahead of time by any central agency what pattern to create.

This example of self-organization whereby a simple structure gives rise to a more complex one, illustrates a very general property of all far-from-equilibrium open systems, namely, that a small change in the environment of a simple system (the small change here being the increase in the temperature gradient) can cause it to self-organize into a complexly organized one.

Development

And, of course, the prime example of structure developing from less structure is the development following conception whereby a single fertilized egg cell develops into a sentient being. Development occurs through self-organization whereby resultant structures are formed as a result of local interactions between cells. To say that development proceeds as a self-organizing process is to say that the elements making up a developing organism, the cells, in a sense, prepare their own structure, bootstrapping themselves up in preparation for the next developmental step. It is to say that molecules and cells in interaction are ultimately responsible for their own eventual structural and functional outcome (which they must be since only one cell, the fertilized zygote, begins the developmental process). The environment (the extracellular matrix) serves predominately to deliver the energy and material to the cellular elements that keep these elements in an open non-equilibrium state. There is a passage from one stable state to another as instabilities are encountered in an existing stable configuration. New order emerges as the system self-organizes to a regime that is stable to the new environment or cellular configuration. Development, then, proceeds through a series of changes of state as stabilities of existing states are continuously tested at each new developmental step.

Here too one major factor in the self-organizing process is the presence of gradients and cellular signals as cells sense and are influenced by the presence of their neighbors. While the genes present in each cell serve to instruct the right proteins to be manufactured during the developmental process, the actual formation of structure is a self-organizing process.

Cells will differentiate into specific types of cells, muscle, skin, etc., depending on which genes are expressed when, and on the presence of already existing specific cell types in the neighborhood. Thus, a cell's spatial position in the environment of other cells plays a fundamental role in its fate. In fact, depending upon a cell's position, it may become part of the embryo proper (from the so-called inner cell mass) or part of the embryo's supporting cast (from the nurturing outer cell layer). This is an illustration of local interactions being important in determining resultant function. There is a limit to how far such local interactions can go, of course. That limit is more or less set by the genetic limits or "know-how" of the organism. Within these limits, however, complex and global patterns are self-organized from the single fertilized egg via communication links between cells and between cells and their environment.

At each stage of development, the developing system owes its stability to the constraining effect it has on the behavior of the individual cells. *In other words, while the individual cells cooperate in the self-organization process of forming the complex structure, they themselves become constrained by the structure they have created through this self-organizing process.* This is generally true of all self-organizing systems, as we shall see later.

Before leaving the developing embryo, we may ask when does the developing embryo acquire mind? But to answer that, we must ask when does the developing embryo acquire brain? Is it when cells differentiate into neural tissue? Is it when the neural tissue invaginates and folds over to form a structure? Is this structure the "brain"? Is it when the tissues that make up the different parts of the brain grow into "recognizable" brain structures? Is it when they become electrically and chemically active? The truth is that we don't know when precisely to call the neural structures brain, and we certainly don't know when in the developmental process "mind" emerges. The mind of an evolving embryo is evolving along with the evolving brain activity. It is a matter of definition as to when we call the evolving mind, mind.

Social Insects and Collective Behavior
Another example that illustrates a self-organizing process is nest-building behavior demonstrated by social insects such as termites. Individual termites collectively build a termite nest that is remarkably elaborate, containing internal structures that have been described as gothic arches and pillars (Deneubourg and Franks 1995; Deneubourg et al. 1987). In other words, a multitude of insects that do not have the end result in mind and have not conceptualized that elaborate structure exhibit collective intelligence. Even though the insect community is made up of individual insects with very limited cognitive ability, yet they as a collective are able to build complex structures through local interactions. These local interactions are via the insects' receptors, which allow them to respond to stimuli emitted by other members of the community as well as to stimuli from the environment. Examples of stimuli from the environment, which may be used to initiate nest building and egg distribution, include irregularities in the ground and temperature gradients. Examples of stimuli from fellow members include pheremones that are laid down. This pheremone diffuses away from the spot where it was deposited, thus creating a concentration gradient that controls building activity.

As we saw earlier, there are several key requirements for self-organization to occur, for example, the system must be open to the environment, be far from equilibrium, and exhibit nonlinear properties. Another is that there be positive feedback. In positive feedback, the elements of the system feed back onto themselves in a way that amplifies any change that occurred by recruiting additional elements to participate in the change. This is called autocatalysis, meaning that the system catalyzes its own growth. In nest building, the positive feedback is such that the more the pheremone deposited previously, the more that will be deposited subsequently (up to a point). Further, in nest building, the interactions are nonlinear, which means that a threshold point may be reached such that the system will take off, that is, bifurcate into a new state. In termite nest building, this occurs when the pheremone that individual termites have laid down reaches a threshold concentration level. At the location where a threshold concentration has been deposited, the termites "decide" to build their nest. Using soil pellets, strips and pillars are constructed, followed by arches that are built between the pillars, in turn, followed by the construction of walls. And each subsequent "pillar" of the nest emerges when a concentration of pheremone exceeds a threshold value, eventually leading to the full structure. As Theraulaz et al. (1998) state, "it may not be necessary to invoke individual complexity to explain nest complexity." As in all self-organizing systems, a complex system (the nest) has emerged out of strictly local interactions. *The coordination of building activity does not depend on the cognitive sophistication of the insects, rather on the evolving structure itself since each change in the structure influences the next building activity.*

Neurons and Emergent Behavior

If we find it difficult to wrap our minds around how complex global structures such as termite nests emerge from simple local interactions between termites, we surely must find it difficult to see how mind arises out of simple interactions between neurons in the brain (Haken 1996; Kahn and Hobson 1993; Szentagothai and Erdi 1989). Yet we know that there is no mind without brain, that brain consists of neurons, and that mind changes when the interactions between neurons change. How does mind emerge out of local interactions between neurons? Is there a self-organizing process by which local interactions between neurons produce patterns, functional structures, and behaviors that eventually lead to mind? We believe that the preceding examples at least suggest the possibility that mind could emerge out of the self-organizing properties of brain activity.

Further, if the answers to these questions are in the affirmative, we do not need to arbitrarily postulate a special neuron or a special set of neurons that act as a director or a choreographer, whose job it is to instruct other neurons how to behave. Instead, the neurons themselves create their own emergent behavior. Individual neurons interact by virtue of their reciprocal and recursive (reentrant) connections with one another. After reaching a functional connection density threshold, which is neither too dense nor too sparse, the individual neurons, like the individual termites, cease acting individually and start participating as part of a group. Local interactions[1] between the neurons determine the outcome of this participation. One outcome is the emergence of a new pattern of activity. The new pattern is not imposed from the outside, but once formed, it constrains the activity of the individual neurons just as each completed structure of the termite nest constrains subsequent behavior of the builders. The newly emergent pattern imposes a constraint on the individual neuronal populations responsible for its creation. This concept is easily visualized by the everyday example of a traffic flow pattern. If the density of automobile traffic is sufficiently high, a traffic flow pattern will self-organize by virtue of interactions occurring within the vehicular traffic stream. This emergent traffic flow pattern will subsequently constrain the movement of those very same automobiles that were responsible for its creation (Kahn et al. 1985).

In the next section, we review some specific neuronal networks in the brain to illustrate how the activity of these neuronal networks affects mind. We return to the discussion of how mind can emerge from an interconnected network of neurons by self-organization in the section "Summary and Concluding Remarks."

Brain Activity and States of Mind

The Dreaming Brain

As we continue to review the brain basis for the emergence of mind, we will first take a look at what dreaming can tell us about the brain–mind connection. The dreaming brain is an example of how changed brain activity can influence mental processing. When we are dreaming, we are "not in our right mind." During the REM stage of sleep, when dreaming is most vivid, specific areas of the brain are functionally disconnected from each other (executive portions of the prefrontal cortex are disconnected from most of the rest of the brain). The result is the very different nature of mental processing during dreaming as compared with waking. Since REM is a unique state, which occurs to every human (and to most animals) and for which the changes in cognition are so pronounced, it serves admirably as a potential test bed for relating changes in cognition to changes in brain activity (Hobson 1988; Nyberg et al. 2000).

Selective Brain Region Activation and
Chemistry Changes During REM Sleep

Electroencephalogram (EEG) studies have shown that the brain when in the REM stage of sleep is hardly asleep in the traditional sense at all, but is as active and its brain-wave output as desynchronized as it is in the wake stage (Hobson 1988, 2001). Later, magnetoencephalogram (MEG) studies (Llinas and Pare 1991) showed not only that the dreaming brain is as active as the wake brain, but also that there is a coherence or synchrony among neuronal populations within the supposed brain-wave desynchronization. This coherence is in the gamma band of frequencies between 20 and 80 Hertz, most often called the 40-Hertz band. Further studies utilizing positron emission tomography (PET) during the REM stage of sleep (Balkin et al. 2002; Braun et al. 1997, 1998; Maquet 2000; Maquet et al. 1996, 2000; Nofzinger et al. 1997, 2002) showed that not only is there as much or more activation of the brain during REM as there is in the wake state, but that there are also very important differences. Specifically, the dorsal lateral prefrontal cortex (DLPFC), which is responsible for executive functions during the wake state, is relatively silent during the REM stage of sleep (Balkin et al. 2002; Braun et al. 1997, 1998; Maquet 2000; Maquet et al. 1996, 2000). On the other hand, the limbic and paralimbic areas, generally associated with emotions, are highly activated during REM. Thus, there is a clear difference as to which brain areas are and are not activated in REM sleep when compared with waking (figure 6.1).

Additionally, the chemistry of the brain radically changes as one goes into the REM stage of sleep. The brain changes from a state of high concentration of the neurotransmitters serotonin and norepinephrine during the wake state to negligible concentrations of these amines during the REM stage of sleep. These neurotransmitters play an important role in our state of mind, for example, their presence affects how attentive we are and how our minds processes information as serotonin and norepinephrine can either attenuate or enhance the response of neurons involved in working memory (Hobson et al. 1999, 2000).

146   David Kahn

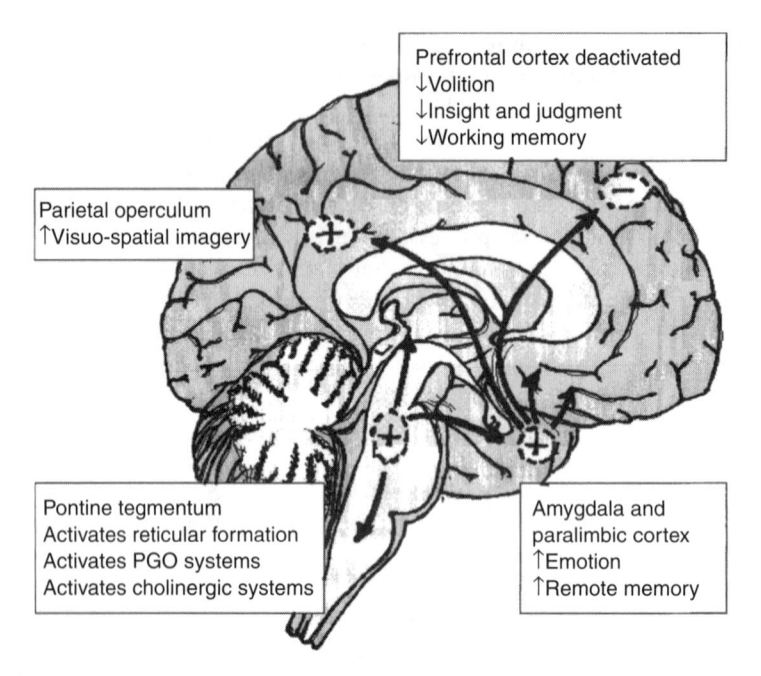

Figure 6.1 Lateral view of the brain showing areas that are activated (↑) and those that are deactivated (↓) in REM compared with waking. From Hobson et al. (2000).

Labels in figure:
- Prefrontal cortex deactivated
  ↓Volition
  ↓Insight and judgment
  ↓Working memory
- Parietal operculum
  ↑Visuo-spatial imagery
- Pontine tegmentum
  Activates reticular formation
  Activates PGO systems
  Activates cholinergic systems
- Amygdala and paralimbic cortex
  ↑Emotion
  ↑Remote memory

What Makes a Dream a Dream and Changes in Brain Functioning

These studies of brain chemistry and selective brain region activation have been supplemented with subjective studies of dreaming. Some researchers (Hobson 1988, 1999; Kahn and Hobson 2003; Kahn et al. 1997, 2002b) have begun to relate the objective measures provided by the EEG, MEG, and brain imaging studies to the subjective measures provided by phenomenological studies in which people are asked to report on their dream experiences. In one such study (Kahn et al. 2000b), it was found that subjects often did not recognize a dream character by appearance or by behavior, but by "just knowing." This was interpreted as indicating that not all areas of the brain could be brought to bear in the recognition process as in the wake state, in accordance with the brain imaging studies that showed that not all areas are active in the REM stage. It was hypothesized that it was this changed and reduced level of communication between regions of the brain that made it difficult to recognize characters in the same way as when awake.

The interpretation of the results of this and other studies is that the disorientation and scrambling of character attributes found in dreams (Hobson et al. 2000; Kahn et al. 2000b, 2002b, 2003) are obligatory manifestations of the changed activity and changed neurophysiology of the dreaming brain.

*Consequences of a Changed Neurophysiology
for the Dreaming Mind*

In general, the changed neurophysiology in REM, for example, the deactivation of executive areas of the prefrontal cortex, causes us to lose control over how the dream progresses. We cannot direct the dream story (except in the relatively rare instances when executive areas of the prefrontal cortex become reactivated during lucid dreaming). Another example is the increased activation of the downstream extrastriate visual cortex and limbic and paralimbic areas, which causes dreams to be high in visual and emotional content (Braun et al. 1997, 1998; Maquet 2000; Maquet et al. 1996). Additionally, there is a reduced functional connectivity between the lateral prefrontal cortex and the precuneus (located in the parietal lobe). This decreased neural activity in the lateral prefrontal–precuneus circuit helps explain the impaired short-term memory and absence of episodic memory in dreams (this absence of episodic memory is why we almost never dream about the exact actual event that took place in waking). The lateral prefrontal–precuneus network is also required for the reconstruction of context and temporal order, it is involved in orchestrating working memory in all the domains, and it plays a critical role in protecting the contents of working memory from distraction (Goldman-Rakic et al. 2000; Mesulam 1998). In short, this reduced neural activity in the lateral prefrontal–precuneus network during the REM stage of sleep compared with waking helps explain many of the unique features of the dreaming mind. These include the lack of access to complete information, our loss of insight, the space and time scrambling of events and characters in the dream, and our inability to recall events of the day. Our ability to think about our future actions, prepare for eventualities, and control what and how we do things are dependent upon a connected functioning awake brain in which the prefrontal cortex is connected with the rest of the brain. And, of course, when awake, we receive input from the outside world so that we can always do a "reality check," something that is not possible while asleep.

Several times every night during REM, we are out of our (awake) minds. We lose the ability to control or willfully guide our actions. On the other hand, it is well to point out that dreams often exhibit creative mental activity. This is because the disconnection of the frontal-parietal areas frees us from many of our learned and over-learned associations and patterns of thinking. This disconnection can lead to the creation of new and sometimes creative mental associations, albeit often bizarre ones. Our minds loosen up as portions of our brains disconnect (Walker et al. 2002).

Changed Brains and Changed Mind

Whereas the changes in brain activity and mental processing during REM and other sleep stages are reversible once we wake up, specific mental functions may be permanently impaired when unfortunate events such as stroke or head trauma affect the functioning of the brain. Some impairments are quite localized, for example, if a region in the human brain where the BA37 and BA19 come together is destroyed, there is an inability to detect visual motion (where BA or Broadman Area is an anatomical labeling scheme). If a region in the mid-portion of the fusiform gyrus (BA37 and BA20) that is specialized for recognition of a face as a face is damaged, we are unable to recognize faces (prosopagnosia). Damage to Wernicke's area (parts of BA22) in the temporal lobe results in an inability to understand words and concepts. Damage to Broca's area (parts of BA44 and BA45) in the frontal lobe results in difficulty to articulate words and naming. Clearly, these examples illustrate the close association between brain activity and mind. Specific kinds of brain activity are required for specific kinds of behaviors such that when the brain regions producing or receiving the brain activity are damaged, the behaviors are compromised.

Changes in Social Functioning

Some consequences on how mind functions when the brain suffers a trauma are of a somewhat different nature. One classical example is illustrated by the change in personality and social functioning that Phineas Gage suffered when a six-foot-long metal tamping rod pierced his face from below and emerged from his forehead. This created a big hole in the front part of his skull and severely damaged the front part of his brain. He actually survived this catastrophic event and, amazingly, his intellectual faculties as measured by standard intelligence tests did not suffer. His personality, however, did change. People reported that they were seeing his personality change from being a likable, considerate, and caring human being to the opposite. We now know that this was due to damage to areas of his ventromedial orbital prefrontal cortices and the connections this region makes with other areas of the brain. Patients with this kind of brain damage have difficulty planning their daily activities and choosing appropriate friends and social activities even though they have normal learning and memory, language, and attention. According to Damasio's somatic marker hypothesis (Damasio 1994), the ventromedial prefrontal cortex (that was destroyed in Phineas Gage) is needed if a decision is to be linked up with emotion and body reaction. The body reaction and emotion come from previous experiences to similar situations. Thus, decision making is dependent on the whole body state. When the ventromedial prefrontal cortex is damaged, however, one is not able to make an appropriate linkage and a disadvantageous choice is often made as there is an inability to reexperience a previously felt negative emotion (Bechara et al. 2000).

Delusional Misidentification

Lesions to other brain areas can have effects on the mind that are also quite dramatic for interpersonal relationships. One such example is the sudden onset of delusional misidentification of people following a traumatic brain injury or after the atrophy of frontal and temporal brain regions, especially in the right hemisphere, which include the limbic areas. One such example of delusional misidentification is called the Capgrass syndrome (Edelstyn and Oyebode 1999; Edelstyn et al. 2001; Oyebode and Sargeant 1996). Capgrass syndrome is characterized by the belief that a familiar person has been replaced by an imposter even though the patient freely admits that the impostor looks like the familiar person. The patient says that the familiar person just doesn't feel like who they are supposed to be. Delusional misidentification symptoms such as the Capgrass syndrome are common in Alzheimer's disease. This suggests that delusional misidentification can also occur as a consequence of the degeneration of brain areas associated with the recognition and updating of memories as occurs in Alzheimer's disease (Forstl et al. 1991, 1994). Once again we see the close association between brain activity and how its loss can lead to dramatic changes in how the mind works, as in this misidentification syndrome.

Another example of delusional misidentification is when a complete stranger is taken to be an old friend (the Frégoli syndrome). Patients with this syndrome also show frontal and temporal lobe atrophy. The syndrome can follow from a traumatic brain injury to the frontal and temporaparietal regions (Feinberg et al. 1999; Joseph and O'Leary 1987). A vivid example of this syndrome is the case of a 71-year-old woman who after suffering brain damage to her right temporal lobe due to an abrupt drop in heart rate mistook her husband for her elder sister. She also misidentified her home, saying it was a rented replica of her real home (Hudson and Grace 2000). She performed normally on tests of verbal intellectual skill and reasoning, though below normal on visual memory tasks. The misidentification was believed to be due to the disruption of connections (brought about by the ischemia) between areas required for the visual recognition of faces and scenes and areas required for memory retrieval; *highlighting the importance of uninterrupted connections in the brain for proper mind functioning.*[2]

Unilateral Neglect

A lesion to another part of the brain has yet a different effect on the mind. Lesions of the right posterior parietal cortex, usually caused by a stroke in the right hemisphere (mainly the supramarginal gyrus, Brodmann's area BA40), are associated with an inability to see what is on the left-hand side of a scene (unilateral neglect). This condition can result in the failure to notice, for example, the position of one's left arm and left leg, or a failure to notice any part of one's body that is on the left side. Such a person may not shave the left side of his face (Halligan et al. 2003). It can also result in the failure to notice things that are within arm's reach on one's left side. A person suffering this change in brain activity pays mind to only part of the world. He pays "no mind" to what is on his left side as that doesn't exist for him; it is all "out of mind."

Brain Stimulation Leading to Out-of-Body Experiences

When the brain is stimulated deliberately, as in the experiments of Penfield and Erickson (1941), it is possible to find out how the mind is affected as a result of electrical stimulation to specific parts of the brain. Penfield wanted to know the function of specific regions of the brain so that he could avoid removing the crucial regions when performing surgery on epileptic patients requiring removal of brain regions to prevent seizures from occurring. When he stimulated one particular area of a patient's brain, the patient reported an out of body experience (OBE). An OBE is when one has the experience of being detached from one's body and viewing it from elsewhere.

More recently, an OBE was reported after electrical stimulation to the vestibular processing areas of a patient's brain (Blanke et al. 2002; Tong 2003). Blanke et al. described a 43-year-old woman who was being evaluated for epilepsy treatment. During stimulation to the angular gyrus in her brain, the patient reported seeing herself lying in bed from above and floating six feet above the bed. Blanke suggests that stimulation of the vestibular and somatosensory body processing areas of the brain led to her OBE. Stimulation administered to a specific region of her brain caused her mind to experience an OBE. In her mind, she was out of her physical body.

Moral Decision Making and the Brain

Lending strong support to the ubiquitous nature of the brain–mind relationship, brain-imaging techniques[3] were used to study brain activity when a subject was asked to make a moral decision. Specific areas of the brain were found to be activated when a moral decision was made, and, further, which areas of the brain responded depended on how strongly the moral decision affected the person personally (Greene 2002). Impersonal moral decision making, that is, decisions based mainly on abstract concepts of right and wrong, was associated with areas normally activated during working memory tasks and decision making, for example, the DLPFC. Subjects responding to personal moral dilemmas (a decision touching them personally and directly), however, had increased activity in areas associated with emotional processing, for example, the medial frontal gyrus and the posterior superior temporal sulcus and the inferior parietal regions, areas thought to integrate emotion into decision making (Greene 2002). Thus, specific areas of the brain respond when thinking about morality, and, further, different areas respond depending on how personally involved one feels in the moral decision.

Religious Experience and the Brain

In further support of the ubiquitous nature of the brain–mind relationship, studies have investigated how brain activity is related to religious thought.

For example, Boyer (2003) suggests that it is possible to develop a cognitive neuroscience of religious beliefs, suggesting that areas of the brain normally involved in human cognition are also specialized for religious thought. Azari et al. (2001) also suggest that religious experience is a cognitive phenomenon. Using PET imaging, they studied self-identified religious subjects and compared them to a nonreligious control group. The self-identified religious subjects read from religious texts and reported when they felt they had reached a religious state. The control group read from secular texts. The frontal–parietal areas of the self-identified religious subjects were activated during religious recitation. The activation of this frontal–parietal circuit is suggestive of a cognitive nonemotional experience. The frontal areas activated included the DLPFC that is associated with conscious monitoring of thought and memory retrieval. The parietal area activated was the precuneus that is important for visual memory and spatial awareness. Interestingly, these are the same areas that, on the contrary, are deactivated during REM stage dreaming, suggesting that a religious experience, at least for these subjects, is very different from a dream experience. The conclusion drawn by Azari et al. is that a religious experience is a cognitive phenomenon as evidenced by the occurrence of brain activity emanating from cognitive structures of the brain.

Matthew Alper (2001) takes a different tack. He argues that the cross-cultural and time immemorial beliefs in gods, soul, and after-life suggest that there is a God part in the brain, that God is an inherited characteristic hardwired in the neurophysiology of the brain. This concept that God is actually hardwired in our brain is likely not to be readily embraced by those who believe that a hallmark of being human is the ability of the human mind to conceive of a God. Nonetheless, Alper argues that there is yet another reason to believe that God is hardwired in the brain, stating that such a hardwired characteristic would have served man by making contemplation of his mortality tolerable. Man being aware of his own death, Alper argues, makes him prone to debilitating depression. Belief in an after-life helps man get through this. In other words, God appeared as an adaptation that allowed man to cope with his awareness of death.

Alper also argues for the brain basis of religious feelings by citing transcranial magnetic stimulation (TMS) studies in which religious, mystical, and spiritual experiences were reported upon electrical stimulation of specific areas of the brain, most notably in the temporal lobe. He argues that the emergence of religious feelings when the temporal lobe of the brain was stimulated is evidence of God existing in the brain.[4] Whether one chooses to believe that this is evidence for the existence of God in the brain, it is certainly evidence for the brain basis of experience, including experiences reported as mystical, religious, and spiritual.

Meditation and the Brain

Specific brain regions are also activated during different kinds of meditation (Lazer et al. 2000; Lou et al. 1999; Newberg et al. 2001). One kind of meditation is based on focused attention. Another kind is based on relaxed acceptance. Not surprisingly, the two kinds of meditation lead to different brain responses because they reflect two different states of mind. Lazar et al. (2000) used functional brain imaging (fMRI) to study the effects of focused kundalini meditation in which the subjects focused their attention on their breathing. This focused attention type of meditation activated the areas involved in attention and arousal—the lateral prefrontal and parietal regions and the anterior cingulate and amygdala. Newberg et al. (2001) also studied focused meditation but here the subjects focused their attention on a visualized image and maintained that focus. Like in the previous study, here too increased activity in the prefrontal cortex was found. This is because focused attention is the basis for this kind of meditation as well.

Using PET, Lou et al. (1999) studied subjects practicing a relaxation type of meditation in which the subject experiences a loss of conscious control, lets go of goals and concerns, and allows thoughts to just come and go (Yoga Nidra). These PET results showed that specific brain areas became activated for specific meditative content. Meditation on the weight of the limbs activated the supplementary motor area; meditation on joy and happiness activated the left hemisphere exclusively, as was to be expected from previous work (Davidson and Irwin 1999). Visual imagery activated the visual cortex and the parietal cortex. There was no prefrontal and no cingulate activity during the meditation, unlike that found in the focused attention kind of meditation studied by Lazar. These two different kinds of meditation studies clearly show that the kind of brain activity produced by meditation depends upon the state of mind during meditation. A relaxed acceptance kind of meditation led to the activation of different brain structures than a focused attention type of meditation. Our brains reflect our state of mind, so it should come as no surprise that meditation would do the same.

Austin (2003) explored yet another meditative experience reporting how losing one's sense of self, or ego, is related to changes in specific brain functions. He described an experience in which there was "no physical or psychic self, but only a kind of anonymous mirror was witnessing the scene . . ." Austin remarks that this special mind state that he reached through his meditation practice, arose as a result of a number of brain and body changes. These changes included changes in the thalamus that allowed it to block external sensory input, and which, in turn, allowed a deeper internal awareness to occur. Clearly, Austin's report also supports the intimate connection that exists between one's brain activity and one's state of mind.

Summary and Concluding Remarks

We have seen evidence for the relationship between brain activity and mind. For example, stroke and trauma patients who suffered brain damage also suffered alterations in how their minds functioned. We also saw that delusional misidentification, unilateral neglect, and changes in social functioning were all related to changes in brain activity. In the section on dreaming, we cited the normal case of REM sleep dreaming where, though there is no anatomical damage to the brain, our mind acts as if there were, at least compared to waking standards. We saw that this was due to the profound changes in brain chemistry and in regional brain activity, which occurred during the REM stages of sleep. We have also seen evidence for how thinking, feeling, meditation, moral decision making, and religious experiences activate specific regions and regional networks of the brain. One obvious conclusion from all this evidence is that there is no mind without brain, whether the mind is engaged in the most mundane task or in the most heartfelt spiritual experience.

We also addressed the more difficult and elusive question of *how* thoughts, emotions, spiritual feelings, and religious experiences arise from brain activity. The brain doesn't consist of experiences like it consists of neurons and other cells, so how does mind emerge? We suggested that the answer to this question might come from an understanding of how complexity arises out of the interaction and self-organization of elements that in themselves have only very simple behaviors. Even though evidence for this fact does not prove that is *how* thoughts, emotions, spiritual feelings, and religious experiences actually arise from brain activity, it does at least lend credence to that possibility. Another important reason for giving credence to this possibility is that self-organization is a fundamental process of nature in that there is a tendency for all nonequilibrium systems to move toward self-organization and collective behavior.

We gave examples of when this happens. In the example of water that was heated from below, we saw that if a sufficiently large temperature difference between the bottom and top surfaces of the water existed, a global hexagonal pattern would spontaneously form on the top surface of the water (the Bénard instability). This global hexagonal pattern emerged as a result of a change from individual to collective movement of the water molecules, which occurred when the temperature gradient reached a threshold level. Each hexagon consisted of billions of individual water molecules that somehow "knew" how to create the hexagonal structure of which they were a part. This is an example of *self-organization*.

In the social insect example, there had to be changing amounts of pheremone to allow structure to build upon itself. But, as we saw, it was the insects themselves who produced the changing amounts of pheremones. They themselves provided for the emergence of a complex nest structure, for the emergence of an elaborate and complex cathedral-like termite nest built without intelligent agency but from the local interactions of millions of termites. Individual termites lost their semi-isolation from the group when they began to act collectively in response to changes in their environment that exceeded threshold levels of stability. Individual behavior became "unstable" and gave way to collective behavior that was stable to the new pheremone-rich environment. How interesting that simplification led to complexity; from the many possible ways in which individual termites could interact to the very few ways in which they could interact. This simplification was due to the emergence of collective behavior that constrained its members' choices.

And embryonic development, which occurred through the process of bootstrapping into ever more complex and differentiated embryonic structures, was accomplished by changes in the environment produced by the cells themselves. Here too, the emergent embryonic structures were a result of local interactions occurring among individual cells. Embryo formation is a dynamic self-organizing process whereby global embryonic structures are formed as a result of local interactions among cells. Once formed, these structures then constrained the subsequent behavior of the individual cells from which they emerged, thereby lending stability to the system.

Haken (1996) has called this process circular causality. In the brain, this is the interplay between neurons and mind. Circular causality allows us to do away with the necessity of postulating a special being or set of beings to initiate mind. Mind, created through the self-organizing properties of brain activity, constrains that very same brain activity that created it. Circular causality insures that the emergent mind constrains the subsequent ways in which the neurons can communicate. The emergent mind is not at the whim of individual neuronal activity even though mind owes its existence to this activity. This provides for stability of the emergent mind. In passing, we note that this may not always work to the best advantage of the individual. For example, an emergent mind that becomes "set" maintains itself by constraining brain activity, thereby preventing other brain activity from which emergent thought patterns and more desirable behaviors might self-organize. If one's mind is made up, this mind set could very well limit communication between individual neurons, which may be necessary to form competing thoughts and beliefs.

Self-organization is not only essential for the *creation* of mind from brain activity, but self-organizing dynamics also help to *dissolve* preexisting mind patterns. Preexisting mind patterns can dissolve when changes occur that are sufficiently large to cause the existing brain activity to become unstable. A transition to a new mind state that is stable in the changed environment may then occur. The process, according to Walter Freeman (2003), happens via chaotic dynamics. Existing thought patterns dissolve when the underlying brain dynamics becomes chaotic. After the existing thought patterns dissolve, the mind is able to take on new learning through subsequent self-organizing brain processes.

In summary, mind emerges out of mindless brain activity. Communication between neurons in the brain takes place through the release of neurotransmitters at synapses between the neurons. This may cause a neighboring neuron to become activated if its membrane potential is sufficiently excited. Nature has assembled a vast and dense network of these simple mindless communication agents, which are both sufficiently interconnected and yet sufficiently independent to allow them to self-organize into what we call mind. Complex patterns of thought and behavior emerge from the tendency of brain activity to self-organize into collective behavior and complexity. Our most profound and spiritual experiences arise out of our own neurons' unique interconnectivity and their propensity to self-organize.

Notes

1. While local interactions between neurons are not difficult to imagine, it is more difficult to see how neurons that are very distant from one another can interact. We know that this does happen as the neurons responsible for vision are located quite distant from those that react to sound and smell. One mechanism by which long-range interaction can be accomplished is via synchronous firing of neural groups whereby neuronal groups are entrained to oscillate collectively. This has been observed, both as collective oscillations in different frequency ranges, but especially the gamma frequency range (20–80 Hertz), and as bursts of synchronous firing by Llinas and Ribary (1993), Engel et al. (1991a,b,c), Gray and Singer (1989), Singer (1989), and Freeman and Barrie (1994), among others. This collective firing helps bring spatially separated neuronal groups together and also to unify the response of multiple neuronal groups to a stimulus.
2. There are advantages, in general, to an interconnected network of brain regions. One important advantage is that many different kinds of functions can be called upon to perform a cognitive task. To recall the identity of a face, visual recall can be supplemented with access to associations that the name brings up, that the recall of specific events brings up, and that the recall of any feelings evoked on seeing the face brings up. As another example, networks that include Wernicke's and Broca's areas allow us to both comprehend and speak our thoughts, and when networks in the visual and auditory areas respond to a given task, sight and sound are added to comprehension.
3. The study of the brain basis of mind received an important input with the advent of the techniques of brain imaging, such as PET and functional magnetic resonance imaging (fMRI) as applied to human cognition. In these techniques, the increase of blood flow or the increase of oxygen utilization in specific brain areas is measured during the performance of a given cognitive task. In this way, we are able to learn about which areas of the brain become oxygen rich during performance of the cognitive task as well which areas become oxygen depleted. This allows speculation on the demands made on different neuronal populations in these brain areas during the performance of the cognitive task. See, e.g., Cabeza and Nyberg (2000) and Frith (1995).
4. In the study cited by Alper (Alper 2001, p. 114), TMS was used to *stimulate* certain regions of the brain. Generally, TMS transiently *disrupts* the function of the neural tissue being stimulated. This allows for identification of the specific neural tissues responsible for a specific behavior. An example of the use of TMS in this way was reported by Fierro et al. (2001). TMS applied over the right frontal cortex and posterior parietal cortex caused a normal subject to experience visuo-spatial neglect similar to that caused by brain damage, as discussed previously in the visual neglect section. For a review of TMS (where it is used in both ways), see Robertson et al. 2003.

References

Alper, M. 2001. *The "GOD" Part of the Brain*. Brooklyn, NY: Rogue Press.
Austin, J.H. 2003. Your self, your brain, and zen. *Cerebrum*, 5, 47–64.
Azari, N.P., Nickel, J., Wunderlich, G., Niedeggen, M., Hefter, H., Tellmann, L., Herzog, H., Stoerig, P., Birnbacher, D., and Seitz, R. 2001. Neural correlates of religious experience. *European Journal of Neuroscience*, 13, 1649–1652.
Balkin, T.J., Braun, A.R., Wesensten, N.J., Jeffries, K., Varga, M., Baldwin, P., Belenky, G., and Herscovitch, P. 2002. The process of awakening: A PET study of regional brain activity patterns mediating the re-establishment of alertness and consciousness. *Brain*, 125, 2308–2319.

Bechara, A., Damasio, H., and Damasio, A.R. 2000. Emotion, decision making and the orbitofrontal cortex. *Cerebral Cortex*, 10, 295–307.
Blanke, O., Origue, S., Landis, T., and Seeck, M. 2002. Stimulating illusory own-body perceptions. *Nature*, 419, 269–270.
Boyer, P. 2003. Religious thought and behaviour as by-products of brain function. *Trends in Cognitive Sciences*, 7, 119–124.
Braun, A.R., Balkin, T.J., Wesensten, N.J., Carson, R.E., Varga, M., Baldwin, P., Selbie, S., Belenky, G., and Herscovitch, P. 1997. Regional cerebral blood flow throughout the sleep–wake cycle. *Brain*, 120, 1173–1197.
Braun, A.R., Balkin, T.J., Wesensten, N.J., Gwadry, F., Carson, R.E., Varga, M., Baldwin, P., Belenky, G., and Herscovitch, P. 1998. Dissociated pattern of activity in visual cortices and their projections during human rapid eye-movement sleep. *Science*, 279, 91–95.
Bressler, S.L. and Kelso, J.A.S. 2001. Cortical coordination dynamics and cognition. *Trends in Cognitive Sciences*, 5, 26–35.
Cabeza, R. and Nyberg, L. 2000. Imaging cognition II: An empirical review of 275 PET and fMRI studies. *Journal of Cognitive Neuroscience*, 12(1), 1–47.
Damasio, A.R. 1994. *Descartes' Error: Emotion, Reason, and the Human Brain*. New York: Grosset/Putnam.
Davidson, R.J. and Irwin, W. 1999. The functional neuroanatomy of emotion and affective style. *Trends in Cognitive Sciences*, 3(1), 11–21.
Deneubourg, J.L. and Franks, N. 1995. Collective control without explicit coding: The case of communal nest excavation. *Journal of Insect Behavior*, 8, 417–432.
Deneubourg, J.L., Goss, S., Pasteels, J.M., Fresneau, D., and Lachaud, J.P. 1987. Self-organization mechanisms in ant societies (II): Learning in foraging and division of labor, in J.M. Pasteels and J.L. Deneubourg (eds.), *From Individual to Collective Behavior in Social Insects*. Basel, Boston: Birkhausser Verlag, pp. 177–196.
Edelstyn, N.M. and Oyebode, F. 1999. A review of the phenomenology and cognitive neuropsychological origins of the Capgras syndrome. *International Journal of Geriatric Psychiatry*, 14(1), 48–59.
Edelstyn, N.M., Oyebode, F., and Barrett, K. 2001. The delusions of Capgras and intermetamorphosis in a patient with right-hemisphere white-matter pathology. *Psychopathology*, 34(6), 299–304.
Engel, A.K., Konig, P., Kreiter, A.K., and Singer, W. 1991a. Interhemispheric synchronization of oscillatory neuronal responses in cat visual cortex. *Science*, 252, 1177–1179.
Engel, A.K., Kreiter, A.K., König, P., and Singer, W. 1991b. Synchronization of oscillatory neuronal responses between striate and extrastriate visual cortical areas of the cat. *Proceedings of the National Academy of Sciences, USA*, 88, 6048–6052.
Engel, A.K., König, P., and Singer, W. 1991c. Direct physiological evidence for scene segmentation by temporal coding. *Proceedings of the National Academy of Sciences, USA*, 88, 9136–9140.
Feinberg, T.E., Eaton, L.A., Roane, D.M., and Giacino, J.T. 1999. Multiple fregoli delusions after traumatic brain injury. *Cortex*, 35, 373–387.
Fierro, B., Brighina, F., Piazza, A., Oliveri, M., and Bisiach, E. 2001. Timing of right parietal and frontal cortex activity in visuo-spatial perception: A TMS study in normal individuals. *Cognitive Neuroscience*, 12, 2605–2607.
Forstl, H., Besthorn, C., Burns, A., Geiger-Kabisch, C., Levy, R., and Sattel, A. 1994. Delusional misidentification in Alzheimer's disease: A summary of clinical and biological aspects. *Psychopathology*, 27, 194–199.
Forstl, H., Burns, A., Jacoby, R., and Levy, R. 1991. Neuroanatomical correlates of clinical misidentification and misperception in senile dementia of the Alzheimer type. *Journal of Clinical Psychiatry*, 52, 268–271.
Freeman, W. 2003. Neurodynamic models of brain in psychiatry. *Neuropsychopharmacology*, 28(SI), 54–63.
Freeman, W.J. 1999. *How Brains Make Up Their Minds*. London: Weidenfeld and Nicolson.
Freeman, W.J. and Barrie, J.M. 1994. Chaotic oscillations and the genesis of meaning in the cerebral cortex, in J. Mervaille and T. Christen (eds.), *Temporal Coding and the Brain*. Berlin: Springer-Verlag.
Frith, C. 1995. Functional imaging and cognitive abnormalities. *The Lancet*, 346, 615–620.
Goldman-Rakic, P.S., O Scalaidhe, S.P., and Chafee, M.V. 2000. Domain specificity in cognitive systems, in M.S. Gazzaniga (ed.), *The New Cognitive Neurosciences*, Cambridge, MA: MIT Press, Chapter 50, pp. 733–742.

Gray, C.M. and Singer, W. 1989. Stimulus-specific neuronal oscillations in orientation columns of cat visual cortex. *Proceedings of the National Academy of Sciences, USA*, 86, 1698–1702.

Greene, J. and Haidt, J. 2002. How (and where) does moral judgment work? *Trends in Cognitive Sciences*, 6(12), 517–523.

Haken, H. 1981. *The Science of Structure: Synergetics*. New York, NY: Van Nostrand Reinhold company.

———. 1996. *Principles of Brain Functioning*. New York, NY: Springer.

———. 2002. *Brain Dynamics. Synchronization and Activity Patterns in Pulse-Coupled Neural Nets with Delays and Noise*. Berlin: Springer.

Halligan, P.W., Gereon, R.F., Marshall, J.C., and Vallar, G. 2003. Spatial cognition: Evidence from visual neglect. *Trends in Cognitive Sciences*, 7, 125–133.

Hobson, J.A. 1988. *The Dreaming Brain*. New York: Basic Books, Inc.

———. 1999. *Dreaming as Delirium: How the Brain Goes Out of Its Mind*. Cambridge, MA: MIT Press.

———. 2001. *Dream Drugstore*. Cambridge, MA: MIT Press.

Hobson, J.A., Pace-Schott, E.F., and Stickgold, R. 1999. Consciousness: Its vicissitudes in waking and sleep—An integration of recent neurophysiological and neuropsychological evidence, in M. Gazzaniga (ed.), *The Cognitive Neurosciences*, 2nd ed. Cambridge, MA: MIT Press.

———. 2000. Toward a cognitive neuroscience of conscious states. *Behavioral and Brain Sciences*, 23, 793–842.

Hudson, A.J. and Grace, G.M. 2000. Misidentification syndromes related to face specific area in the fusiform gyrus. *Journal of Neurology, Neurosurgery and Psychiatry*, 69, 645–648.

Joseph, A.B. and O'Leary, D.H. 1987. Anterior cortical atrophy in Fregoli syndrome. *Journal of Clinical Psychiatry*, 48, 409–411.

Kahn, D. and Hobson, J.A. 1993. Self-organization theory of dreaming. *Dreaming*, 3, 151–178.

———. 2003. State dependence of character perception: Implausibility differences in dreaming and waking consciousness. *Journal of Consciousness Studies*, 10, 57–68.

Kahn, D., Krippner, S., and Combs, A. 2000a. Dreaming and the self-organizing brain. *Journal of Consciousness Studies*, 7, 4–11.

———. 2002a. Dreaming as a function of chaos-like stochastic processes in the self-organizing brain. *Non-Linear Dynamics, Psychology, and Life Sciences*, 6, 311–321.

Kahn, D., Pace-Schott, E.F., and Hobson, J.A. 1997. Consciousness in waking and dreaming: The roles of neuronal oscillation and neuromodulation in determining similarities and differences. *Neuroscience*, 78, 13–38.

———. 2002b. Emotion and cognition: Feeling and character identification in dreaming. *Consciousness and Cognition*, 11, 34–50.

Kahn, D., Stickgold, R., Pace-Schott, E.F., and Hobson, J.A. 2000b. Dreaming and waking consciousness: A character recognition study. *Journal of Sleep Research*, 9, 317–325.

Kahn, D.L., De Palma, A., and Deneubourg, J.L. 1985. Noisy demand and mode choice. *Transportation Research*, 19, 143–153.

Kelso, J.A.S. 1995. *Dynamic Patterns. The Self-Organization of the Brain and Behavior*. Cambridge, MA: MIT Press.

Lazar, S.W., Bush, G., Gollub, R., Fricchione, G.L., Khalsa, G., and Benson, H. 2000. Functional brain mapping of the relaxation response and meditation. *NeuroReport*, 11, 1581–1585.

Llinas, R.R. and Pare, D. 1991. Of dreaming and wakefulness. *Neuroscience*, 44, 521–535.

Llinas, R. and Ribary, U. 1993. Coherent 40 Hz oscillation characterizes dream state in humans. *Proceedings of the National Academy of Sciences, USA*, 90, 2078–2081.

Lou, H.C., Kjaer, T.W., Friberg, L., Wildschiodtz, G., Holm, S., and Nowak, M. 1999. A ^{15}O-H$_2$O PET study of meditation and the resting state of normal consciousness. *Human Brain Mapping*, 7, 98–105.

Maquet, P. 2000. Functional neuroimaging of normal human sleep by positron emission tomography. *Journal of Sleep Research*, 9, 207–231.

Maquet, P., Laureys, S., Peigneux, P., Fuchs, S., Petiau, C., Phillips, C., Aerts, J., DelFiore, G., Degueldre, C., Meulemans, T., Luxen, A., Franck, G., Van Der Linden, M., Smith, C., and Cleeremans, A. 2000. Experience-dependent changes in cerebral activation during human REM sleep. *Nature Neuroscience*, 3(8), 831–835.

Maquet, P., Peteres, J.M., Aerts, J., Delfiore, G., Degueldre, C., Luxen, A., and Franck, G. 1996. Functional neuroanatomy of human rapid-eye-movement sleep and dreaming. *Nature*, 383, 163.

Mesulam, M.M. 1998. From sensation to cognition. *Brain*, 121, 1013–1052.
Newberg, A., Alavi, A., Baime, M., Pourdehnad, M., Santanna, J., and d'Aquili, E. 2001. The measurement of regional cerebral blood flow during the complex cognitive task of meditation: A preliminary SPECT study. *Psychiatry Research: Neuroimaging Section*, 106, 113–122.
Nicolis, G. and Prigogine, I. 1977. *Self-Organization in Non-Equilibrium Systems*. New York: John Wiley.
———. 1989. *Exploring Complexity*. New York: W.H. Freeman and Company.
Nofzinger, E.A., Buysse, D.J., Miewald, J.M., Meltzer, C.C., Price, J.C., Sembrat, R.C., Ombao, H., Reynolds, C.F., III, Monk, T.H., Hall, M., Kupfer, D.J., and Moore, R.Y. 2002. Human regional cerebral glucose metabolism during non-rapid eye movement sleep in relation to waking. *Brain*, 125, 1105–1115.
Nofzinger, E.A., Mintun, M.A., Wiseman, M.B., Kupfer, D.J., and Moore, R.Y. 1997. Forebrain activation in REM sleep: An FDG PET study. *Brain Research*, 770, 192–201.
Nyberg, L., Persson, J., Habib, R., Tulving, E., McIntosh, A., Cabeza, R., and Houle, S. 2000. Large scale neurocognitive networks underlying episodic memory. *Journal of Cognitive Neuroscience*, 12, 163–173.
Oyebode, F. and Sargeant, R. 1996. Delusional misidentification syndromes: A descriptive study. *Psychopathology*, 29, 209–214.
Penfield, W. and Erickson, T.C. 1941. *Epilepsy and Cerebral Localization*, Spring Field, IL: Charles C. Thomas.
Robertson, E.M., Théoret, H., and Pascual-Leone, A. 2003. Studies in cognition: The problems solved and created by transcranial magnetic stimulation. *Journal of Cognitive Neuroscience*, 15(7), 948–990.
Scott, A. 2004. Reductionism revisited. *Journal of Consciousness Studies*, 11(2), 51–68.
Singer, W. 1989. Search for coherence: A basic principle of cortical self-organization. *Concepts in Neuroscience*, 1, 1–11.
Szentagothai, J. and Erdi, P. 1989. Self-organization in the nervous system. *Journal of Social and Biological Structures*, 12, 367–384.
Theraulaz, G., Bonabeau, E., and Deneubourg, J.L. 1998. The origin of nest complexity in social insects. *Complexity*, 3(6), 15–25.
Tong, F. 2003. Out-of-body experienced: From Penfield to present. *Trends in Cognitive Sciences*, 7, 104–106.
Tononi, G. and Edelman, G.M. 1998. Consciousness and complexity. *Science*, 282, 1846–1851.
Tononi, G., Edelman, G.M., and Sporns, O. 1998. Complexity and coherency: Integrating information in the brain. *Trends in Cognitive Sciences*, 2, 474–484.
Walker, M.P., Liston, C., Hobson, J.A., and Stickgold, R. 2002. Cognitive flexibility across the sleep–wake cycle: REM-sleep enhancement of anagram problem solving. *Cognitive Brain Research*, 14, 317–324.

CHAPTER SEVEN

Converting: Toward a Cognitive Theory of Religious Change

PATRICIA M. DAVIS AND LEWIS R. RAMBO

Introduction

The term conversion may be used to refer to a range of phenomena that encompasses an individual's internal experience, the growth in spiritual maturity of the individual, and outward public affiliation. While there would seem to be a distinction between the unique inward experience and the outward change of affiliation, these processes often work together and are mutually reinforcing. A more illuminating distinction is between the use of the word "conversion" to signify a discrete *event* or an ongoing *process*.

Conversion involves the personal adoption of, and investment of faith in, a particular group of religious rituals, relationships, roles, and rhetoric, which together serves as a system of cognitive meaning for the individual. It assumes the individual is converting *from* a meaning system as well as *to* a meaning system. Conversion is thus usually the process of disruption of the existing meaning system, which results in disorientation, and eventual reorientation to a new or revised meaning system. Religious teachings may contribute to disequilibrium, facilitating and guiding transformation.

In this chapter, we will review the salient research on conversion and on cognitive psychology. We will then apply the methodology of cognitive linguistics to Christian conversion language. We will use the research in cognitive linguistics and psychology that has recently shown that metaphor is an important way of thinking and exists as one of the cognitive structures. The importance of "primary metaphors," that is, metaphors based on early childhood experiences that form the experiential basis for comprehension, will be explored.

Then, we will review the biblical scholarship on the original ancient Hebrew, Greek, and Latin words used in primary Christian texts to discuss conversion. In many cases, the scholarship shows that contemporary translations have substituted conceptual words for words with literal meanings, which are actually primary metaphors—the very building blocks of conceptual thought. By examining the literal language used in the ancient texts for primary metaphors, we will develop an understanding of the underlying thought structure.

We will show that the literal meanings for the Christian concept of conversion refer to the primary experience of turning and walking in a type of A PURPOSEFUL LIFE IS A JOURNEY metaphor. These concepts expressed are understandable across cultures because of the universal embodied experience of physical orientation and locomotion. Interestingly, the words are used in a subversive way, emphasizing disruption of orientation, bodily reorientation, and stepping. And, rather than using the more obvious word choices for planned journeys, the ancient texts sometimes choose words that emphasize the step-by-step, moment-to-moment quality of walking.

This analysis of the language of Christian conversion is consistent with the disruption and disorientation we have identified as a necessary intermediation between two meaning systems. However, the language may also imply a type of conversion from a fixed meaning system to a new life orientation open to continual meaning system adjustment.

Conversion Literature Review

This section of the chapter provides a brief review of some key concepts and theories in the academic field of conversion studies. Approaches from the fields of anthropology and theology are included with psychological approaches when they relate to the psychological processes of conversion.

Wallace: Mazeway Reformulation—Cultural Breakdown to Breakthrough

The anthropologist Anthony F.C. Wallace (2003) has proposed the concept of "mazeway reformulation," the view that a culture responds to potential collapse by a process of reconstitution. He posited that when a culture's meaning system is threatened, the existing core myths, rituals, and symbols are broken down and reformulated to revitalize the culture. This process occurs through a key individual who has a vision or conversion experience that modifies the old way to adapt to the current situation. The individual's new vision is spread through disciples who both communicate and adapt the original new vision, thus creating a cultural revitalization.

While Wallace's focus was on the cultural process, his model assumes a psychological dimension to the renewal of culture. Cultural renewal is possible because myths, rituals, and symbols are deeply embedded in the individual psyche and are capable of change, development, and transformation. Wallace termed these cognitive patterns of cultural meaning "mazeways."

Smith: Jungian Perspective on the Creation of the Self

Based in Jungian psychology, Curtis Smith (1990) has posited a theory of conversion that emphasizes religions as systems of symbols. His archetypal theory of conversion asserts that the human psyche is structured with shared fundamental patterns. As an individual human is confronted with a universal experience, a profound need is triggered within the person's psyche. Religious symbols provide the archetypal meaning for the experience, fulfilling the psyche's need. Thus, conversion occurs because the psyche's existing symbol system does not support the psyche in the current experience, and the new religion provides the symbol system that fulfills the psyche's need.

McFague: Religious Language Initiates Conversion through Psychic Disruption

Based on the analysis of parables by John Dominic Crossan, Sallie McFague (1978) proposed that conversion is a process of orientation, disruption, and reorientation with the parables serving to initiate conversion through disorientation. Crossan had proposed that the Christian parables—the stories told by Jesus in the Bible—should be viewed not as sources of comfort, consolation, and peace, but rather as sources of discomfort, disruption, and vulnerability. He suggested the metaphor of life as existing on a raft in the open ocean—one's stability and security is limited and something greater than we can understand is just beyond. He also suggested we view the challenge of Christian parables to our imagination as comparable to being at the edge of the raft exposed to the open ocean expanse (Crossan, 1975, pp. 41–46).

McFague embraced the raft metaphor and further argued that in authentic Christian conversion, the reorientation was not characterized "by a new stable worldview, but by a radical vulnerability" to a lifelong process of reorientation, of life on the edge of the raft (McFague 1978, pp. 258–259). Thus, she proposed that through the parables, the Christian conversion "must be a lifelong journey to pass over to a new orientation to reality" in which every day is lived in "a strange world, a new world in which to live and love" (p. 268).

Metzner: Universal Metaphors Used to Describe Transformation

The psychologist Ralph Metzner observed recurring metaphors used to describe spiritual transformation across cultures and religions. He identified ten universal metaphors: metamorphosis, awakening, uncovering, being liberated, purification by fire, wholeness from fragmentation, outbound journeys, return journeys, rebirth, and unfolding. Metzner proposed that the examination of the use of these metaphors might be helpful in stimulating spiritual transformation.

Metzner treated outbound journeys and return journeys as separate but related metaphors. However, he recognized that in actual use by spiritual authorities, the description often included both outbound and return metaphoric components (1986, p. 106), and that the return journey metaphor presupposed an outward component (p. 132). He identified wandering and labyrinths as cases of the outward journey metaphor.

Morrison: Medieval Christian Conversion as Life Process

A historian of hermeneutics in Western culture, Karl Morrison came to see the understanding of conversion as a central theme. In his research on the meaning of conversion in the medieval period, he found that Christian conversion was understood quite differently than it is commonly understood today. Then it was understood to be a lifelong inward process of the heart turning toward Christ and subsequent reformation.

Morrison notes that the notion of a slow process of conversion rather than a sudden change is present in the Christian New Testament in both the metaphors used and in the example of the convert Paul. The metaphors of "germination, growth and fecundity are not metaphors of sudden change" (1992, p. 38). Similarly, although Paul is often cited as the exemplar of sudden conversion,

> Paul realized the consequences of his vocation slowly, not only in the days immediately following it, but also in the fourteen years that he spent in Arabia before he undertook his apostolate and in the tentative beginnings that eventually directed him to the Gentiles. (p. 38)

Rambo: Conversion as a Multi-Stage Process

Rambo has reviewed contemporary studies of conversion in the social science literature encompassing psychology, sociology, and anthropology. He presents conversion as a dynamic interaction that takes place in a set of interlocking systems. Rambo proposes a seven-stage conversion process, which incorporates both the convert's internal psychology and the external environment. He notes that conversion is accomplished by the individual's needs and aspirations, but also by the group into which the person is converted and the greater social matrix. Rambo's seven stages are: context, crisis, quest, encounter, interaction, commitment, and consequences.

Rambo distinguishes between conversion to a group and commitment to a path of transformation:

> In fact, I would argue that people who convert and remain the same are not really on a spiritual path of transformation. They have enshrined the conversion as a sacred moment and relive the event over and over again, but it has little power to transform their lives. Change is persistent and important and continuous, and most religious traditions expect and foster change by providing ideology and techniques for the ongoing development and maturation of their members. (1993, p. 163)

Cognitive Science Literature Review

This section of the chapter provides a brief review of some key concepts and theories in the academic field of mind–brain science, which may be applicable to the study of conversion.

Ramachandran and Persinger: Temporal Lobes and God Consciousness

V.S. Ramachandran and Michael Persinger have discussed the neurophysiological basis for religious experience. They have independently concluded that the temporal lobes of the brain are the site of religious experience. Ramachandran says that epileptic seizures in the temporal lobes are often accompanied by religious experiences. He posits the possibility that repeated experiences may permanently alter the pathways of the brain:

> These seizures—and visitations—last only a few seconds each time. But these brief temporal lobe storms can sometimes permanently alter the patient's personality so that even between seizures he is different from other people. No one knows why this happens, but it's as though the repeated electrical bursts inside the patient's brain (the frequent passage of massive volleys of nerve impulses within the limbic system) permanently "facilitate" certain pathways or may even open new channels, much as water from a storm might pour downhill, opening new rivulets, furrows, and passages along a hillside. This process, called kindling, might permanently alter—and sometimes enrich—the patient's inner emotional life (Ramachandran and Blakeslee 1998, p. 179).

Persinger (1983, 1993) has carried out a program of research to document and explore the relationship between stimulation of the temporal lobe structures and types of religious experience.

Freeman: Brain Structure, Chaos, and the Scrooge Effect

The neurobiologist Walter Freeman has researched the brain structure of learning and the acquisition of new material that is inconsistent with previously learned material. Freeman has concluded that a mechanism for unlearning, dissolution of existing structure, is necessary before new material can be incorporated into the structure. Chaos must be induced in order for the brain to replace the existing structure with a new structure that incorporates the anomalous information. The process of learning new anomalous material is thus one of order, chaos, and then new order.

Freeman has related this learning process to normal socialization processes, as well as to the techniques for behavior modification in religious conversions. He labeled it the "Scrooge Effect": "The dissolution prepares for new learning by self-organization, where-by the pre-existing life history of an individual is transiently weakened, even melted down, so that new structure can grow that is not logically consistent with all that came before" (2003, p. S59).

Freeman identifies biological techniques for inducing dissolution including isolation, severe physical exercise, fasting, lack of sleep, and the induction of powerful emotional states.

Varela, Thompson, and Rosch: Buddhism and Cognitive Science

Francisco Varela, Evan Thompson, and Eleanor Rosch have presented the understandings of the cognitive structure of the brain developed by cognitive science in juxtaposition with descriptions of Buddhist mindfulness/awareness practice (Varela). Both approaches result in the ultimate conclusion of a lack of ultimate foundations for either the individual's mind or its objects, the world. Their conclusion is reminiscent of McFague's conclusion regarding the parables in its letting go of a clinging to absolute ground:

> When we widen our horizon to include transformative approaches to experience, especially those concerned not with escape from the world or the discovery of some hidden, true self but with releasing the everyday world from the clutches of the grasping mind and its desire for absolute ground, we gain a sense of perspective on the world that might be brought forth by learning to embody groundlessness as compassion in a scientific culture. (1993, p. 254)

Lakoff and Sweetser: Cognitive Linguistics

Historically, metaphor has been seen as a trope or embellishment to language. However, late-twentieth-century work in linguistics and cognitive science has shown that, rather than an embellishment, metaphor is an important way that people think. The "mind contains an enormous system of general conceptual metaphors—ways of understanding relatively abstract concepts in terms of those that are more concrete" (Lakoff 2001, p. 265).

We draw from our experiences of sensorimotor physical movement to reason conceptually about abstract things. Primary metaphors are experientially grounded mappings from the sensorimotor domain to the domain of subjective judgment. George Lakoff and Mark Johnson explain:

> Primary metaphors are part of the cognitive unconscious. We acquire them automatically and unconsciously via the normal process of neural learning and may be unaware that we have them. We have no choice in this process. When the embodied experiences in the world are universal, then the corresponding primary metaphors are universally acquired. This explains the widespread occurrence around the world of a great many primary metaphors. (1999, p. 56)

Thus, primary metaphors are actually part of the mind—neural connections between sensorimotor experience and subjective experience.

Research regarding the primary metaphor KNOWING IS SEEING demonstrates both early childhood, acquisition and general cross-cultural directionality from sensorimotor experience to conceptual experience. Chris Johnson (1999) has shown the original conflation of knowing and seeing in the language use of the child "Shem." In early childhood, the use of a phrase like "I see what you mean," would refer to a physical case of visually perceiving and thus comprehending the caretaker's meaning regarding a physical object. Later, a phrase like "I see what you mean" would also be used as a metaphorical expression of comprehension with no actual visual component.

Eve Sweetser's etymological research (1990) has shown the general historical linguistic trend that words for seeing tend to later acquire the meaning of knowing as well.

In metaphor theory, the word "domain" is used to describe a category of experience. The "source" domain is the more concrete or familiar area of experience. The "target" domain is the less familiar or abstract concept, the goal or target of understanding. The mind "maps" from the familiar source domain onto the target domain. Mappings are tightly structured and these correspondences are a fixed part of our conceptual systems.

Some other examples of primary metaphors cited by Lakoff and Johnson (1999) include:

- STATES ARE LOCATIONS;
- CHANGE IS MOTION;
- ACTIONS ARE SELF-PROPELLED MOTIONS;
- PURPOSES ARE DESTINATIONS.

In the notation of metaphor theory, the abstract concept, the target domain, is listed on the left, and the source domain, the concrete sensorimotor experience, is listed on the right. Thus, in the examples above, the mind maps from its primary sensorimotor experience:

- of locations to understand abstract states,
- of motion to understand change,
- of self-propelled motion to understand action, and
- of destinations to understand purposes.

We build up complex metaphors from our primary metaphors. Lakoff and Johnson use the metaphor A PURPOSEFUL LIFE IS A JOURNEY to illustrate the idea of complex metaphors (Lakoff and Johnson 1999, pp. 60–63). Spatial location and physical movement are inherent in this classic example. This complex metaphor A PURPOSEFUL LIFE IS A JOURNEY includes two of the primary metaphors listed earlier: PURPOSES ARE DESTINATIONS and ACTIONS ARE MOTIONS. The complex metaphor also includes the fact that a long trip to a series of destinations is a journey.

Unlike primary metaphors, complex metaphors may be culture specific. Lakoff and Johnson propose the A PURPOSEFUL LIFE IS A JOURNEY metaphor as an example that is pervasive in contemporary Western culture but is not universal. Lakoff explains that, along with primary metaphors and facts, this metaphor requires the cultural belief that people are supposed to have life purposes.

They propose that the experience of a journey with a traveler, destinations, and an itinerary is mapped onto the concept of a purposeful life as follows:

- A PERSON LIVING A LIFE IS A TRAVELER;
- LIFE GOALS ARE DESTINATIONS;
- A LIFE PLAN IS AN ITINERARY;

Here LIFE GOALS ARE DESTINATIONS is a special case of the primary metaphor referred to earlier, PURPOSES ARE DESTINATIONS. This metaphor carries with it our cultural expectations about journeys and maps them onto life purpose.

> A journey requires planning a route to your destinations. Journeys may have obstacles, and you should try to anticipate them. You should provide yourself with what you need for your journey. As a prudent traveler, you should have an itinerary indicating where you are supposed to be at what times and where to go next. You should always know where you are and where you are going next. (Lakoff and Johnson 1999, p. 62)

When these cultural judgments about journeys are mapped onto the concept of PURPOSEFUL LIFE, a series of powerful inferences, judgmental "shoulds," are created:

- If a journey requires planning, then one's life requires planning.
- If one should anticipate obstacles on a journey, then one should anticipate obstacles in life.
- If one should bring provisions on a journey, then one should provide for oneself in life.
- If a journey requires an itinerary, then life requires a planned set of goals.

These are not logical conclusions from primary sensorimotor experience, but mappings from a culturally based complex metaphor.

The Language of Christian Conversion

Metaphor is the way people think at the deepest level. The methodology of metaphor theory has recently been applied to religious concepts. Edward Slingerland (2004) has used metaphor theory as a methodology for comparative religion to understand Chinese Confucian moral concepts. Mary

Therese DesCamp and Eve Sweetser (2005) have used metaphor theory to explore the cognitive understanding of God in Christian and Hebrew texts.

The methodology of metaphor theory may also be used to explore the Christian concept of conversion. One of our primary experiences is our physical embodied location in space. This experience is applied through metaphor to help us understand many important concepts including time, love relationships, and life purposes. The Christian language of conversion is an expression of a spatial location metaphor. The words used for concepts of Christian religious conversion and new lifestyle are really expressions of primary sensorimotor experience. They literally express the experiences of turning and walking along a path.

The original literal meanings of the ancient Hebrew, Greek, and Latin words used to discuss conversion are actually about spatial reorientation. All three ancient words carry some connotation or possibility that the turning is a returning toward an earlier or original position.

Conversion is the English word used to translate the concept from the Hebrew Bible expressed in Hebrew as *sub*. This Hebrew word literally means "to turn." "The verb sub, translates the idea of changing route; to come again, to retrace one's steps. In a religious context, it means to be turned away from what is bad and to be turned toward God" (Leon-Defour 1967, p. 486). An example that clearly illustrates this usage is from Jeremiah: "It may be when the house of Judah hears of all the disasters that I intend to do to them, all of them may *turn* from their evil ways and, so that I may forgive their iniquity and their sin" (Jer. 36:3, NRSV, p. OT 488). Isaiah uses the journey in the Exodus from Egypt and the return to the Promised Land as a redeemed people as the metaphor for group conversion: "I have swept away your transgressions like a cloud, and your sins like mist; *return* to me, for I have redeemed you" (Isa. 44:22, NRSV, p. OT 927). In the Greek version of the Hebrew Bible, the Septuagint, *sub* is translated as *epistrephein*, which also means literally to turn.

The Christian religion incorporates the Hebrew Bible as its "Old Testament" and calls the new Christian religious material the "New Testament." The New Testament, originally in Greek, continued to use *epistrephein*. When the New Testament was translated into Latin, the word used was *conversio*, again rooted in the concept of directional rotation. In this following example from Acts, the turning from is both literal and metaphorical because Paul's speech occurs outside the temple of Zeus where the people wish to sacrifice after Paul healed a man with crippled feet: "Friends, why are you doing this? We are mortals just like you, and we bring you good news, that you should *turn* from these worthless things to the living God, who made the heaven and the earth and the sea and all that is in them" (Acts 14:15, NRSV, p. NT 182). In the following example from the Epistles of Peter, the metaphor of a shepherd and sheep is used to describe Christian conversion: "For you were going astray like sheep, but now you have *returned* to the shepherd and guardian of your souls" (1 Pet. 2:25, NRSV, p. NT 340).

Corollary to the spatial location literal meaning of conversion, the words to discuss Christian lifestyle are about self-locomotion. The literal meaning of life conduct is walking along a path or way.

> References to walking in the ways of God or his law would make up a small anthology of biblical verses (there are over a hundred references in Psalms and the book of Proverbs alone). Specimen passages include the following: "He leads me in right paths" (Ps. 23:3, NRSV); "I will instruct you and teach you the way you should go" (Ps. 32:8, NRSV); "to those who go the right way I will show the salvation of God" (Ps. 50:23, NRSV); "you search out my path . . . and are acquainted with all my ways" (Ps. 139:3, NRSV). (Ryken 1998, p. 630)

It is noteworthy that one of the Psalms referenced above is the familiar Psalm 23, which begins "The Lord is my shepherd, I shall not want" (Ps. 23:1, NRSV, p. OT 691). Thus, the quotation from 1st Peter above, which refers to the sheep returning to the shepherd, is a cultural reference that connects "return" to known imagery of "right paths."

In the New Testament, both the Gospel of John and the Epistles make frequent use of the "walk" and "way" to refer to the Christian life (Banks 1987).

Marilyn Harran's summary of St. Augustine's conversion language is also instructive:

> Man may be either turned to God by humble piety or turned away from Him by lifting up the horn of false liberty. Once man is converted to God, he does not simply occupy a stable, unchanging position, but begins to travel the arduous road of pilgrimage. Conversio transforms peregrinato from an undirected movement of aimless wandering into a directed movement of hopeful pilgrimage toward rest in God. (1983, p. 26)

The Christian A PURPOSEFUL LIFE IS A JOURNEY Metaphor

The Christian language of conversion is based on a spatial metaphor of directional rotation. Can we break down this language into metaphoric structure? It would seem to be a specific case of the A PURPOSEFUL LIFE IS A JOURNEY metaphor, that is, A PURPOSEFUL LIFE IS WALKING.

> As always, symbolic meanings grow out of the physical phenomenon. Walking on a path involves choosing to enter on the path and to pursue it in a given direction, progress toward a destination, making wise rather than foolish choices along the way, taking care for safety and not getting lost, and arriving at a goal. (Ryken 1998, p. 630)

The questions then become the where, how, and with whom of the metaphor submappings. It is important to note that the A PURPOSEFUL

LIFE IS WALKING metaphor is not always used in a consistent manner. While it is always true that the more abstract concept of one's life is mapped onto the concrete experience of walking, the submappings may differ:

LIFE CHOICES ARE DIRECTIONAL CHOICES
LIFE CHOICES ARE PATHS (OR WAYS)
STYLE OF LIFE IS STYLE OF WALKING
GOD IS THE COMPANION
GOD IS THE LEADER
GOD IS THE GOAL

It is also noteworthy that Jesus' instructions to his apostles sometimes directly contradict the cultural expectations about journeys from the A PURPOSEFUL LIFE IS A JOURNEY metaphor:

Take no gold, or silver, or copper in your belts, no bag for your journey, or two tunics, or sandals, or a staff; for laborers deserve their food. (Matt. 10:9–10, NRSV)

When they hand you over, do not worry about how you are to speak or what your are to say: for what your are to say will be given to you at that time; for it is not you who speak, but the Spirit of your Father speaking through you. (Matt 10:19–20, NRSV)

He said to his disciples, "Therefore I tell you, do not worry about your life, what you will eat, or about your body, what you will wear. For life is more than food, and the body more than clothing. Consider the ravens, they neither sow nor reap, and yet God feeds them. Of how much more value are you than the birds! And can any of you by worrying add a single hour to the span of your life? If then you are not able to do so small a thing as that, why do you worry about the rest? (Luke 12: 22–27, NSRV)

Applying these precepts, the cultural inferences from the A PURPOSEFUL LIFE IS A JOURNEY metaphor are reversed:

- A journey requires planning becomes do not plan;
- One should anticipate obstacles on a journey becomes do not anticipate;
- One should bring provisions on a journey becomes do not bring provisions;
- A journey requires an itinerary becomes do not use an itinerary.

A Straight and Narrow Path or a Labyrinth?

The language of Christian conversion may be advising temporary disorientation in order to reorient the individual to a new meaning system.

Alternatively, the language may be suggesting a permanent disorientation, that is, a continual state of openness to change. This may be expressed in the comparison of walking a straight and narrow path and walking a twisting, turning maze or labyrinth.

Conversion as reorientation to walking a straight and narrow path would be comparable to Lakoff's description of the A PURPOSEFUL LIFE IS A JOURNEY metaphor with his cultural inferences. One knows the rules. One knows where one is going. One can assess one's progress and the progress of others. These concepts are frequently expressed by converts to Christianity. This kind of conversion is a subtype of the A PURPOSEFUL LIFE IS A JOURNEY metaphor, which may be called A CHRISTIAN LIFE IS WALKING A STRAIGHT AND NARROW PATH. This is a subtype of the A PURPOSEFUL LIFE IS A JOURNEY metaphor where a CHRISTIAN LIFE is a subtype of a PURPOSEFUL LIFE and WALKING A STRAIGHT AND NARROW PATH is a subtype of a JOURNEY.

However, we have found in primary Christian texts another subtype of the A PURPOSEFUL LIFE IS A JOURNEY metaphor, which may be called the A CHRISTIAN LIFE IS WALKING A LABYRINTH metaphor. When walking a labyrinth, one is on a path, however, one is required to continually turn and continue along a new orientation. Here, once again, a CHRISTIAN LIFE is a subtype of a PURPOSEFUL LIFE. The subtype of JOURNEY is WALKING A LABYRINTH.

The use of the labyrinth for this metaphor is chosen based on the language we have analyzed. However, it is also consistent with medieval Christian symbolism. In her historical study of labyrinths, Penelope Doob presents the important distinction between multicursal and unicursal labyrinths. In multicursal labyrinths, there are "wrong" choices; not all paths lead to the goal. In unicursal labyrinths, there may be many twists and turns, but there is only one path; so to continue is to eventually reach the goal. The image of the unicursal labyrinth was an important symbol in medieval Christianity. They were frequently carved into placed baptismal fonts and inlaid in the floors of cathedrals (1990, pp. 117–133). Doob's analysis concludes that the church was teaching that, with Christ, the confusing and unstable turnings of life, the baffling complexities, are actually superbly articulated designs by God. We can add that the unicursal center represented God and eternal life as the goal.

The labyrinth metaphor represents the twisting, turning journey of lifelong Christian conversion. This type of imagery is present not only in the medieval universal labyrinth motif, but also in the language of Christian mystics. For example, in his discussion of journey metaphors, Metzner referred to the sixtenenth-century Christian saint Theresa of Avila's description of what seems to be a labyrinth-type journey through the inner castle:

> In it, the human soul is likened to a crystalline castle surrounded by seven concentric "mansions." The innermost circle is that of the King of Glory, the indwelling Spirit who is the Sun, the Light that is the center of Being. Each mansion has many rooms, many doors, passages,

gardens, mazes, fountains: these are the many aspects and facets of our nature, which, according to Teresa, the soul must be permitted to explore. (1986, p. 122)

It is interesting also to note that Wallace chose mazeway as metaphoric language to describe how the processes of the mind itself are formulated and reformulated.

The A CHRISTIAN LIFE IS WALKING A LABYRINTH metaphor may recur in the Christian tradition because its submappings and inferences closely fit the experience it is meant to express. The word inferences is used to discuss the results that occur in both the metaphor and the more conceptual experience the metaphor is used to understand. Some important inferences for the medieval unicursal labyrinth are:

1. In the physical experience of a labyrinth, one cannot assess one's progress toward one's goal. In walking the path of the labyrinth, one may be closer to the center at one point, and then out at the edge soon after. Being close to the center does not necessarily indicate one is near the end of the journey. The journey twists in and out before the end is reached. This has a disturbing corollary for Christian spiritual growth. One may need to face away and walk away from God as perceived in order to continue on the path and eventually reach God.
2. As one walks the labyrinth, one cannot assess one's progress in relation to others. The other who now appears at the outer edge may be closer to the end of the journey to the center than one is. Similarly, Christian teaching emphasizes the inability to judge the other.
3. As one walks the labyrinth, one must be willing to turn in the opposite direction, again and again. It is correct to walk north, then it is correct to walk south. It is only not correct to refuse to turn or to make one's own path. The corollary in Christian spiritual growth may be that one must be open to about-faces in religious matters. What was the obvious and appropriate course of action may no longer be so. One may have been following the direction of the Holy Spirit and may continue to do so, even (or especially) as one reverses course. In the labyrinth, each reverse in course is actually leading us closer to the center.

The raft metaphor of Crossan and McFague would be very similar to the medieval unicursal labyrinth if it were a type of river rafting trip rather than a raft on the open ocean. We can imagine a first-time river rafting trip on a twisting route to the sea. This would preserve the quality of life on the edge of the raft, but place it within an unknown but structured experience with a definite conclusion. Many of the inferences for the medieval unicursal labyrinth would transfer to such a raft trip experience.

The A CHRISTIAN LIFE IS WALKING A LABYRINTH metaphor may be understood to be about converting rather than about conversion.

In conversion, one has reoriented to a fixed meaning system. In the A CHRISTIAN LIFE IS WALKING A LABYRINTH metaphor, one is in a continual process of reorientation, continually converting.

Conclusions

Conversion may be defined as an initial God consciousness experience, as a commitment to a particular set of meaning-making beliefs, or as an ongoing life experience. We have examined the language in primary Christian texts to study the metaphors used in discussing conversion. Lakoff and Sweetser have shown that metaphors are the way humans make meaning. By examining metaphors, we can understand the basic conceptual structures used to reason. Thus, our examination of the metaphors used to discuss conversion illuminates the conceptual structures used to reason about this spiritual experience.

We have found that the A PURPOSEFUL LIFE IS A JOURNEY metaphor is used repeatedly in discussing conversion. However, in the Christian texts, this metaphor has two distinct subcases with very different inferences for the understanding of the conversion experience.

In the A CHRISTIAN LIFE IS WALKING A STRAIGHT AND NARROW PATH metaphor, conversion is the temporary experience of being placed on the path. The conversion experience, with its disruption of one's meaning system, is a temporary experience necessary for of the reorientation to a new or revised meaning system. Conversion is the precursor to a permanent meaning system, which can be described metaphorically as a STRAIGHT AND NARROW PATH. The inferences from this metaphoric model are confidence, certainty, and a refusal to deviate.

The second Christian subtype of the A PURPOSEFUL LIFE IS A JOURNEY metaphor, which we found in our examination of the embodied metaphor in the actual words used in Christian texts to describe conversion, is A CHRISTIAN LIFE IS WALKING A LABYRINTH. In this metaphor system, conversion is an ongoing process of continual reorientation. Conversion as an ongoing process is similarly described through McFague's examination of the Christian parables, and Varela, Thompson, and Rosch's examination of Buddhist mindfulness. The inferences from this metaphoric model are humility and an openness to change.

References

Banks, R. 1987. "Walking" as a metaphor of the Christian life: The origins of a significant Pauline usage, in E. Conrad (ed.), *Perspectives on Language and Text*. Winona Lake: Eisenbrauns, pp. 303–313.

Crossan, J. 1975. *The Dark Interval: Towards a Theology of Story*. Niles, IL: Argus Communications.

DesCamp, M. and Sweetser, E. 2005. Metaphors for God: Why and how do our choices matter for humans? The application of contemporary cognitive linguistics research to the debate on God and metaphor. *Pastoral Psychology*, 53, 207–38.

Doob, P. 1990. *The Idea of the Labyrinth from Classical Antiquity Through the Middle Ages*. Ithaca: Cornell University Press.
Freeman, W. 2003 Neurodynamic models of brain in psychiatry. *Neuropsychopharmacology*, 28, S54–S63.
Harran, M. 1983. *Luther on Conversion: The Early Years*. Ithaca: Cornell University Press.
Johnson, C. 1999. Metaphor vs. conflation in the acquisition of polysemy: The case of SEE in M.K. Hiraga, C. Sinha, and S. Wilcox (eds.), *Cultural, Psychological and Typological Issues in Cognitive Linguistics*. Amsterdam: Benjamins, pp. 155–169.
Lakoff, G. 2001. How metaphors structure dreams: The theory of conceptual metaphor applied to dream analysis, in K. Bulkeley (ed.), *Dreams: A Reader on the Religious, Cultural and Psychological Dimensions of Dreaming*. New York: Palgrave, pp. 265–284.
Lakoff, G. and Johnson, M. 1999. *Philosophy in the Flesh: The Embodied Mind and its Challenge to Western Thought*. New York: Basic Books.
Leon-Oefour, X. 1967. *Dictionary of Biblical Theology*. Paris: Oesclée.
McFague, S. 1978. Conversion: Life on the edge of the raft. *Interpretation*, 32, 255–268.
Metzner, R. 1986. *Opening to Inner Light: The Transformation of Human Nature and Consciousness*. Los Angeles: Jeremy P. Tatcher.
Morrison, K. 1992. *Understanding Conversion*. New Haven, CT Yale University Press.
Persinger, M. 1983. Religious and mystical experiences as artifacts of temporal lobe function: A general hypothesis. *Perceptual and Motor Skills*, 57, 1255–1262.
———. 1993. Paranormal and religious beliefs may be mediated differentially by subcortical and cortical phenomenological processes of the temporal (limbic) lobes. *Perceptual and Motor Skills*, 76, 247–251.
Ramachandran, V. and Blakeslee, S. 1998. *Phantoms in the Brain*. New York: William Morrow.
Rambo, L. 1993. *Understanding Religious Conversion*. New Haven: Yale University Press.
Ryken, L. (ed.) 1998. *Dictionary of Biblical Imagery*. Downers Grove: InterVarsity Press.
Slingerland, E. 2004. Conceptual metaphor theory as a methodology for comparative religion. *Journal of the American Academy of Religion*, 72 (1), 1–31.
Smith, C. 1990. Religion and crisis in Jungian analysis. *Counseling and Values*, 34, 177–185.
Sweetser, E. 1990. *From Etymology to Pragmatics: Metaphorical and Cultural Aspects of Semantic Structure*. Cambridge: Cambridge University Press.
Varela, F., Thompson, E., and Rosch, E. 1993. *The Embodied Mind: Cognitive Science and Human Experience*. Cambridge, MA: MIT Press.
Wallace, A.F.C. 2003. Revitalization movements, in R. Grumet (ed.), *Revitalizations and Mazeways: Essays on Culture Change*, vol 1. Lincoln: University of Nebraska Press, pp. 9–29.

CHAPTER EIGHT

Cognitive Science and Christian Theology

CHARLENE P.E. BURNS

> Cognitive science challenges our complacent theological claims about human nature and the human relation to God. It challenges but need not threaten, and if we listen closely and think deeply, our theological understanding will be the richer for it.
>
> —Peterson 1997, p. 627

Concerns about the impact of scientific discoveries on Christian theology have been a reality for at least 400 years. From Copernicus and Galileo came the first major challenges to humanity as the imago dei and its home as the center of God's purpose-filled creation. Subsequent discoveries compelled theologians to rethink everything from the meaning of miracles to the creation story itself. Again and again, doomsayers and scientific reductionists decreed the demise of religion under the burden of scientific "truth." Yet religion survived and theologians carried on the task of interpreting doctrine in light of a rapidly changing world. But today, say some, theology might finally collapse under the weight of science. This is so because the cognitive sciences offer a challenge to so many aspects of theological reflection. As science uncovers the ways in which neurochemistry affects the mental life, serious questions arise about what it means to say we have a soul, that humanity is created in the image of God, that we sin yet have free will, that we can encounter the divine through prayer, mystical experience, and revelation, even to say that there is a personal God.

During the past two decades a great deal of theological attention has been focused on the questions posed by the cognitive sciences. The issues are complex both scientifically and theologically. This is so in large part because "cognitive sciences" is an umbrella term used at present to designate a variety of disciplines, ranging from neuroscience to philosophy to evolutionary biology and psychology. A complete exploration is obviously far beyond the scope of this essay, but we can benefit from a survey of some of the major issues. Because the most immediate challenge is to theological anthropology, the bulk of this essay deals with the questions raised and possibilities offered by the so-called embodied mind for theologies of the soul and personhood. Our point of entry into this discussion is then a survey of recent thought on consciousness.

The Problem of Consciousness

Consciousness is the most profound problem for the cognitive sciences, as evidenced by how difficult it is to define. The literature is inconsistent but one thing is clear—no matter how it is defined, consensus is that consciousness is central to understanding human existence. Even this simplest definition, "an organism's awareness of its own self and surroundings," indicates its position as the source of all that sets humanity apart from other creatures. Societies, religions, conscience, art, music, science are all only possible because we are conscious beings (Damasio 1999).

The problem of explaining consciousness, sometimes called the brain–mind or the mind–body problem, occurs on at least two levels. Philosopher David Chalmers calls these the "easy" and "hard" problems of consciousness. The "easy problem" is that dealing with phenomena that can be explained in terms of functions and abilities. This includes phenomena like attention, purposeful control of behavior, the ability to categorize stimuli, to report on mental states, or to integrate information. Chalmers advocates calling these phenomena "awareness." These are easy problems because the cognitive sciences can tell us much about how they happen.

The "hard problem" of consciousness is that of experience itself: "What makes the hard problem hard . . . is that it goes beyond problems about the performance of functions . . . even when we have explained the performance of all the cognitive and behavioral functions in the vicinity of experience—perceptual discrimination, categorization, internal access, verbal report—there may still remain a further unanswered question: why is the performance of these functions accompanied by experience?" (1995, p. 200). The hard problem is the problem of "qualia" or phenomenal qualities of experience, which cannot be explained in terms of abilities or functions. This includes the sensory richness of human life experience: color, taste, smell, sadness, joy, the quality of physical pain. Functionally, it is possible to explain how the brain processes scent, but no amount of physical information about a perfume, for example, or about the neurochemistry of smell can actually capture its smell (Edwards 1997).

A great deal has been discovered in recent decades regarding the easy questions, but until neuroscience answers the hard problem of consciousness, the task is incomplete. Even so, some claim that in having demonstrated the physiological basis for the mental life, what it means to be a conscious human being is explained. We are nothing but a "pack of neurons" (Crick 1994) and all religious experience is nothing but the by-product of the human mind, a "complex biological machine" that manages "to give airy nothing a local habitation and a name" (Boyer 2001, p. 330).

This extreme form of materialism (where only physical things are real) is clearly incompatible with Christian theological claims, since it means that a human person is nothing but a physical organism whose mental and spiritual life will eventually be completely understood by science. These contentions end in the ontological reduction of religious claims—since spiritual experience and ideas of God are products of neurochemistry, they do not actually exist. Sigmund Freud's theory that religion is an illusion based in projection of infantile wishes for protection and unresolved oedipal issues is one early example of this approach. More recently, Canadian neuropsychologist Michael Persinger claims that because electrical stimulation of a certain part of the brain can elicit in the subject a "sensed presence," this demonstrates that religious experiences of the divine are nothing more than by-products of neurochemistry in the temporal lobe (1987, p. 4). Further, some claim that because they are biological phenomena, mental states are located in the brain. But this isn't true; the neurons involved in generation of thoughts are in the brain, but the thoughts themselves do not have a location or occupy space (Edwards 1997).

A complex entity like the brain certainly must be broken down into its smallest constituent parts in order to be explored. This does not, however, necessitate a materialist interpretation of the data. Reductive materialism is not tenable for many scientists because of the hard problem of qualia mentioned earlier. Dreams are a good example of the hard problem of consciousness. If materialism is right, then there ought to be a perfect correlation between brain states and dreams such that the neurologist could determine what someone is dreaming, not just that one is. This hard problem indicates there is a kind of privacy to our inner thought world that isn't overcome by science. For these reasons, some theorists advocate a more moderate form of materialism in which the findings of cognitive science are acknowledged as true without claiming they invalidate religious experience. In this way, religious experience is epistemologically but not ontologically reduced to neurochemistry and psychological experience. The newly emerging field of "neurotheology" is an example (Peterson 2003).

Neurotheology

Neurotheology is a term coined by James Ashbrook in 1984 to describe his attempt to integrate neuropsychological research and theology (1984, p. 331). In a 1997 book, *The Humanizing Brain: Where Religion and Neuroscience Meet*, he and Carol Rausch Albright read theology through the neurosciences. While they operate from an epistemological reductionist standpoint in which they accept that "the workings of the brain correspond with people's understanding of the divine," they also claim a realist position regarding religious ideas; we have these ideas "because they have survival value, and . . . they contain elements of truth" (pp. xx–xxi). Ashbrook and Albright claim that the human brain itself is revelatory of information about

God. They argue that since the human brain is preprogrammed for pattern and face recognition, we cannot help but anthropomorphize the universe and our ideas of God, and because our brains are hardwired to seek meaning, God is experienced as Logos-Word or Reason Itself (pp. 21–23).

Ashbrook and Albright argue that brain function mirrors the divine. They base this in a theory of brain function wherein three layers of evolutionary development can be seen: the reptilian or the responsive brain, the paleo-mammalian brain, and the rational brain or neocortex. The reptilian brain is thought to be the most ancient. It generates responses like territoriality and hierarchical thinking. Ashbrook and Albright relate this to religious experience by maintaining that it is reflected in claims to a Holy Land and in descriptions of an "Almighty" God. The second layer or paleomammalian brain is implicated in emotion; therefore, God is understood in the Christian tradition to be a loving, nurturing deity. The most recently evolved layer, the rational brain, is concerned with meaning-making and organization, thus God is Reason Itself, the Ground of Being (Ashbrook and Albright 1997).

Two physicians, Eugene d'Aquili and Andrew Newberg, in *The Mystical Mind: Probing the Biology of Religious Experience* (1999), explain that these assertions have not been established through "a substantive integration" of neuropsychology and theology. For d'Aquili and Newberg, neurotheology refers to the study of "how the mind/brain functions in terms of humankind's relation to God or ultimate reality." They offer a model for understanding how it is that God (referred to as "pure consciousness" or "absolute unitary being") is generated by the brain and yet at the same time is generative of material reality (pp. 5, 18).

To prove the validity of the model, d'Aquili and Newberg used Single Photon Emission Computed Tomography (SPECT) to monitor a small group of Tibetan Buddhist monks in meditation and Catholic Franciscan nuns in prayer. SPECT imaging measures blood flow to the brain, and studies have demonstrated that blood flow increases to areas of the brain when they are activated. In the meditating Tibetan monks, blood flow increased to the areas associated with attention and decreased to the orientation areas of the parietal lobe, which inform us about the position and movement of the body. When the nuns prayed, a more verbal activity than Buddhist meditation, the parietal lobes showed the same decrease in blood flow as in the monks, while the areas associated with language showed increased flow. This is taken to support the proposed model that an altered state of consciousness, Absolute Unitary Being (AUB), is generated when areas of the brain experience alterations in blood flow. When areas of the brain that function to maintain self–other distinction are not stimulated, a sense of "oneness" ought to ensue. At the same time, heightened awareness would result from stimulation of areas related to increased concentration (pp. 118–119). According to the researchers, AUB is a state in which the practitioner "loses all awareness of discrete limited being and of the passage of time." If associated with positive emotion, it tends to be interpreted as the experience of God; if associated with neutral emotion, it tends to be interpreted as an encounter with the nameless Absolute or with Nirvana (pp. 109–110).

D'Aquili and Newberg elaborated these findings into a theory of brain function in which they conceptualize the brain in terms of "cognitive operators." They examine seven of these functional components, which they say integrate sense perception, emotion, and thought so as to produce what we call the mind. The operators most pertinent to religious experience are the "holistic" and "causal" ones. The holistic operator does just what its name implies—it generates a perception of the overall reality or gestalt of an experience. It would function in religion to engender a sense of unity or oneness of/with the divine. The causal operator, aptly named, guides the brain's tendency to see causal relations among events. According to d'Aquili and Newberg, it is the driving force behind religion, science, philosophy, myth-making, and all other human explanatory endeavors. It explains the evolution of religions that speak of a first cause, or God.

Although the cognitive operators are functional components of the mind, it is important to note that the researchers insist they are representations of the way the mind acts on input to the brain and each one of them requires the entire brain. The holistic operator, for example, is generated primarily from the right parietal lobe, but many other parts of the brain interact to finally produce the thoughts, emotions, and behaviors that are its end result. D'Aquili and Newberg's claims rest on somewhat more scientific ground than Ashbrook and Albright's, but they advocate caution in generalizing on the basis of findings. They claim only to have demonstrated a connection between neurophysiology and religious experience. Yet, in the final chapter of the book, they say that neurotheology offers the possibility of a "megatheology," which would contain claims acceptable to all religious traditions and provide a basis upon which to build genuine ecumenical dialogue (p. 210).

The major criticisms of neurotheological approaches from the standpoint of science revolve around the starting point. All four of the theorists discussed earlier assume the reality of the divine. They use cognitive and neurosciences to illuminate what is for them the unquestioned divine–human relationship. This is valid for theological reflection, but is a questionable starting point for something that purports to be a science. For this and other reasons, some call neurotheology a "quasi-science" (Feit 2003). D'Aquili and Newberg say they offer models for understanding how it is that God or AUB is generated by the brain and yet at the same time is generative of material reality. But legitimately speaking, from the scientific standpoint, all that can be said is that reports of encounters with the divine or a sensation of oneness with creation accompany certain brain states, which are also correlated with motivation, the emotions, and sexual acts. Nothing can be said about the source of material reality itself.

Neurotheology's conflation of "God" and AUB is itself quite problematic. Buddhism is a nontheistic religion; there is no God and material creation is eternal, without beginning. The ultimate goal of practitioners is to achieve nirvana, a state that cannot be described since one who achieves it ceases to exist in any normal sense of the term. Nirvana is sometimes described as a state of supreme bliss, but the word literally means "to extinguish," and so it is a mistake to claim much comparative ground between Christian and Buddhist afterlife expectations. The Christian God concept is of a personal deity who not only created material reality, but also sustains it and will at some future point radically transform it into the *baseleia*—a realm of actual existence that perfectly reflects the divine. To claim to have made discoveries about God through the study of Buddhists who are meditating with the goal of being "extinguished" as a life force is at minimum disrespectful of Buddhist teachings and at worst a Western monotheistic arrogance.

Another problem arising from the conflation of God and AUB is that the research can be used to support diametrically opposite conclusions. The data might lead us to say that religious experience is merely one of many brain states. The fact that certain areas of the brain are activated during meditation, prayer, and ritual practice actually tells us nothing about the reality of God. Or, oddly enough, we might reach the opposite conclusion. Since there is no distinction between AUB and God, it also makes sense to conclude that human consciousness and God are indistinguishable. In fact, Newberg sometimes sounds as though his AUB is actually Tillich's God, the ground of Being: "So, if Absolute Unitary Being truly is more real than subjective or objective reality—more real, that is, than the external world and the subjective awareness of the self—then the self and the world must be contained within, and perhaps created by the reality of Absolute Unitary Being" (Newberg et al. 2001, p. 155).

Fraser Watts, Starbridge lecturer in Theology and the Natural Sciences at the University of Cambridge, is a psychologist and ordained Anglican minister who has extensively studied these issues. His work helps to illuminate the methodological problems that currently plague neurotheology. Watts urges caution in regard to d'Aquili's theory of cognitive operators, since these are not standard terms in neuroscience (Russell et al. 1999). In d'Aquili and Newberg's work, the methodological difficulties compound because two scientists have developed a useful but highly speculative scientific framework and then used it to speak of theological concerns. One problematic outcome is that they apply the theory to all forms of mystical experience, even though it was developed through study only of intentionally generated experiences. Reports of spontaneous mystical encounters are relatively common, and there is no way to determine whether or not the unplanned-for experience differs from the structured forms they studied. Religious experience, particularly of the mystical type, is deeply subjective. We only know it exists because of reports from believers of states of altered consciousness. These states sometimes arise in the midst of intentional religious activity, like prayer, and sometimes unexpectedly, seemingly from outside the self. Proving that changes in neurological blood flow are always associated with mystical experiences is likely well neigh impossible.

Further, it is not clear just how fundamental mystical experience is to religious practice. And finally, "religion" is a widely variant phenomenon across individuals and cultures and so it is not clear how applicable the findings of neurotheology are across the board. Watts argues that because religion is such a complex phenomenon, any theory of it must be multilevel and mindful of the entire structural system of human cognition. We do not yet have such a theory.

Additional problems stem from this confusion of methodologies because (1) neurotheology is an amalgam of disciplines that may not be entirely compatible and (2) none of these thinkers is primarily a theologian, even though they make theological assertions.[1] All four "neurotheologians" are seeking integration of scientific and theological claims, something that clearly calls for methodological caution. Ian Barbour has written extensively on method in religion and science dialogue and his now-seminal typology of methods for bringing religion and science fruitfully together will help illustrate this point. Barbour (1990) describes three methods for incorporating scientific discoveries and theory into theological reflection: a natural theology, a systematic synthesis, or a theology of nature.

Natural theology begins with the claims of science and moves toward theological doctrine. Arguments for the validity of religion and existence of God are based on human reason—a move that obviates the need for special revelation. Problems for Christianity tend to arise when making use of this approach since much of the theological core of the faith is based precisely in the special revelation of God in Jesus of Nazareth.

A systematic synthesis of religion and science can be quite powerful, but requires a metaphysics to serve as the explanatory scaffolding. Process Theology is perhaps the most successful synthesis yet achieved, but it is clearly not without problems. Many theologians have incorporated some of the insights gained through attempts to incorporate the Process metaphysic into Christian thought, but as a stand-alone theological system, it has fallen out of favor, in part because the Process synthesis, like natural theologies, runs into trouble in the face of claims to special revelation.[2]

Ilia Delio, O.S.F., a professor of Church History, argues that without philosophical exploration of the nature of reality and God's relation to it, the neurotheologians have committed a "fallacy of misplaced contingency," wherein God's reality is contingent upon one's experience of God. In this framework, we have a "new anthropic principle—that is, in order for God to exist, we must exist, because it is we humans who give a conscious voice to the existence of God" (2003, p. 578).[3]

A theology of nature is less ambitious than synthesis. It begins with religious belief and incorporates only those well-supported scientific claims that have an impact on theological assertions. In doing theology of nature, theologians do not appeal to science as proof of theology in the way natural theologies do. Here, science functions as a tool to help rethink traditional doctrines. Since religious belief is primary, in this approach it is best to avoid highly tentative theory in favor of generally accepted scientific claims.

This prevents having to constantly rework theology whenever speculative theories are discarded or proven incorrect by scientists (Barbour 1990).

Ashbrook and Albright start with a natural theology approach but end with what seems to be a synthesis since there is a latent metaphysics at work. The underlying metaphysics is a sort of panpsychism in which reality itself is not so much conscious as it is humanlike. This explains why one is left with the sense that they have appropriated science to serve as proof of religion and composed a "confession of faith written in a scientific vein."[4]

Methodological criticisms aside, it cannot be denied that mental states are intimately related to neurochemistry. Even though the hard problems of consciousness (and by extension, of religious experience) have not been solved, theology must grapple with the implications of the facts of science. Our personalities and emotions, indeed consciousness itself, are in some sense by-products of the activity of the physical brain. This means that traditional religious claims, particularly radically dualistic ones, which identify the person with a soul that exists independently of the body, must be reconsidered. This area has thus far received the most attention from theologians and biblical scholars and to it we now turn.

Neuroscience and the Christian Soul

"Soul" in traditional Christian thought is a concept that has functioned in at least two ways: to designate the enduring facet of the self (that which in some sense survives death), and as a label for the aspect of the human that is accountable for moral choices and capable of communion with God (Brown et al. 1998). Traditional western conceptualizations of the soul and life after death have been under attack for many years and from many directions. Enlightenment critics of the seventeenth and eighteenth centuries declared belief in the soul as an entity that survives physical death superstitious and unscientific. The nineteenth century brought with it Marxist humanism and its insistence that all religious teachings are oppressive tools of the powerful designed to placate the masses with promises of fulfillment in a life to come. Psychoanalysts reduced the soul to the psyche and determined that belief in its immortality is merely the product of psychological defense against the terror of death. Some twentieth-century feminists and environmentalists insisted that belief in a survivable soul is destructive in its dualist deprecation of material creation. Present-day neuroscientists and philosophers of mind maintain that there is no such thing as a soul; for them, the mental life of the human is a by-product of neurochemical and electrical activity in the brain.

The strongest recent challenge to dualist anthropologies is the materialist framework found in philosophy of mind and many cognitive scientists' writings. Epiphenomenalism is probably the most popular form of materialism, with its claim that mental states are nothing more than by-products of brain events. The epiphenomenal mind is only a consequence of neural firing, so there can be no such thing as a mind or soul apart from the body. In other words, mental states are completely reduced to the physical. Just as psychoanalysis once reduced religion to the neurotic projection of infantile wish fulfillment, some scientists today have reduced the person to neurochemistry.

The problem for theology is that of accepting the truths of science without compromising the theological truth that we are embodied spiritual creatures. The assumption among philosophers and scientists regarding religious ideas of the soul seems to be that if the "Cartesians can no longer live in their disembodied souls," neither can the Christians consider their souls to be immortal (Peters 1999, p. 305), although there is no universal consensus even among neuroscientists. For example, three Nobel Laureate neuroscientists have come to three divergent conclusions: John Eccles believes dualism to be the logical explanatory model for the body/mind problem, while Francis Crick advocates an extreme reductive materialism, and Roger Sperry a nonreductive materialism (Brown et al. 1998). This lack of consensus among scientists is an important point to remember throughout the following discussion.

Popular Western thought has long insisted that the human being is in some sense more than just a material body. There has, however, never been consensus on the number and kind of "ontological ingredients" it takes to make a person (Cooper 2000, p. 9). The human as a trichotomy (body, soul, and spirit), dichotomy (body and soul), and monism (body–soul unity) have all been proposed. For most of Christian history, dualist language of one form or another has tended to predominate, even though a strong dualism of material body and separable, inherently immortal soul or spirit is alien to the scriptures (Cullman 1958). Christian teachings on the afterlife do not require body/soul dualism but they do insist on a future resurrection of the body. As we shall see, this tendency to speak of a dual human nature is driven by the claim that the essence of the person in some sense resides or rests with Christ prior to the final transformative bodily resurrection of all.

Theological difficulties arise when reductionist claims lead theorists to confuse the two functions of the soul concept noted earlier. For many, anything that implies the soul lacks substantiality or independence from the body necessarily eliminates the religious significance of moral accountability and the possibility of life after death. If this were so, it would certainly be theologically problematic since one central aspect of Christian doctrine includes the promise of life after death in communion with God. Not quite so obvious is the fact that it is somewhat more theologically consistent with both scripture and tradition to accept the claims of science that insist on the embodied nature of the human self and that our embodiment is itself morally significant.

Recently, Christian theologians interested in science and religion dialogue have begun to reexamine the scriptures and explore ways of speaking about the soul that are true to the tradition and at the same time scientifically and philosophically coherent. Theologians and biblical scholars under the auspices of the Center for Theology and the Natural Sciences and the Vatican Observatory Foundation have done most of this work. These "sophisticated conciliationists," as philosopher-of-mind Owen Flanagan (2002) calls them, offer a philosophical framework of nonreductive physicalism in place of dualism.

Cognitive Science and the Solipsistic Soul

Nonreductive physicalism is the philosophical position that all causation of functional properties is physical even though the functional cannot be reduced to the physical. The position is predicated on acceptance that this is preferable to eliminative or reductive materialism, which defines the person as nothing but a physical being, all of whose experiences—mental, spiritual, or otherwise—will eventually be explained by science. On this philosophical basis, the sophisticated conciliationists define the person as "a physical organism whose complex functioning, both in society and in relation to God, gives rise to 'higher' human capacities such as morality and spirituality" (Brown et al. 1998). It is possible, say the conciliationists, to reject the nonmaterial soul and, at the same time, speak coherently of consciousness and spiritual capacities, if we appeal to the scientific and philosophical concepts of supervenience and emergence. In this framework, the soul and consciousness become circumstantially supervenient emergent properties of the brain, which exert causal influence on the body in a top–down direction.

Supervenience has been given a variety of definitions in philosophy of mind, some of which border on contradiction. In general, though, it describes a relation of dependence between two sets of properties. Put simply in the context of the mind–body problem, strong supervenience means that a physical base property, P, for mental property, M, guarantees the occurrence of M; if something brings about P, it necessarily brings about M as well—always and everywhere (Kim 1996). Strong supervenience is the interpretation advocated by epiphenomenalists; mental events are just a special kind of physical event. Obviously, this interpretation challenges claims to spirituality.

Philosophical theologian Nancey Murphy (1999) advocates a weaker form, called circumstantial supervenience. She argues that this interpretation of supervenience avoids causal reductionism by recognizing that: (1) there are multiple characteristics that make for goodness (multiple realizability); (2) mental properties are what they are not just by reason of the neurochemistry that brings them about, but also by reason of their function.

Circumstantial supervenience allows for the possibility that identical events occurring under different circumstances might not produce the same outcome. The moral quality of "goodness" illustrates the point: goodness supervenes on a set of characteristics. To use St. Francis as a case in point, his goodness supervenes on characteristics like celibacy, charity, desire to do no harm, and so on. Goodness is a noncausal consequence of having these qualities. But these traits do not exhaust the possibilities for goodness. There are other traits St. Francis did not display that in some circumstances call for the label of goodness. Some traits, like celibacy, only contribute to goodness under certain conditions or in certain circumstances. If St. Francis had been married, his wife may well have found his vow of celibacy problematic (Murphy 1999).

To claim that the soul is an emergent property is to say that an exhaustive description of the underlying physical state of the human is necessary but not sufficient for explaining the soul's existence. An emergent property is an unanticipated outcome or by-product, a "something more": whereas normally a + b should = c, if the relation between a and b is one of emergence, a + b = c + 1 (Murphy and Ellis 1996). Advocates say that supervenience helps explain causation in the direction of physical to mental, but experience tells us that mind/body causation is not a one-way street. "Bottom up" causation in firing neurons and neurochemical transfer across synapses is certainly necessary, as far as we now know, for any mental or physical event. But evidence indicates that causation flows from the top down as well. This has been most clearly demonstrated in perception studies. How we perceive sensory input is determined both by the stimulus itself and by individual expectations regarding the stimulus. We very often see what we expect to see rather than what is really there.

Nonreductive physicalists say this means the physical determines initial emergence of the mental, but does not fully determine the outcome of the mental after its emergence. If the mental life is an emergent feature of the complex biological structure of the brain's interaction with its environment, a person is a physical being whose multifaceted functioning gives rise to higher human faculties like the moral and spiritual. The soul is a property, a quality, or a phenomenon, not a substance: the physical brain causes but does not determine it.

While this approach might overcome some of the difficulties faced by theologians who want to speak intelligibly about the soul in light of neuroscientific discoveries, it does not solve all our problems. It may, in fact, concede too much too soon to science. There are philosophical, practical, and theological issues at stake here and the territory is not well mapped. Huston Smith says that these theologians are earnest Christians who do not see themselves as surrendering to the scientific worldview and yet they are doing just that. The nonreductive physicalist's God is "(1) the world's first and final cause, who (2) works in history by controlling the way particles jump in the indeterminacy that physicists allow them" (2001, pp. 75–77). The nonreductive physicalist's soul is (1) the higher level cognition that emerges out of and supervenes upon the physical brain and (2) is in some sense capable of freedom and moral accountability.

Although nonreductive physicalism is attractive to those who hope to further the conversation between religionists and scientists, in reality it does not add much to the conversation, nor does it advance an ontological explanation of mind or soul. In a very real sense, we are all nonreductive physicalists when it comes to the things of everyday life. Donald Wacome (2004) uses pizza to illustrate this point. He says that a nonreductive physicalist interpretation of the self is no more revelatory than the claim that we cannot adequately explain what a pizza is by talking about its quantum structure. And it is not at all clear how, even with circumstantial supervenience, this frees the soul from the bonds of physical causation. In attempting to free theology of an immaterial soul, the nonreductive physicalists have intensified the problem of freedom and the will.

Theologically, we ought not reduce all aspects of human life to the physical: scripture and tradition recognize that we are clearly embodied creatures, and at the same time something more. Nor ought we overly spiritualize. One reason strongly dualist accounts are problematic is that they inherently tend toward overemphasis on the survival of the spiritual to the detriment of the physical. Taken too far, dualist interpretations end in Gnostic claims that the physical world is a prison from which we must escape. Theology must seek the middle ground here, to recognize that our urge to speak dualistically about brain/mind, body/soul is a reflection of the intuition that subjective experience is not reducible to the physical. If it turns out that the person is nothing more than an epiphenomenon of neurochemistry, all sorts of theological problems ensue. What is the moral status of someone who suffers from Alzheimer's disease, severe brain damage, or congenital cognitive disability if that which makes us persons is nothing more than neurochemistry?

Persons are more than just brain states and neurochemical reactions, but it is not true that only the spiritual matters. Major advances have been made in understanding how the brain and nervous system function; scientists have found that specific areas of the brain are directly involved in physical sensation and in some types of emotional response. But no one has yet identified any physical location that, when stimulated, causes a person to make decisions and choices. Humans clearly have a mental capacity that in some sense operates extra-neurologically (Jewett 1996).

The conciliationists have given primacy to a scientific worldview that necessitates a materialist metaphysics as they hope to preserve ideas arising from within a thought world that necessitates less determinate language. And perhaps it is here that something like a category mistake has been made by those eager to vindicate Christian theology in the face of the challenges of cognitive science. It seems to me that the nonreductive conciliationists are themselves engaged in a reductionist project. In attempting to make theology consonant with cognitive science, they are implying that theological language is reducible to the language of science (which I do not believe to be the case). In attempting to communicate on scientific terms, the conciliationists have allowed the conflation of mind, consciousness, and soul. Theologically speaking, these terms do not refer to the same aspect of human life.

Assume, for the case of argument, that the language of theology were reducible to that of the neurosciences. This would mean that theological terms could be replaced by neuroscientific terms without loss of meaning. In this understanding, saying that language about the soul is in principle replaceable with language about brain function is not to claim that the soul does not exist; all this says is that the language of science is referentially adequate to express theological concepts. So far so good. However, since any physicalism necessarily entails the claim that there is no immaterial aspect of the human, all forms of physicalism, not just reductionisms, cause problems for talk about soul or self. Some would argue here that a modest physicalism allows the position that all knowledge can be expressed as statements about physical objects and activities without making any ontological or metaphysical claims. Still, it is hard to see how this really solves anything since the mind is a material entity and we are nonetheless left with ontological implications.

On a practical level, the language and arguments here are esoteric, not easily understood by the nonphilosophically minded. Besides, it is possible to reach the same goal through careful use of scripture and tradition and without conceding too much to what actually is nascent scientific theory. Since the theory and data of cognitive science can and have been used to support diametrically opposed philosophical positions on the brain/mind question, theologians ought to adopt the "principle of parsimony" in responding to scientific claims. Even Nancey Murphy admits, "no amount of evidence from neuroscience can prove a physicalist view of the mental" (Brown et al. 1998, p. 139). By entering the brain/mind fray from the standpoint of the individual mind, the conciliationists (inadvertently, I think) accept a basic premise of science that compromises important theological points. That premise is that in order to understand what it means to be a person, science must be made primary. This "natural theology" method requires that the object of study be the individual mind in isolation from others, a necessarily solipsistic enterprise. No solipsistic enterprise can adequately support the communal imperative of the two Great Commandments, that we love God and neighbor with the whole self, physical and spiritual and mental. Jesus said we attain eternal life in this way: "You shall love the Lord your God with all your heart, and with all your soul, and with all your strength, and with all your mind; and your neighbor as yourself."[5]

A "theology of nature" methodology, wherein we start with scripture and theological anthropology, makes it possible to hold in tension the tenets of Christian faith and the discoveries of science without granting too much to as yet unproven scientific theory. It allows us to reach conclusions that capture the gist of nonreductive physicalism and eliminate dualistic errors that have crept into Christian thought over the centuries without compromising theological priorities. I want to offer some constructive reflections on this, but first a brief survey of the history of Christian deliberations on the soul is necessary.

The Embodied Christian Soul

The Christian notion of the soul is a relational and moral concept, the product of evolving thought traceable in the biblical texts. In the creation story (Gen. 2:7), the human being is a "breathing 'living' clod of earth" whose life-breath comes from God (Gillman 1995). Adam isn't made of dust; he is dust out of which God has brought life.

The general scholarly consensus is that in the Hebrew Bible the person is a unity of body and soul, what John Cooper (2000) calls a functional holism in which there is a duality of ingredients. We do not "have" souls; the human is human only as body-and-soul. The Hebrew terms for soul and spirit do not refer to any sort of immaterial entity. We are clearly embodied beings. Higher human cognitive facilities like moral decision making, loving, and praying are attributed not only to the spirit or soul, but also to the "gut" and the heart. The heart is in fact the seat of the human conscience in ancient Hebrew thought (Gorsky 1999). A detailed examination of the Hebrew words and their various interpretations goes beyond the scope of this chapter, and there is some question as to the usefulness of this enterprise for uncovering theological anthropology in the scriptures anyway. We can say, however, that there is nothing in the terms themselves that unquestionably supports dualist anthropologies.

The Hebrew texts do not offer much in the way of explanation about what happens to us after death, but some general comments can be made here. The person survives in some sense, although whether consciously or unconsciously is unclear. Sheol is the resting place for all who die—a realm that seems to be cut off from God. It is not Hell—one's moral standing has nothing to do with it. It functions as a place of waiting for all who die (Cooper 2000).

Moral responsibility is a communal, not individual, concept at this time; subsequent generations are thought to bear liability for the sins of the present. In Exodus 20:5 and Deuteronomy 5:9, for example, children are punished by God for the sins of the parents "to the third and fourth generation." In later texts, this idea is replaced by individual accountability for sin. Jeremiah and Ezekiel tell us that "all shall die for their own sins" (Jer. 31:29–30), and "it is only the person who sins that shall die" (Ezek. 18:2–4). As cultural ideas of individual responsibility took shape within the Jewish community, theological ideas of the human being expanded to allow space for the individual to stand in direct and personal relationship with God, and the soul concept that evolves alongside the individualization of moral responsibility is an expression of this. So in summary, we can say that the predominant view of the human in the Hebrew Bible is that of a psychosomatic unity for which life continues in some sense after death in the shadowy realm of Sheol until such time as God acts to bring about a return to bodily existence on this earth that will have been transformed by God.

During the intertestamental period, beliefs about life after death take on more clarity and we see the development of a variety of views. The whole range of beliefs appears within Judaism during these centuries: the Sadducees apparently believed in annihilation while others accepted continued existence in a kind of sleep state until the future resurrection. On the other end of the spectrum, the Pharisees seem to have adapted the Greek idea of a separate soul to Jewish thought (Cooper 2000).

One way in which the body/soul unity and its essential tie to morality is underscored in both Christian canonical testaments is through the metaphor of the heart. As noted above, in the Hebrew Bible, the heart is the seat of conscience and is the organ with which we ought to think, whereas the emotions (also firmly seated in the body) are "located" in the gut. The heart is the "place" of relation to God: hardening the heart separates us from God, and having a new heart signifies spiritual rebirth. In the New Testament, heart is synonymous with the inner being or total self. It is the ethical center and seat of memory. This "cardiac anthropology" is clearly a holism: the literal physical organ is the center of spiritual and psychic integration of the person in relation to God (Ware 2002).[6]

Evaluating theological anthropology of the New Testament has been made more knotty by the fact that in the Gospels, soul language functions primarily in the context of teachings on salvation. It is not a technical term in a theological or philosophical sense, yet attempts have been made to read these meanings into the texts. Paul uses a variety of terms that some have taken to mean we are a trichotomy of body, soul, and spirit; body and soul are part of the natural human and spirit in this interpretation is some sort of commodity that is added to the individual upon conversion to the faith.[7] But since Paul's surviving letters are only a small sample of all he most likely did write, and they are "occasional" pieces, written to small communities to address particular and personal concerns, theological caution is in order. Attempts to construct systematic theological doctrine from his writings have often run into difficulty because what we now have of his thought is far from systematic.

Complicating the picture is the fact that in the New Testament, beliefs in the resurrection of the body and in an intermediate state after death receive more attention than in the earlier texts. Since the idea of an intermediate state does appear in the New Testament, attempts to eliminate it from the discussion are misguided; the idea is there, but there is no speculation as to the nature of the interim state. All we can glean from the texts themselves is (1) some sort of intervening status was expected for the dead-but-not-yet-resurrected and (2) this state represents some sort of union with Christ (Cullman 1958).

The belief in a transitional circumstance ought not be news to us, since some sort of shadowy existence apart from God had long been a part of Judaism, and the earliest Christians were, after all, mostly Jews. John Cooper argues that some type of dualism is required by the worldview implicit in this teaching, since to be with Christ, the "core personhood" must continue on beyond death of the body. "All we know from Scripture is that in God's providence human beings can exist in fellowship with Christ without earthly bodies . . . any philosophical anthropology which accounts for this possibility must necessarily be dualistic." He advocates a "holistic dualism" in which body and soul function as a unity and yet are separable at death (2000, p. 163).[8]

Joel Green argues that, while there are diverse views of the person in the scriptures, the dominant theme is an "ontological monism" that precludes any sort of disembodied existence. Belief in an intermediate state after death does not, he says, necessitate an immaterial and separable soul. Since the New Testament texts are concerned first and foremost with soteriology—that Jesus, as the Messiah, has brought about the conditions for the salvation of humankind—extraction of theological anthropologies from the texts must be done in light of this. When body and soul are contrasted in the scriptures, it is in the context of soteriological value: that which upholds the spiritual versus that which upholds the selfish use of material reality. What seems to have mattered more to the New Testament writers than clearly developed theological anthropologies was the person's relationship to others and to God (Brown et al. 1998).

In *Minding God: Theology and the Cognitive Sciences*, Greg Peterson says that the spirit "emerges out of the activities of the mind/brain, which in turn are intimately connected to the body. A spiritual transformation, therefore, is in some sense also a biological one. Soteriology must therefore include the whole person." He goes on to say that this "seems strange" since biology belongs to the world of science and talk of spirit to theologians and ministers (2003, p. 94). Biblical scholarship, however, shows that originally Christian soteriological ideas did include the whole person. The clearest illustration of this is Jesus' own resurrection and ascension. Christ really died. If his soul had been immortal and separable, what was the point of the resurrection? And, as Adrian Thatcher so drolly put it, "What then is the ascension [if dualism is correct]? A highly visual way of saying cheerio? It is, rather, the return of the transformed, transfigured, glorified, yet still embodied, Christ to the Father" (1987, p. 184).

Many present-day academics suffer from "dualophobia" (Cooper 2000, p. 27). It is clear that human nature is portrayed in the scriptures as dual—made up of both a physical body and an aspect with the capability of communion with God such that it can survive beyond death of the body. This does not, however, necessitate a Platonic or Cartesian dualism of substances that are different in essence. Dualophobics sometimes wrongly proclaim that the traditions' earliest theologians were dualists through-and-through, when in fact the separable soul becomes doctrinal only during the Middle Ages when teachings on purgatory and the sale of indulgences made elaboration necessary.

Early Christian theologians extensively examined the difference between the Platonist's inherently immortal soul and the scriptural soul's immortality as a gift of God's grace. For example, Justin Martyr (ca.100–165 CE) discusses the non-Christian nature of a Platonic soul. He is perfectly clear on this: "I pay no regard to Plato" in speaking of the soul, for it "lives not as being itself life, but as the partaking of life" because "God wills it to live . . . for it is not the property of the soul to have life in itself" (Wolfson 1993, p. 305).[9] The scriptural imperative that God alone possesses immortality (1 Tim. 6:17) was unquestioned in the early decades of Christian theology.

During the third through fifth centuries, theologians worked out their thought in conversation with the predominant philosophies and cosmologies of the day. In the third century, Tertullian was a materialist. He taught traducianism—the doctrine that souls are passed from parents to offspring through the physical reproductive act. His soul was a "corporeal substance," not of matter but of the breath of God, formed along with the body at birth. The mind, he said in the *Treatise on the Soul* (1978), is in the soul and life ceases when the soul and body separate.

In the next century, Augustine of Hippo became one of the most influential theologians in Christian history. Augustine's thought on the soul, like so much of his theology, changed over time as he moved from Gnostic to neo-Platonic philosopher to Christian theologian. But in order to be true to the intent of this article, attention is given only to his later thought. In *City of God*, he says that the human is "constituted by body and soul together . . . the soul is not the whole man; it is the better part of man, and the body is not the whole man; it is the lower part of him. It is the conjunction of the two parts that is entitled to the name of 'man' " (1980, XIII.24). The human is a combination of body and soul, the relationship of which is inexplicable, a "miraculous combination" wrought by God (XXII.24).

This variance of opinion continues into the Middle Ages. Before Thomas Aquinas and his rediscovery of Aristotelian philosophy, neo-Platonism held sway. In the thirteenth century, Aquinas incorporated Aristotelian metaphysics into theology and gave what would become the orthodox Catholic interpretation of the soul as the form of the body. He said the rational soul (which only humans have) is a simple substance, meaning it has a spiritual nature, which is incomplete until united with the body. Aquinas thought that the human embryonic soul was "vegetative" like that of other living creatures, having only the capacity for survival. Once the embryo develops sufficiently, through a special act of creation, God replaces the vegetative with a rational soul. The rational soul makes possible the higher cognitive functions like abstract thinking and willing. This rational soul is the "form" of the body, meaning that the soul is that which makes it possible for the material body to actualize its potential to become a thinking, loving human being (Aquinas 1964). In the words of one twentieth-century Protestant theologian, the soul is the life of the body, as a result of the action of the divine creative spirit (Pannenberg 1985).

In the seventeenth century, Descartes recovered Platonic dualism with a vengeance. The so-called Cartesian revolution made the division between purely spiritual soul and material body complete. For him, there were two kinds of reality, "thinking" substance (the mind, angels) and "extended" substance (material things). He said, "the soul by which I am what I am, is entirely distinct from body, and is even more easy to know than the later; and if body were not, the soul would not cease to be what it is" (1970, p. 101). Descartes offered a neat solution for his day, a kind of "interactionism" that said the soul is in the pineal gland of the brain. The two distinct realities of soul and body influence one another though that gland. Exactly how this happened, he was not able to say. Later Cartesian thinkers struggled with this problem and found the only satisfactory explanation in the claim that interaction is made possible by a special act of God. Drawing such a hard line between soul and body created the mind/body problem, which foreshadowed the difficulties that plague theologians, cognitive scientists, and philosophers today.

What's a Theologian to Do?

The solipsistic pull we noted in attempts to incorporate neuroscientific claims into theology is a problem for Christian theology. This is so because although the gift of the incarnation is salvation for the individual person, it is meaningful only in the context of relationship. This truth begins with God's own self. God as Trinity is an expression of the intuition that person-in-relation is the very existence of God. God's being is itself an act of communion in the begetting by the Father of the Son and in the bringing forth of the Holy Spirit (Zizioulas 1985).[10] God became incarnate in order to make real to us the imperative that we must love God and one another. That which matters most is individuals-in-relation. We are each of us held individually accountable for our choices but that which makes those choices meaningful is the manner in which we live in relation to others and to God. In our examination of scientific developments, we have seen that these teachings are challenged by the clearly physical nature of human cognition. Some of the assertions of cognitive science threaten to reduce all of this to the neurochemistry of the individual brain. Many theologians respect the importance of scientific advances and at the same time refuse to be painted into a strongly physicalist corner. A way forward comes from another area of the cognitive sciences—that which focuses not on the individual mind but on human cultures. Cultural psychology and anthropology offer intriguing clues to the uniqueness of the human, and a potent language for advancement of a theology of nature approach to the soul.

Nearly all human mental abilities are found to some degree in other species, but the capacity for formation of elaborate cultures and complex relationships seems to be uniquely ours. Humans are actually very similar to other primates in terms of the biology of cognition. Where we differ seems to be in the realm of relationship; we identify with "conspecifics," others like us, more deeply than other primates. Nonhuman primates are quite creative and intelligent but in learning they do not rely as much on other primates as do we. A chimpanzee, for example, will see the outcome of another chimp's behavior and try to replicate it but apparently cannot understand the connection between the other primate's behavior and the outcome. Human beings, even as young as two years, focus more on the strategy used by other humans, and this indicates a different mental process. We understand other humans to be intentional agents with a mental life like our own and we try to understand things from the other's standpoint. The chimp sees the outcome of another's behavior and strives to get there from within her own mental world, without attention given to the tactics used by others. Nonhuman primates are certainly causal and intentional beings but they do not seem to understand the world in those terms (Tomasello 1999). Human beings are much more dependent upon others than our primate cousins: "The human brain is the only brain in the biosphere whose potential cannot be realized on its own" (Donald 2001, p. 324). And so cultures have evolved in order to support and sustain the human species.

Our development as a species is a dual process of biological and cultural evolution; our minds a hybrid outcome of biology and culture. The human capacity for interpersonal relatedness, which gives rise to complex cultures, may itself be an emergent property of evolution, but the changes that most clearly set us apart may be more correctly seen as historical and ontogenetic, not purely biological. Evolutionary biology, anthropology, and psychology tell us that the dramatic evolution of cognitive skills in the human cannot have occurred without the simultaneous development of culture. Our brains have developed in ways that assume the existence of an external means for pooling cognitive resources. By means of symbol systems like language, human minds can learn "not just from the other [as other species do] but through the other" (Tomasello 1999, p. 6). Cultures are ecologies that function as storehouses, which make it possible for humans to share learning, language, and customs over generations. The "symbolic technologies" of our cultures "liberate consciousness from the limitations of the brain's biological memory systems" and enable us to break the bonds of biology (Donald 2001, pp. 305–306).

Perhaps now we can agree with the "sophisticated conciliationists" that the soul is not an ethereal entity that exists apart from the body. It is rather "the net sum of those encounters in which embodied humans relate to and commune with God (who is spirit) or with one another in a manner that reaches deeply into the essence of our creaturely, historical, and communal selves" (Brown et al. 1998, p. 101). Following Aquinas to some extent, we can say that the soul is the organizational structure that makes the individual person possible, and culture is the organizational structure that makes humanity itself possible. Individual minds become persons in and through relationship to one another in cultures and to God in spiritual communion. As Jurgen Moltman says, a person comes to be in the "resonance-field of relationships of I-you-we . . . The 'person' emerges through the call of God" (Brown et al. 1998, p. 225).

When we understand the soul in this way, the deeply felt intuitions that humans are transcendent beings and that the spiritual is really real are preserved. Further, the ethical is emphasized rather than lost, as critics of physicalist theories of the person fear. The final status of the human being, whether it be as a soul that survives death or a no-self that hopes to escape the cycle of rebirth, is always tied to morality. In the interpretation of soul offered here, the soul is that which is most true about the person. It is the locus of God's action within the individual, the place where divine Wisdom enters in and makes of us "friends of God and prophets" (Wisd. 7:27). The soul is not so much consciousness or even the self as it is the seat of relationship.

In Christian theology, the centrality of relationship is vital. This is the insight of Trinitarian theology: God's own being is an act of communion. Humanity, the imago dei, reflects this reality in that communion is likewise constitutive of what it means to be a person. In and through cultures, humanity pools its resources and thereby fulfills its potential. In communion with other minds, we learn what it means to be created in the image of God and we are made capable of living that reality.

Culture functions for us as the "community of memory" (Bellah 1985). Our souls are held in communal memory, during life and afterward. This reading overcomes problems that arise for physicalist interpretations in the face of Alzheimer's disease and other brain disorders. Materialist interpretations cannot avoid the implication that a person suffering from brain damage is in some sense less-than-human. A theology of cultures added to our reflections on soul allows us to say that culture is the guardian of our souls. In and through cultures, we learn moral behavior even when role models fail. In and through cultures, the souls of those who suffer from diseases that challenge all our assumptions about what it means to be a person are preserved in communities of memory.

In this interpretation, the biblical insistence that immortality is a gift from God is preserved, since the soul is both "within" the individual and "outside" in the community. Survival of death is then both "objective immortality," in the sense that what immediately survives death is the cumulative effects of one's life on the cultural and communal memory, and "subjective immortality," which is the gift given in God's own time by God to the individual believer. We have also preserved the Biblical ethical imperative of communal responsibility and full personhood in and through relationship to one another and to God.

The Christian idea of the afterlife for the soul finds its fullest expression and fulfillment in eschatological hope. Much of what Jesus said and did was not new to Judaism, although the emphasis he placed on caring for others without regard to social status was quite striking. What set Christianity apart from Judaism was (and is) belief in the resurrection of Jesus from death. The promise of life after death for the Christian can only be understood in the context of Jesus' resurrection, and the resurrection can only be understood within the framework of Jesus' teachings about the world to come, or the Kingdom of God. And what is the "Kingdom of God" if not an ethical, relational concept?

Although interpreted individualistically in both scientific and popular circles, the Christian concept of the soul is a relational, moral perception. Its origin is in the Hebrew Bible's understanding of the human being as a creation of the one God. The life-breath comes from God, is a gift of love, and is best understood in terms of relationship. The soul—as the sum total of who each of us has been, is, and will be in this earthly life—is first and foremost an ethical, future-oriented concept. It is a product of the relational character of human being. What is most important about it is its function as that which joins us "to other individuals and to our community, and to God" (Brown et al. 1998, p. 222). We are spiritual beings. What this means can be illuminated to some extent by the sciences. But in the end, it may well be that the very neurological and cultural realities that make our spirituality possible are the same realities that give us our limitations (Teske 1996).

Notes

1. Carol Rausch Albright and James Ashbrook both studied theology but their careers have focused on other aspects of religious life and thought.
2. See Burns (2002, especially pp. 76–81) for a discussion of some insoluble problems for Christology that arise within the Process system. See Bracken (2004, pp. 161–174) for a revised model of Process Theology, which he claims provides an adequate philosophical basis for belief in the soul and life after death.
3. There are philosophical systems compatible with these early claims of neurotheology—Hegel's vision of the world process as the coming-to-be of God as Spirit suggests itself here. His attempt at an absolute idealism (in which nature and Geist/spirit are not reduced to one another, nor are they held apart dualistically) might be fruitful were one to attempt a systematic synthesis.
4. Review comment by Paul MacLean, quoted on book cover.
5. Luke 10:25–28; cf. Deut. 6:5; Lev. 19:18; Matt. 22:36–40; Mark 12:28–34; Gal. 5:14; Jas. 2:8.
6. There are many passages that refer to heart in this sense. A few include: Deut. 6:5; Ezek. 11:9, 18:31; Prov. 4:23, 23:26; Ps. 64:6; Matt. 15:19, 22:37; Rom. 1:24, 8:27.
7. Trichotomy is sometimes used to make a distinction between the soul's rational intellectual capacity and the spiritual capacity for relation to God. This distinction allows us to say that other forms of life have souls while at the same time preserving the status of human as imago dei.
8. Cooper makes a distinction between what he calls ontological holism and functional holism. He argues convincingly that the Hebrew Bible does not entail the former, in which neither the body nor the soul has ontological standing; only the body–soul totality is a human being. This form of thought is very similar to monism and is certainly precluded by the New Testament. Functional holism sees the human as an integrated functional system, not a compound of separate parts: "the parts do not operate independently within the whole, and . . . would not necessarily continue to have all the same properties and functions if the whole were broken up . . . An organism is a prime example . . . organs can survive separation from their organisms" (pp. 45–46). For Cooper, an intermediate state necessarily presupposes a dualism of some sort. Luke 16:19–31 is his support text. Jesus tells a parable in response to Pharisees' accusations that he has violated Jewish law. He contrasts the rich man who goes to Hades/Sheol after death and the poor Lazarus, who finds himself carried away by angels to be "with Abraham."
9. Wolfson shows that while it is true that the Patristic soul is "essentially Platonic" in that it was thought to be incorporeal, "these early Christian theologians did not attribute immortality and resurrection to the nature of the soul: they are only ours by the will and action of God" (pp. 320–331).
10. John Zizioulas argues that Christian Trinitarian theology gave the world a radically new understanding of "person." Before the Patristic theologians, it had no ontological content in Greek or Roman thought.

References

Aquinas, T. 1964. *Summa Theologia*, Book I, Questions 75 and 76. Blackfriars ed. New York: McGraw-Hill.

Ashbrook, J. 1984. Neurotheology: The working brain and the work of theology. *Zygon: Journal of Religion and Science*, 19, 331–350.

Ashbrook, J. and Albright, C. 1997. *The Humanizing Brain: Where Religion and Neuroscience Meet.* Cleveland, OH: Pilgrim Press.

Augustine, A. 1980. *City of God.* H. Bettenson (trans.). New York: Penguin Books.

Barbour, I. 1990. *Religion in an Age of Science.* San Francisco: Harper San Francisco.

Bellah, R. 1985. *Habits of the Heart: Individualism and Commitment in American Life.* Berkeley, CA: University of California Press.

Boyer, P. 2001. *Religion Explained: The Evolutionary Origins of Religious Thought.* New York: Basic Books.

Bracken, J.A. 2004. Reconsidering fundamental issues: Emergent monism and the classical doctrine of the soul. *Zygon: Journal of Religion and Science*, 39, 161–174.

Brown, W.S., Murphy, N., and Malony, H.N. (eds.). 1998. *Whatever Happened to the Soul? Scientific and the Theological Portraits of Human Nature.* Minneapolis, MN: Fortress Press.

Burns, C.P.E. 2002. *Divine Becoming: Rethinking Jesus and Incarnation.* Minneapolis, MN: Fortress Press.

Chalmers, D.J. 1995. Facing up to the problem of consciousness. *Journal of Consciousness Studies*, 2, 200–209.

Cooper, J.W. 2000. *Body, Soul, and Life Everlasting: Biblical Anthropology and the Monism–Dualism Debate.* Grand Rapids: William B. Eerdmans.

Crick, F. 1994. *The Astonishing Hypothesis: The Scientific Search for a Soul.* London: Simon and Schuster.

Cullman, O. 1958. *Immortality of the Soul or Resurrection of the Dead?* London: Epworth Press.

Damasio, A. 1999. *The Feeling of What Happens: Body and Emotion in the Making of Consciousness.* San Diego, CA: Harcourt.

D'Aquili, E. and Newberg, A. 1999. *The Mystical Mind: Probing the Biology of Religious Experience.* Minneapolis, MN: Fortress Press.

Delio, I. 2003. Brain science and the biology of belief: A theological response. *Zygon: Journal of Religion and Science*, 38, 573–585.

Descartes, R. 1637. *The Philosophical Works of Descartes.* vol. 1. (1970).

Donald, M. 2001. *A Mind so Rare: The Evolution of Human Consciousness.* New York: W.W. Norton & Co.

Edwards, P. (ed.) 1999. *Immortality.* Amherst, NY: Prometheus Books.

Elizabeth S. Haldane and Ross G.R.T. (trans.). New York: Cambridge University Press.

Feit, J.S. 2003. *Probing Neurotheology's Brain, or Critiquing an Emerging Quasi-Science.* Paper presented to the Critical Theory and Discourses on Religion Selection, American Academy of Religion, 2003 Annual Convention, Atlanta, GA, November 22–25, 2003.

Flanagan, O. 2002. *The Problem of the Soul: Two Visions of the Mind and How to Reconcile Them.* New York: Basic Books.

Gillman, N. 1995. *The Death of Death: Resurrection and Immortality in Jewish Thought.* Woodstock, VT: Jewish Lights Publications.

Gorsky, J. 1999. Conscience in Jewish thought, in Jayne Hoose (ed.), *Conscience in World Religions.* Notre Dame, IN: University of Notre Dame, pp. 129–154.

Jewett, P.K. 1996. *Who We Are: Our Dignity as Human.* Marguerite Shuster (ed.). Grand Rapids: William B. Eerdmans.

Kim, J. 1996. *Mind in a Physical World: An Essay on the Mind–Body Problem and Mental Causation.* Cambridge, MA: MIT Press.

Murphy, N. (Winter 1999). Downward causation and why the mental matters. *CTNS Bulletin*, 19 (1), 13.

Murphy, N. and Ellis, G. 1996. *On the Moral Nature of the Universe.* Minneapolis, MN: Fortress Press.

Newberg, Andrew, Eugene D'Aquili, and Vince Rause. 2001. *Why God Won't Go Away: Brain Science and the Biology of Belief.* New York: Ballantine.

Pannenberg, W. 1985. *Anthropology in Theological Perspective.* Matthew J. O'Connell (trans.). Philadelphia: Westminster Press.

Persinger, M. 1987. *Neuropsychological Bases of God Beliefs.* New York: Praeger Books.

Peters, Ted 1999. Resurrection of the very embodied soul? Russell et al. (eds.), p. 38.

Peterson, G. 2003. *Minding God: Theology and the Congnitive Sciences* (Minneapolis: Fortress Press).

Peterson, G.R. 1997. Cognitive science: What one needs to know. *Zygon: Journal of Religion and Science*, 32, 615–627.

Russell, R.J., Murphy, N., Meyering, T.C., and Arbib, A.E. (eds.) 1999. *Neuroscience and the Person: Scientific Perspectives on Divine Action*. Notre Dame, IN: University of Notre Dame.

Smith, H. 2001. *Why Religion Matters: The Fate of the Human Spirit in an Age of Disbelief*. San Francisco: Harper San Francisco.

Tertullian. 1978. A treatise on the soul, in Alexander Roberts and James Donaldson (eds.), *Ante-Nicene Fathers. The Writings of the Fathers Down to A.D. 325*. Grand Rapids: William B. Eerdmans.

Teske, J. 1996. The spiritual limits of neuropsychological life. *Zygon: Journal of Religion and Science*, 31, 209–234.

Thatcher, A. 1987. Christian theism and the concept of a person, in A. Peacocke and Grant Gillett (eds.), *Persons and Personality: A Contemporary Inquiry*. Oxford: Basil Blackwell, pp. 180–196.

Tomasello, M. 1999. *The Cultural Origins of Human Cognition*. Cambridge, MA: Harvard University.

Wacome, D.H. 2004. Reductionism's demise: Cold comfort. *Zygon: Journal of Religion and Science*, 39, 321–338.

Ware, K. 2002. How Do We Enter the Heart? *Paths to the Heart: Sufism and the Christian East* (especially pp. 2–23). James Cutsinger (ed.), Bloomington, IN: World Wisdom.

Wolfson, H.A. 1993. Immortality and resurrection in the philosophy of the church fathers, in Everett Ferguson, David Scholer, and Paul Finney (eds.), *Doctrines of Human Nature, Sin, and Salvation in the Early Church, vol. X, Studies in Early Christianity*. New York: Garland Books, pp. 301–336.

Zizioulas, J.D. 1985. *Being as Communion*. Crestwood, NY: St. Vladimir's Press.

CHAPTER NINE

Overcoming an Impoverished Ontology: Candrakīrti and the Mind–Brain Problem

RICHARD K. PAYNE

Personal Preface

In my mid-twenties, I spent a couple of months of one summer working alone, constructing a retaining wall made of broken concrete block. While doing this work, I took to contemplating, as if it were a koan, the question, "If there is no self, then what is reincarnated?"

Introduction: Caveat

There is no one, single Buddhist view. Buddhism is more than two-and-a-half millennia old, has moved out of its Indian homeland into a variety of markedly distinct religious cultures, and retains canonic writings in over half a dozen languages. The lineages of thought and practice are as fully diverse as one might imagine they would be in such circumstances. Even when it comes to such teachings as the nonexistence of the self (anātman), which an overwhelming majority of Buddhists would understand as central to their religion, there is a wide range of interpretations and explanations. As a consequence, it would be inconsequential to present some generalized Buddhist view as if it were the Buddhist view. To avoid this, I have selected one particular Buddhist thinker—the medieval Indian Buddhist religious philosopher Candrakīrti—as a basis from which to develop a contemporary understanding of the teaching of the nonexistence of the self and its contribution to one part of the contemporary, cognitive scientific understanding of human existence—debates over the relation of mind and brain. At the same time, it would be misleading, if not dishonest, to present some specific Buddhist view as if it alone were authoritative.[1] For example, Vasubandhu—Candrakīrti's near contemporary—while agreeing on the basic principle of the nonexistence of the self, made a rather different argument regarding what the self actually is from the one made by Candrakīrti (Duerlinger 2003).

In the following paragraphs, we will first introduce four of the key concepts of Buddhist discourse on the self: no self (anātman), psychophysical aggregate (skandha), emptiness (śūnyatā), and interdependence (pratītyasamutpāda). The similarity of contemporary views of the self and the Buddhist analyses described raises the question of whether or not Buddhist thought can make any additional contribution to the current impasse in discussions of the relation between mind and brain. In order to answer this question, we will focus more closely on a series of refutations of the self that are found in one of Candrakīrti's works. What Candrakīrti says about the self will then be adapted to the mind, and applied to the current discussions attempting to determine what the relation between mind and brain is. It should also be made clear from the outset that the argument here is a philosophic one intended to contribute to cognitive science—a capacious domain, one which is accepting of a wide range of disciplinary approaches.

From No Self to Emptiness

ANĀTMAN: The Claim That There is "No Self" (anātman) and What is Being Denied

It is fairly widely recognized in the contemporary Western discourse on Buddhism that from its earliest recorded form, Buddhism has as one of its consistent themes the denial of the self (ātman). This is usually, however, taken out of its historical, religious, social, and cultural contexts. From its very beginnings, Buddhist thought has not been created in the splendid isolation of mystical experience. Rather, it has been the consequence of disagreement and debate. The polemic character of Buddhist thought—like every religious tradition—necessitates understanding who its interlocutors were in order to fully understand the claims that are asserted. It is fundamentally anachronistic to simply take the assertion of emptiness of the self and place it into the psychologized contemporary discourse. A dehistoricized and decontextualized denial of the self comes to be discussed as if it unproblematically meant the same thing in the context of fifth century BC India as it means in the Euro-American context at the beginning of the twenty-first century. Understanding the doctrine of no self in a psychological context may be a necessary step in making Buddhist thought relevant in our own world, but it can only meaningfully be accomplished after having stretched ourselves to incorporate the context out of which the denial of the self originated. If we wish to understand the Buddhist denial of the self, we need to ask, "What exactly is it that is being denied?"

Decontextualized, the assertion that there is no self is commonly taken as a psychological claim.[2] One such psychological interpretation is based on a distinction between the "little self" and the "big self" (or Self), which has its own roots in the conflation of late-nineteenth- and early-twentieth-century esoteric and Perennialist conceptions of the self with psychological conceptions. For example, Steven Collins points out that the Perennialist religious philosopher R. Zaehner's

> view of Buddhism, conditioned by a Jungian-influenced sensibility to other religions, tended always to speak of the denial of self as merely "the elimination of ego." This formulation leads the way for him, as for so many others, to suggest that there is a Self, or Real Self behind the (small) self or ego. Thus he speaks of "the Buddhist convention of using the word 'Not-Self' to mean something other than the Ego which has direct experience of both the subjective self and of objective phenomena"; and declares that "the Buddha... recognizes that there is an eternal being transcending time, space and change; and this is the beginning of religion. Moreover the Hindus, overwhelmingly, and the Buddhists when they are off their guard, speak of this eternal being as the 'self.' " (1982, p. 9)

In this psychological interpretation of the meaning of the teaching of anātman, the denial is understood to be directed at the personal, psychological, or phenomenal self—that is, the self-conception every person has of being the center of subjective, first-person experience. This is expressed by phrases such as "the self is an illusion," which are not uncommon in the Western discourse on Buddhism. To focus on this one point for a moment, such an expression is a mistaken appropriation of one of a set of metaphors employed to describe the impermanence of the self in which it is said that the self is *like* a magical illusion, a cloud, a dream, bubbles on a stream. Nāgārjuna, the founder of the Madhyamaka school of thought, says in stanza 66 of a work called *Seventy Stanzas Explaining How Phenomena are Empty of Inherent Existence*: "Produced phenomena are similar to a village of gandharvas,[3] an illusion, a hair net in the eyes, foam, a bubble, an emanation, a dream, and a circle of light produced by a whirling firebrand" (Komito 1987, p. 94). The point is not that they are merely illusory, but rather that, being created, all of these things are impermanent and continue to exist only so long as conditions for their existence continue. The same applies to the self.[4] To understand this more fully, however, it is necessary to place the denial in its own intellectual context, prior to attempting to understand its significance for our contemporary discussions of the nature of mind, brain, and self.

The period in which Buddhism originated is generally considered to have been a particularly vital and creative one in the history of Indian religious thought. It was in this period that the Upaniṣads were compiled, reflecting a wide variety of religious speculations on the nature of the self. A variety of metaphoric images are used to express these speculations.

One of these metaphors is the essence at the center of a seed, which makes the seed grow into the kind of plant that it does. This is employed in a dialogue between the sage Āruni and his son Śvetaketu. After having asked his son to bring him a fig and splitting it apart to reveal the seeds, the son is directed to split one of the seeds. Asked what he sees there, Śvetaketu replies, "Nothing." The father then explains, "Verily, my dear, that finest essence which you do not perceive—verily, my dear, from that finest essence this great sacred fig tree thus arises. Believe me, my dear, . . . that which is the finest essence—this whole world has that as its self. That is Reality. That is Ātman. That art thou (*tat tvam asi*), Śvetaketu" (Zimmer 1951, p. 336).

Another metaphor is that there are five "sheaths" covering the true, essential self. These five sheaths are: the physical body (anna-maya-kośa), which is the base for waking consciousness; the vital force or breath (prāṇa), the mind and senses (manas), and the comprehension or understanding (vijñāna; these middle three constituting the subtle body that is the base for dream consciousness); and bliss (ānanda), which is the basis for dreamless sleep. This final is the causal body of ignorance (avidyā), concealing the true self (p. 415).

A third metaphor equates the self with the breath. "The term *ātman*, which attained such great significance in the *Upaniṣads*, meant primarily 'breath' in the *Ṛg Veda*, though in later hymns it came to mean 'vital spirit.' Upon the death of a person the individual *ātman* is said to mingle with the wind, the breath of the gods" (Reat 1990, p. 59). And a fourth metaphor is the monistic equation of the true or absolute self (ātman) with absolute being (Brahman).[5] In general, then, the conception of the self that emerges is one in which the self is in some way permanent, eternal, absolute, or unchanging.[6] It is also simultaneously universal and individual (p. 180). The view is that there is an essence and that it can be known.

Richard Gombrich has summarized this situation, saying "The brahminical scriptures of the Buddha's day, the Brāhmaṇas and the early Upaniṣads, were mainly concerned with a search for the essences of things: of man, of sacrifice, of the universe. Indeed, brahminical philosophy continued in this essentialist mode down the centuries" (1996, pp. 3–4). The essentializing character of brahminical thought applied not only to the personal self, but also to the existence of all things. It was in dialogue with brahmins promoting these kinds of views that the Buddha denied that persons and all other existing things had a self or essence of any kind (p. 33).

It is, in other words, this permanent, eternal, absolute, or unchanging conception of the self that is being denied. In Peter Harvey's close textual examination of what is being denied, it was a concept of the self as "an unconditioned, permanent, totally happy 'I,' which is self-aware, in total control of itself, a truly autonomous agent, with an inherent substantial essence, the true nature of an individual person" (1995, p. 51). It is this complex of conceptions of the self that is being denied. Instead, human beings are ongoing processes, continually changing and lacking any essence; they are systems that may be analyzed in a variety of ways. One of these analytic systems is that of the five psychophysical aggregates.

SKANDHAS: *The Set of Five Psychophysical Aggregates into which Early Buddhist Scholastic Philosophy (Abhidharma) Reduces or Dissolves the Self*

Perhaps the most frequently employed analysis of human existence found in Indian Buddhism is into five "heaps" or aggregates (S. skandha; P. khandha).[7] These are described as the constituents of a person. However, this should not be taken to mean that they are in any way themselves permanent, absolute, eternal, or unchanging. They are themselves further analyzable into their own constituent elements, called dharmas, and it is the grouping of these into coherent wholes that constitute the aggregates.

The five psychophysical aggregates are fairly consistently rendered as matter (S. rūpa, P. rūpa), sensations (S. vedanā, P. vedanā), perceptions (S. saṃjñā, P. saññā), mental formations or volitional factors (S. saṃskāra, P. saṃkhārā), and consciousness (S. vijñāna, P. viññāṇa). Sue Hamilton explicates the meaning of these five as aggregates by reference to the traditional metaphor of a cart:

> in just the same way that a cart is made up of various bits and pieces that, when assembled, we *call* a cart, but there is no separate independent thing that *is* the cart, so each human being is an assembly of these five *khandhas* which gives rise to the convention of a living being, but there is no separate independent thing that is the self of that person. The *khandhas* are thus most often described as being aggregates: a human being consists of a group of five aggregate parts, none of which, individually or collectively is one's self. (2000, p. 27)

Since the aggregates themselves are subject to change—unstable, impermanent—they cannot themselves be taken to be the self (Gombrich 1996, p. 41). The term "aggregate," however, may mislead one into thinking that the person is seen as some sort of random collection—an unorganized "heap" of things. As the metaphor of the cart indicates, the aggregation forms an organized, functioning whole—a system. This interpretation of the aggregates as an understanding of human existence as a functional system corresponds with recent understandings of the Buddha as having been more concerned with how things work than with what they are or are not (Hamilton 2000, p. 21; and Gombrich 1996, p. 27).

ŚŪNYATĀ and PRATĪTYASAMUTPĀDA:
Emptiness and Interdependence as Synonyms

Like all religio-philosophic systems, Buddhist thought does not remain static. While early Buddhist thought focused largely on the emptiness of persons, analyzing them into the psychophysical aggregates (skandhas) and elements (dharmas), later Buddhist thought—Madhyamaka—turned its attention to the emptiness of things. Here, "things" (a term often maligned in philosophic discourse) refers to any and all existing things, that is, entities that have some efficacy—both phenomenal objects of perception, such as tables and chairs, and subtle objects such as the psychophysical elements, and the dharmas (or in modern Western discourse such subtle objects as atoms and subatomic particles). Most importantly, for foundations of Buddhist philosophy at that time, Madhyamaka critiqued those very dharmas that had previously been understood as the irreducible elements of existence. The analysis of human conscious experience into these elements allowed for a complete description without reference to a personal self. In going beyond this approach, Madhyamaka critiqued the earlier view as one that treated the dharmas as metaphysical absolutes.

The Madhyamaka view of the nature of the existence of both persons and things can be summarized by the phrase "emptiness (śūnyatā) is identical with interdependence (pratītyasamutpada)." Emptiness refers to the absence of any permanent, eternal, absolute, or unchanging essence (or self in the broader sense indicated by the term ātman). In other words, all existing things, whether persons or objects, are empty of a self or essence that has any status as a metaphysical absolute.

Interdependence, also translated as dependent arising or dependent co-arising, refers to the idea that everything that exists does so as the result of causes, that there is nothing that exists autonomously or independently of other things. In other words, there is nothing that is *sui generis*. Causal relations serve to interconnect all things—a view expressed by the contemporary Buddhist teacher Thich Nhat Hahn in his image of the sheet of paper, in which are present the sunshine that grew the trees that provided the pulp for the paper as well as the food that fed the lumberjack who cut them.

In the foundational Madhyamaka work Nāgārjuna's *Mūlamadhyamakakārikā*, this relation is expressed in a pair of stanzas:

> 18. Whatever is dependently co-arisen
>   That is explained to be emptiness.
>   That, being a dependent origination,
>   Is itself the middle way.
> 19. Something that is not dependently arisen,
>   Such a thing does not exist.
>   Therefore a nonempty thing
>   Does not exist.
>   (Garfield 1995, p. 69)

Emptiness and interdependence are known in Buddhist terminology as the "two truths," and despite the common tendency in Western discussions to interpret them hierarchically, they are simply synonymous—different ways of expressing the same conception of the existence of things.

In this all too brief a summary, I am emphasizing the continuity of Buddhist thought, the shifts of emphasis that I have characterized as anātman-skandhas-śūnyatā are, from the perspective that I am presenting here, differing expressions of the same fundamental concept that changed as Buddhist thought developed over time—a refinement and reexpression, rather than in any way a fundamental change.

Congruence between Buddhist and Contemporary Views of the Self

Many contemporary discussions of the personal or psychological self have called into question the traditional Western conceptions of the self as a permanent, unitary, unchanging source of identity. A few examples of this contemporary critique can be found in cognitive linguistics, cognitive neuroscience, anthropology, and cultural psychology.

George Lakoff has examined the metaphoric structures of ordinary language as a means of uncovering cognitive structures, arguing that such metaphors are not simply decorative flourishes, but rather foundational to human thought. His work in cognitive linguistics has established both the importance of metaphoric thinking and its basis in the embodied character of human being in the world.

In his essay "The Internal Structure of the Self," Lakoff claims that cognitive linguistics "can give us some evidence as to how we conceptualize the internal structure of the Self" (1997, p. 93). While explicit discussions of the self in Western literature have tended to build on conceptions of a unitary soul as immortal, immaterial, and immutable, what Lakoff discovers in his examination of the metaphors underlying the ordinary ways of talking about the self is a fundamentally divided conception.

One of the principal metaphors "is reminiscent of the traditional Western model of the transcendental ego" (p. 93). Lakoff uses the terms Self and Subject to talk about this model of the self. This metaphor carries the ethics of Western culture implicitly within itself. In discussing the modality of self-reflection, Lakoff says,

> You, the Subject, the locus of consciousness, rationality, and judgment, are looking at your Self, the locus of your needs, desires, and passions. Our culture tells us that the Subject, our locus of consciousness and reason, *should* be in control of our Self, so that our desires and passions do not get out of hand and lead us to harm others. Our culture also tells us that there is a way, a single way, we really are, and an objective viewpoint from which we could see who we really are, if only we could reach that place. (p. 94)

Despite the frequency of this metaphor of Self and Subject, Lakoff does not find a coherent or consistent system of "conventional metaphors for conceptualizing the internal structure of the Self" (p. 93). Looking beyond the principal Self–Subject metaphor, he finds that "the rest of the system does not define a single consistent Subject, but rather an inconsistent collection of many different Subjects" (pp. 93–94).

According to Lakoff, this does not, however, mean that such conceptions are entirely arbitrary.

> Though we do not have a single monolithic unified conception of the Self, perhaps each conception that works for us does reflect some reality. In short, it may be the case that the collection of conceptions of the Self that have evolved to serve us well in our unconscious conceptual systems may give us the best account of what the Self is that we can get using the kind of conceptual resources available to us in our unconscious conceptual systems. (p. 110)

Lakoff's use of cognitive linguistics to analyze the understandings of the self embedded in ordinary language reveals a diversity of such conceptions. Cognitive neuroscience seems to also reveal a similar diversity, though by simply asserting a definition, such diversity can appear to be brought under control.

Joseph LeDoux, a cognitive neuroscientist, discusses a variety of approaches to the question of the self, including logical, philosophic, and psychological. He concludes this, saying that in his own view,

> the self is the totality of what an organism is physically, biologically, psychologically, socially and culturally. Though it is a unit, it is not unitary. It includes things that we know and things that we do not know, things that others know about us that we do not realize. It includes features that we express and hide, and some that we simply don't call upon. It includes what we would like to be as well as what we hope we never become. (p. 31)

In sharp contrast, we find Thomas Metzinger, another cognitive neuroscientist, stating that the main thesis of his magisterial *Being No One*

> is that no such things as selves exist in the world: Nobody ever *was* or *had* a self. All that ever existed were conscious self-models that could not be recognized *as* models. The phenomenal self is not a thing, but a process—and the subjective experience of *being someone* emerges if a conscious information-processing system operates under a transparent self-model. (2003, p. 1)

Both anthropological and psychological literature evidence a similar movement toward the idea that the "self" is a construct. For example, Clifford Geertz notes that "the Western conception of the person as a bounded, unique, more or less integrated motivational and cognitive universe; a dynamic center of awareness, emotion, judgment, and action organized into a distinctive whole and set contrastively both against other such wholes and against a social and natural background is, however incorrigible it may seem to us, a rather peculiar idea within the context of the world's cultures" (Collins 1982, 2).

An example of this cultural malleability of the self is the work of Jeannette Mageo. Based on her fieldwork in Samoa, Mageo theorizes two ways in which the self is conceptualized—egocentric and sociocentric. Egocentric conceptions of the self are those in which the self is seen as distinct and separate from others, an individual and autonomous self. In contrast, sociocentric selves are defined by their relations with others, their place in a family network or social hierarchy. Mageo does not hypothesize these as mutually exclusive categories, but rather as two ends of a spectrum.

Such self-conceptions are enforced by the moral discourse of a society, but in doing so, it reveals that the dominant social modality predominates through the suppression of the other modality, making it everpresent as a problem. For example, in the egocentric society of the United States, sociocentric behaviors such as being caught up in the group enthusiasm of a football game or a rock concert—or a lynch mob—may be negatively described as instances of "losing oneself."

Mageo distinguishes ontological premises, about the way in which the self exists from moral premises, about the way the self should be. "In the United States a learned pride in independence converts the ontological premise that persons are separate into a moral premise that they should be. In Samoa a learned pride in service to superiors converts the ontological premise that people are role players within a larger group into a moral premise that they should be" (1998, p. 21).

Indeed, the anthropological literature amply evidences that different societies have different conceptions of the self. The corollary to this then is that there is no objective referent for the term self, but rather that the self is a social construct. We will expand on these views later, presenting a view in which social constructs do have actual existence in the realm of what we will call the intersubjective.

While Mageo outlines two organizational structures, Alan Roland, a psychoanalyst, compares the formations of the self in India, Japan, and the United States, and discerns "three overarching or supraordinate organizations of the self: the familial self, the individualized self, and the spiritual self" (1988, p. 6). The familial self allows people to function in hierarchical social relations, both in families *per se* and in other social groups. In contrast to an orientation on hierarchy, the individualized self as found in the United States enables the individual to work in fluid social situations, one in which autonomy is both allowed and expected. The spiritual self is highly private, can coexist with the other kinds of selves, and "is realized and experienced to varying extents by a very limited number of persons through a variety of spiritual disciplines" (p. 9).

The literature of anthropology and cross-cultural psychoanalysis cited here demonstrates the social construction of the self at two levels. First, its express intent is to demonstrate how the self is constructed differently in different societies. At a second, reflexive level, however, the very fact that these different disciplines can formulate the idea that the self is not a unitary, autonomous entity, but rather something that is socially constructed in a variety of differing forms, demonstrates that very social construction—this time in the context of an academic metadiscourse.

In one way of thinking about the congruence between contemporary conceptions of the self as a social construct and the Buddhist denial of the self, this "proves" the truth of Buddhism. It is never the case, however, that such congruence between a scientific theory and a religious system proves the truth of the latter, as the claims regarding such religious matters as salvation or awakening can never be proven true in this way. Conversely, and equally easily, in another possible way of thinking about this congruence of conceptions of the self, it proves the irrelevance of Buddhism. In this interpretation, Buddhist conceptions of the mind are a kind of antiquated gesture toward understanding which—now that we have science—can be discarded as having been passed beyond.

Despite these various critiques, the common or naive conception of the self that continues to be found in contemporary culture remains dualistic and essentialized. Mind, soul, spirit, or self as a metaphysical essence—eternal, absolute, unchanging, permanent—that is fundamentally distinct from matter, body, or brain. The heritage of Western culture is deeply imbued with such conceptions of the self. The specific formulations of this dualism can be identified in Platonic, neo-Platonic, Christian, and Cartesian sources. Given the similarity between Buddhist critiques of the self as ātman and the contemporary formulation of the self as social construct, which serves as a critique of an essentialized dualistic conception of the self, it might seem that Buddhism has nothing further to contribute.

However, as suggested earlier, Buddhist critiques of the self are not limited to the personal or psychological self. It includes a critique of all metaphysical absolutes that is of any essence described as absolute, eternal, unchanging, or permanent. This formulation is most clearly expressed in the Madhyamaka philosophy originating with Nāgārjuna, and further developed by Candrakīrti.

Candrakīrti on the Self

Candrakīrti (ca. seventh century CE) was a medieval Indian Buddhist philosopher considered today to be one of the most important proponents of the Middle Way system of thought (Madhyamaka), established by Nāgārjuna (late second century CE). Candrakīrti is credited with developing an interpretation of Madhayamaka that limited argumentation to the use of *reductio ad absurdum* argument forms (prasaṅga, hence Prasaṅgika). One of his works is known as *Entry into the Middle Way* (*Madhyamakāvatāra*), an expository work structured around the ten stages of the Bodhisattva Path, which begins with the perfection of compassion and culminates in full awakening.

The longest section in *Entry into the Middle Way* is the sixth one, in which Candrakīrti discusses the perfection of wisdom. This section is devoted to the central teaching of the Madhyamaka school, emptiness (śūnyatā). In the course of his discussion, Candrakīrti refutes five different theories regarding the personal self. These are that the personal self exists intrinsically (pudgala), that it differs from the five psychophysical aggregates, that it is the same as the five psychophysical aggregates, that it is a composite of all of the psychophysical aggregates, and that it is the same as the body (Huntington and Wangchen 1989, pp. 171–174).

Candrakīrti attributes the idea of an intrinsically existing self to non-Buddhist philosophers, while acknowledging that there are a variety of views of the self within the various non-Buddhist schools. According to Candrakīrti, non-Buddhists hold a view of the self as "eternal, inactive, without qualities, a nonagent, and the partaker" of all objects of knowledge (p. 171). He dismisses this kind of a self as having no more actual, objective existence than "the son of a barren woman." The latter is a commonly used expression in Buddhist philosophy to indicate something that—despite being something that one can talk about—has no actual, objective existence. The point of this argument is that non-Buddhists hypothesize a self that has no origin, and that anything that has no origin does not exist. As Jamgön Mipham (b. 1846, d. 1912), a famed premodern Tibetan Buddhist exegete, explains in his commentary on Candrakīrti's *Entry into the Middle Way*, "The characteristics of the self, as expounded in the non-Buddhist treatises . . . are disproved by the very argument of 'no origin' that they themselves advance. If the basis (in this case the self) has no existence, its characteristics likewise have no existence" (Padmakara, p. 284). In this way, the argument of the non-Buddhists is turned back on itself, the kind of argumentation that Candrakīrti propounds.

The second of Candrakīrti's refutations is directed against the idea that there can be any self established on the basis of anything other than the psychophysical aggregates. In other words, there is nothing other than the five psychophysical aggregates to be found in human existence, nothing in addition to those five that can be "considered to be the cognitive basis for clinging to an 'I' " (Huntington and Wangchen 1989, p. 172). Mipham comments that if there were a self that existed separately from the aggregates, then one would be able "to apprehend it independently of the aggregates, whereas in fact this never happens" (Padmakara, p. 284). Candrakīrti next considers three understandings of the existence of the self, which he presents as variants of one another. The self may be the same as the psychophysical aggregates, part of the aggregates, or simply the single aggregate of consciousness. According to Mipham, these different positions are held by different groups of Buddhists who interpret the Buddha's statement that those who think of themselves with the thought of being a separate, independently existing self are only referring to the five psychophysical aggregates (p. 285). Candrakīrti rejects the first on the grounds that, "If the self *is* the psychophysical aggregates, then there would have to be a plurality of selves, since there is a plurality of aggregates" (Huntington and Wangchen 1989, p. 172). He then goes on to discuss the contradictions inherent in these variant conceptions of the self as being based on the aggregates.[8] The next conception of the self to be refuted is that it is identical with the composite of the five aggregates, that is, all five of them taken together constitute the self in the same way as when the constituent parts of a carriage are properly assembled that is exactly what is meant by carriage. However, as Mipham puts it, such a composite entity "lacks real, substantial existence, [and therefore] the collection of aggregates is not the self" (Padmakara, p. 291).

The next view that Candrakīrti refutes is that the self is identical with the form aggregate, that is, that which possesses the body. He points out that such a view is self-contradictory in that according to such a view, the self is the agent that causes the body to act, and that it is incoherent to suggest the agent is identical with the object upon which it causes an effect. The final view of the self that he refutes is that it is something about which one cannot express "whether it is identical to or different from [the aggregates], permanent or impermanent, or anything else" (Huntington and Wangchen 1989, p. 175). Despite being inexpressible, the self does, however, actually exist. In Candrakīrti's view, however, anything that actually exists must be expressible.

Having refuted all of these views, Candrakīrti makes it clear that in his opinion the self is only a "dependent designation" (Skt. prajñaptir upādāya). He employs the now-familiar analogy of the carriage. Just as a carriage does not exist except as a designation dependent upon the parts that make it up, so also the self does not exist except as a designation dependent upon the parts that make it up, such as the five psychophysical aggregates. As Candrakīrti says:

> The self is not a real, existent thing, and thus it is not constant,
> And it is not inconstant, for it has no birth or ending.
> Attributes like permanence do not apply to it,
> And it is not, nor is it other than, the aggregates.
> (§ 163. Padmakara, p. 91)

In Candrakīrti's presentation, we can clearly see the ontological character of the treatment of the self in the Buddhist denial of the self. The self has no metaphysical status as permanent, eternal, absolute, or unchanging, rather, it is simply a conventional expression used to identify the constituent parts of a human being as an ongoing process, and depending on those parts. As James Duerlinger has expressed it, Candrakīrti "thinks that first-person singular reference to ourselves does not depend upon a reference to something that ultimately exists. This does not mean that he thinks that 'I' is not a referring expression. Rather, it means that it refers to a mentally constructed 'I' and to nothing else" (2003, p. 2). This fits well with the idea that the self is a mental construct that is significantly conditioned by social conceptions of it.

Since it does not refer to any essentially existing thing, self (and mind, soul, spirit) can be used to mean a variety of things depending upon who is doing the defining.

Extending Candrakīrti's suggestion that the self is a mental construct to include its social character, creates an opportunity to go beyond the dualistic conceptions discussed earlier. Dualistic conceptions of the self can be described as suffering from an impoverished ontology, a two-term ontology. This impoverished ontology distinguishes mind (which we will henceforth use as a cover term for all related concepts, e.g., self, consciousness, soul, spirit) from brain (which we will henceforth use as a cover term for all of its related concepts, e.g., body, matter). As a consequence of this impoverished ontology, first-person experience is consistently—and as we hope to demonstrate later, mistakenly—identified with mind, while brain is identified as the other term, the physical existence contrasting with the mental.

Monistic solutions to the inadequacies of the impoverished ontology fail to capture adequately the character of first-person experience.[9] For example, the eliminativist position claims that by taking a physical base as fully explanatory, it overcomes the Cartesian dualism of *res cogitans* and *res extensa*. However, under the representationalist version of eliminativism, it inadvertently reintroduces a different kind of dualism—that between the external world and the world as represented inside the head.

Rather than attempting to resolve the "mind-brain problem" within the confines of a two-term ontology, following Candrakīrti's lead we can consider a three-term ontology. The three terms of such an ontology would be objective, subjective, and intersubjective.

This usage of objective and subjective is not to be confused with the contemporary popular usage in which the terms mean unbiased and biased, respectively. Instead, objective is used here to refer to those things that are part of the public space, things that we may observe jointly—a table, a chair, a sunset. Such things are objects, and hence objective in that we all may observe them and agree that we are doing so, admittedly from our own perspectives—they are objects for all of us. Subjective things, however, are entirely private, specifically first-person experiences such as my pain. Such first-person experiences are limited to one subject, they are not available objectively; they only exist subjectively, as part of my subjective experience. It is important to stress at this point that both objective and subjective entities actually exist.

Intersubjective things also actually exist, that is, they have real effects, but their existence is dependent upon human culture, society, and language.[10] The real efficacy of such intersubjective entities is evidenced by the fact that people will fight and die for causes, for ideals; how uncountably many are the real, objective deaths that have been brought about by the intersubjectively existing entity patriotism? In such an ontological schema, brain is objective, first-person experience is subjective, and mind (soul, person, consciousness, self) is intersubjective, that is, it exists as a socially, culturally, and linguistically constructed entity.[11]

Another way of approaching this is to distinguish between the ways in which we can define our terms; we can define both brain and first-person experience ostensively, that is, we can point out what a brain is, show pictures of one, and so on; similarly, we can say to someone with whom we are visiting a zoo, "Do you see the penguins?" and then point out that the visual experience of the penguins is a first-person experience, especially if it is the case that they can see where the penguins are and I can't. The mind, however, cannot be defined ostensively, it can only be defined stipulatively—"What I mean when I say mind is . . ." In another example, a table exists objectively, our ideas about the style of a table and its integration with the style of other pieces of furniture exist intersubjectively, while the pain experienced upon stubbing one's toe on the table exists subjectively.[12] This is why when someone attempts to claim that the self or mind is an eternal, absolute, unchanging, or permanent something or other that is separate from the brain (or its processes, or the central nervous system, or anything physically determined at all), they are not pointing at any objectively existing entity, but rather stipulatively defining how they use the term.

While the objective and subjective as used here are informed by Kantian conceptions, the intersubjective has a background in phenomenology. It plays an important role in the development of the sociological phenomenology of Alfred Schutz, Thomas Luckmann, and Peter Berger, as well as the psychological phenomenology of Maurice Merleau-Ponty (Dillon 1988; Merleau-Ponty 1962; Schutz and Luckmann 1973). In particular, the work of Berger and Luckmann, *The Social Construction of Reality* (1966), seems to have initiated much of the contemporary focus on social constructivism.

While the explicitly philosophic dimension of these ideas may not be evident, it does seem to form at least part of the background to such developments in cognitive science as the work on "distributed cognition," in which, for example, problem solving is recognized as a social or intersubjective activity, rather than one that occurs only within the confines of a single brain. The point being made here, however, is not that the mind actually exists in the social realm, but rather that since it can be defined in this way, it is a social construct, actually existing as an intersubjective entity.

In other words, the arguments over whether to primarily identify the mind with the brain or with first-person experience are irresolvable because they are based on a fundamental confusion, a category mistake. This is an adaptation of the view developed by Gilbert Ryle in his classic work, *The Concept of Mind* (1949/2002). Ryle is famous for his application of the idea of category mistake to the concept of mind (pp. 16–17). The logic of this is that there is nothing that can be called the mind that is other than or in addition to the workings of the mind, any more than there is something that can be called a university that is other than or in addition to students, professors, libraries, classrooms, and so on. Ryle, however, seems intent on dismissing the concept of mind entirely, or at least as a locative—"The phrase 'in the mind' can and should always be dispensed with" (p. 40). What I am pointing out here, however, is that the concept of mind may be employed so variously, including as a locative, because it is an intersubjective entity. There is no goal of "purifying" discourse here, but rather simply an attempt to sort out the confusion resulting from identifying subjectively existing first-person experience with intersubjectively existing mind.

This three-fold ontology corresponds to the three-fold system developed by Eve Sweetser in her work in cognitive linguistics, indicating that the three ontological categories that I have suggested are already implicitly recognized in our speech. Sweetser argues that semantics is "inherently structured by our multi-leveled cultural understanding of language and thought" (1990, p. 21). The three domains that she identifies are physical, social, and epistemic. The physical refers to the way in which language can express something about the world, a description or "model of the world" (p. 21). This corresponds to what I have been referring to as the objective. The social level is that in which language can itself be an action, "an act in the world being described" (p. 21). This social level is the intersubjective. And the epistemic refers to the way in which language can be used to refer to "a premise or conclusion in our world of reasoning" (p. 21). The subjective ontological domain then corresponds to Sweetser's epistemic.

Sweetser's work also indicates how it is that these different domains can come to be confused. However frequently conceptual structures from one domain come to be applied in another, this does not

> imply that physical, social, and epistemic barriers have something objectively in common, at however abstract a level. My idea is rather that our *experience* of these domains shares a limited amount of common structure, which is what allows a successful metaphoric mapping between the relevant aspects of the three domains. The mapping itself, then further structures our understanding of the more abstract domains in terms of our (more directly experientially based) understanding of the more concrete domains. (p. 59)

In terms of our discussion here, this implies that the while the three ontological domains remain distinct, how, for example, the mind is defined (either explicitly or implicitly) directly influences—"structures" in Sweetser's terminology given earlier—how first-person experience and the brain are understood.

Mind and Brain

One specific instance of the difficulty of discussing the relation between mind and brain under a two-term ontology is found in Jeffrey M. Schwartz and Sharon Begley's *The Mind and the Brain* (2002). As a psychiatrist, Schwartz has had success with patients suffering from obsessive–compulsive disorder (OCD). His treatment protocols were developed in light of mindfulness meditation (S. smṛti, P. sati), and involve not only learning to focus attention but also a didactic element about the neurophysiology of OCD.[13] On the basis of his success, he developed what can only be considered a dualist theory that the mind—in the form of attention and will—directly effects the brain. According to Schwartz, the mind makes changes in the structure of the brain. Although he focuses on neuroplasticity (the changes in how neurons are organized), plasticity "is built into the fabric of brain and mind at multiple levels, from the molecular to the cognitive" (Black 2000, p. 119).

Schwartz's argument is that:

1. attention creates mental force, a concept he qualifies as distinct from the standard four forces of contemporary physics (Schwartz and Begley 2002, p. 318);
2. mental force is able to influence neuroplasticity, that is, changes in the structural organization and allocation of resources (i.e., blood flow) to the brain;
3. mental force is able to do this by virtue of quantum effects,[14] specifically that just as observation at molecular level (e.g., ammonia molecule [p. 352]) fixes a particular quantum state, attention fixes a particular state of the brain (e.g., the desire to weed the garden instead of counting cans in the pantry);
4. because we know that attention affects neuroplasticity, mental force (the link between the two) is real.

Schwartz's theory returns us to a dualistic conception of mind and brain as existing independently of one another and having effects upon each other. The model is obviously akin to Descartes', although Schwartz has changed the locus of interaction from the pineal gland to quantum effects in the synapses. Schwartz's argument also seems to be at least implicitly informed by vitalist arguments of the eighteenth century (Braddon-Mitchell and Jackson 1996, p. 4). Vitalism argued that there must be a separately existing life force (e.g., Bergson's *elan vital*) because organic chemistry could not be explained in terms of inorganic. Schwartz has simply revised the terms from life force to mental force, but is still making what is fundamentally an argument from ignorance—almost always considered to be a fallacy.

From comments Schwartz makes regarding his view of the deleterious consequences of materialism (Schwartz and Begley 2002, pp. 257–258), one suspects that the appeal of his reinterpretation of the vitalist argument in terms of directed mental force is similar to that it held for some "neo-vitalists" at the end of the nineteenth century—it appears to support the claims of religion. As Peter J. Bowler expresses the religious appeal of neo-vitalism, "If life was a distinct nonmaterial force, its introduction into the material world would presumably have to be supernatural" (2001, p. 161).

By hypothesizing separate, independent existence of mind and brain, Schwartz draws conclusions that are based on a unidirectional understanding of causality—either mind effects brain, or brain effects mind.[15] Two sets of experimental results that he discusses, however, are not nearly as clear cut as this. The first are a set of experiments with monkeys whose sensory neural connection between, for example, an arm and the brain has been severed, or what is technically called "deafferented" (Schwartz and Begley 2002, Chapter 4). These experiments showed the plasticity of cortical organization. Rather than a fixed association between the sensory input of the arm and a specific area of the brain, the area associated with the now deafferented arm would be taken over by neighboring functions, such as sensory inputs from the face. If Schwartz's theory is to hold good, it must provide an explanation for the cortical plasticity displayed by these monkeys. Presumably, he would say that the changes in cortical organization result from the monkeys paying attention, exercising their "directed mental force" in such a fashion that quantum effects at the synaptic level are altered, creating new connections, and thereby altering the cortical mapping of functions. It is unclear, however, what it means for monkeys to be paying attention or exercising their will in such a case. Cortical reorganization is a constantly ongoing process, one that is a "dynamic balance within systems" (Kaas 2000, p. 232). In other words, no matter what the deafferented monkeys do, there will be cortical reorganization. The primary question then is the mechanism that brings this about. It would seem counter to Occam's Razor to add the concepts of attention, will, and directed mental force to established explanations that rely on the electrochemical character of the nervous system. Stimuli, including learning to focus one's thoughts, produce electrochemical effects that are critical to neuronal growth and survival, and the development of axonal connections [see, e.g., the discussion of receptor tyrosine kinases, Zhou and Black (2000, p. 215); and also, the discussion of excitatory transmitter glutamate, LeDoux (2002, p. 81)]. It is a standard criteria in the evaluation of scientific theories that the simpler is the better. This is the force of Occam's Razor and the reason that a theory that unnecessarily includes quantum effects is less likely to be true than one that doesn't. While plasticity does not provide evidence of a nonphysical, mental force, what it does support is an understanding of the brain as a complex ongoing process, one that is interdependent with the world. According to Joaquín M. Fuster, "Experience begins to play its structural, network-building role early in ontogeny, and that role persists throughout life" (2003, p. 39). Thus, cortical reorganization does not simply result from the mind producing effects on the brain, but rather out of the non-dual, interdependent relations between the environment (including the social environment), the body (including the brain), and first person experience.

Similarly, in the therapeutic applications Schwartz describes, it is not simply the exercise of the "will" ("I will myself to go out and weed the garden, and not count the cans in the pantry again.") that is involved in the actual situation. In addition to the didactic component of the training, there are a lot of social stimuli and, one must assume, reinforcement involved in the actual training process (Schwartz and Begley 2002, Chapter 2). Rhetorically, one might pose the following questions: If it were simply an exercise of the will that produces therapeutically beneficial outcomes, then why should there be any relapse? Didn't those clients try hard enough? And wouldn't this in turn simply be another version of "it's all in your head"?

In keeping with both Ryle and Madhyamaka critiques of reified concepts (Ryle's category mistakes), an additional set of questions can be raised. Where is this will when it is not being exercised? Is the will something separate from acts of willing (Ryle 1949, pp. 62–82)? Likewise, one may ask the following: Where does the mind go when it is not thinking? Is the mind something separate from the thinking? Where does consciousness go when one is not aware? Is consciousness something separate from the awareness?

In contrast to Schwartz's dualism of mind and brain with its unidirectional conception of causality, a non-dual, interdependent conception of the relation between first-person experience and the brain is open to the complexity of the actual situations—the plurality of causal factors involved. Adding Candrakīrti's understanding of the socially constructed character of such concepts then also allows us to avoid reifying will, mind, consciousness as objectively existing entities separate from the set of activities referred to by each of these concepts. The single answer to the rhetorical questions above is that will, mind, and consciousness exist as intersubjective entities.

Conclusion

The "mystery" of consciousness results (in [large] part) from its mystification. This mystification is embedded in the way it is conceived. When conceived of as an entity of some kind that is present in all of our moments of conscious awareness and separate from those moments of consciousness, it remains beyond the scope of any scientific explanation.

However, this conception of consciousness may not be the best or the most appropriate one. In Western philosophy of mind, from Brentano onward, there has been a view of consciousness as intentional. That is, that consciousness is always consciousness of something. (The something in question may, of course, have a variety of ontological statuses.) This is similar to views of mind found in Buddhist psychology as well.

From this perspective, consciousness seems far less mysterious an entity. If electrical stimulation of a part of the brain can make one aware of a memory, and if being aware of that memory is being conscious of it, then that particular conscious moment is explained by reference to the electrical stimulation of a portion of the brain. There need be nothing else—no some sort of something somewhere that hasn't been explained

In other words, there is no mystery to be explained. To those who ask "Why this experienced sense of interiority?," the only answer can be "Because that's the way it is."

While working on this essay, I came across a note I made to myself several years ago. That note simply read "Is mind–body dualism a cultural artefact?" Having worked through the issues discussed here, I am now convinced that the answer to my own question is "Yes."

Notes

1. Of course, Buddhists actually do assert one particular view as authoritative, arguing from the (presumptive) superiority of their own tradition.
2. This important distinction was made by Harvey Aronson (1998; see also Aronson 2004).
3. Literally, "fragrance eaters," ethereal heavenly musicians who are thought to live by consuming fragrances (Keown 2003; s.v. "gandharva").
4. While this is not the place to detail the history of the Western misunderstandings of Buddhist concepts, we can simply note in passing that Schopenhauer's use of the authority of the exotic to support his own views seems to have been foundational (see Halbfass 1988, pp. 105–120; see also, Droit 1997).
5. The monist view is generally agreed upon as central to the Upaniṣadic conceptions of the self: "Die grosse Leistung der Philosophie der Upaniṣads ist die Lehre von der identität des als unvergänglicher Kern des Individuums erkannten Ātman mit dem Brahman" (Bechert and von Simson 1993, p. 105). However, in the later, medieval developments of Vedānta, the monistic interpretation is only one of three interpretations of the relation between ātman and brahman found. In addition to the monism of Advaita Vedanta, there was also a dualist view (Dvaita), and a qualified monism (Viśiṣṭādvaita). The monist interpretation is that most familiar to Western audiences probably because of the dominance of Advaita Vedanta in the representation of Indian thought. There are another five major schools besides Vedanta (Nyāya, Yoga, etc.), which also had their own distinct psychological theories.
6. This formula—"permanent, eternal, absolute, or unchanging"—is not canonic, but rather my own way of characterizing the essentialist conception of the self that Buddhism denies.
7. In some of the older English language discussions, this term is rendered "confections," but in the literal sense of something "put together," rather than in the colloquial use meaning candy.
8. Detailing Candrakīrti's arguments would require additional explanations of such abstruse Buddhist philosophical concepts as momentariness, karma, and disagreements over the status of the self at nirvana, all of which would take us far beyond the scope of what is needed here. The interested reader is referred to Mipham's commentary.
9. This has been extensively discussed by several authors (see, e.g., Chalmers 1996; Wallace 2000; Siewert 1998).
10. This is an extension of Austin's notion of the performative function of language, or what Searle calls speech acts. Speech acts have their efficacy in the intersubjective realm (Searle 1997).
11. This is broader than what Ian Hacking means when he speaks of "interactive kinds" in contrast to "natural kinds" (1999, p. 59).
12. These philosophic reflections are not offered as particularly profound, but rather as simply a useful terminology by which to talk about certain things that do have real consequences, such as the ideas of democracy and human rights, and that therefore—according to Buddhist thought—do actually exist, but do not exist in the same way as dogs and newspapers do. For a much more adequate and philosophically sophisticated treatment of these issues, see Hacking 2002.
13. At various points, Schwartz claims that his theory is consistent with Buddhist thought. For example,

    > Volition, or Karma, is the force that provides the causal efficacy that keeps the cosmos running. According to the Buddha's timeless law of Dependent Origination [i.e., interdependence], it is because of volition that consciousness keeps arising throughout endless world cycles. And it is certainly true that in Buddhist philosophy one's choice is not determined by anything in the physical, material world. Volition is, instead, determined by such ineffable qualia as the state of one's mind and the quality of one's attention: wise or unwise, mindful or unmindful (Schwartz and Begley 2002, p. 294).

While what may be called "mentalistic" interpretations of Buddhism, such as this one presented by Schwartz, are common in the rhetoric of Buddhist modernism, it does not represent a universally accepted (or even, perhaps a majority) view of Buddhist teachings. Most critically, not all Buddhist thinkers agree that karma—being the actions of body, speech, and mind—is solely mental, or that the mental exists independently from the physical. For example, the very traditional and widespread description of the processual nature of human existence known as the "twelve links in the chain of causation" (ignorance, mental formations, consciousness, mind and body, senses, contact, sensation, craving, attachment, becoming, birth, and suffering) is not causally initiated by volition (Mitchell 2002, pp. 40–42). Rather, the twelve are arranged in an unending circle, the *discussion* of which may begin with any one of them.

14. Schwartz's appeal to quantum effects is at the least a strained one. It depends upon a very problematic analogy between the "observations" of molecular, atomic, and subatomic phenomena, and the "attention" paid to one particular thought rather than another. This is an argument by analogy based on the assertion of the similarity of observation and attention, but one that leaves out several important differences, e.g., observation in quantum physics involves the addition of energy to the item being observed—human beings cannot simply "attend" to such phenomena, there has to be something that goes into the items and comes back out to allow for an observation of this kind. This particular aspect of quantum physics would, therefore, not seem to breach the closure principle (nonphysical phenomena, e.g., thoughts, cannot have any physical effect, i.e., the physical realm is closed to nonphysical influences), while Schwartz's theory does [on the closure principle and the issues of causal relations between the mental and the physical, see Wallace (2000, p. 81)]. While space here does not allow full treatment, two additional problems of Schwartz's work may be noted. First is his repeatedly typifying the dominant scientific view of the brain as deterministic. However, there is a very strong interpretation of scientific "laws" as statements regarding probabilities, and not all probabilities are the consequence of quantum physics. Second is the underlying conflation of a materialistic (only physical existence is real) and naturalistic conceptions. Naturalistic explanations of the mind need not be reductionistic in the way materialistic (or eliminativist) ones would be; they would simply avoid supernatural explanations [see Flanagan (2002)].

15. Schwartz talks about causality being bidirectional, "the arrow of causation relating brain and mind must be bidirectional" (Schwartz and Begley 2002, p. 95). However, he is using this term to refer to two unidirectional causal relations, which are different from the concept of a non-dual interdependence that I am proposing here based on the thought of Nāgārjuna and Candrakīrti.

References

Aronson, Harvey B. 1998. Review of Jeffrey B. Rubin, *Psychotherapy and buddhism: toward an integration*. *Journal of Buddhist Ethics*.

———. 2004. *Buddhist Practice on Western Ground: Reconciling Eastern Ideals and Western Psychology*. Boston: Shambhala Publications.

Bechert, H. and von Simson, G. (eds.) 1993. *Einführung in die Indologie: Stand, Methoden, Aufgaben*, 2nd ed. Darmstadt: Wissenschafliche Buchgesellschaft.

Berger, P. and Luckmann, T. 1966. *The Social Construction of Reality: A Treatise in the Sociology of Knowledge*. New York: Doubleday.

Black, I.B. 2000. "Introduction" to Section II. Plasticity, in Michael S. Gazzaniga et al. (eds.), *The New Cognitive Neurosciences*, 2nd ed. Cambridge and London: MIT Press.

Bowler, P.J. 2001. *Reconciling Science and Religion: The Debate in Early-Twentieth-Century Britain*. Chicago and London: University of Chicago Press.

Braddon-Mitchell, D. and Jackson, F. 1996. *Philosophy of Mind and Cognition*. Oxford: Blackwell Publishers.

Chalmers, D.J. 1996. *The Conscious Mind: In Search of a Fundamental Theory*. New York and Oxford: Oxford University Press.

Collins, S. 1982. *Selfless Persons: Imagery and thought in Theravāda Buddhism*. Cambridge: Cambridge University Press.

Dillon, M.C. 1988. *Merleau-Ponty's Ontology*. Evanston, IL: Northwestern University Press.

Droit, R.P. 1997. *Le Culte du Néant: Les Philosophes et le Bouddha*. Paris: Éditions du Seuil.
Duerlinger, J. 2003. *Indian Buddhist Theories of Persons: Vasubandhu's "Refutation of the Theory of a Self."* London and New York: RoutledgeCurzon.
Flanagan, O. 2002. *The Problem of the Soul: Two Visions of Mind and How to Reconcile Them*. New York: Basic Books.
Fuster, J.M. 2003. *Cortex and Mind: Unifying Cognition*. New York: Oxford University Press.
Garfield, J.L. (trans.) 1995. *The Fundamental Wisdom of the Middle Way: Nāgārjuna's Mūlamadhyamakakārikā*. New York and Oxford: Oxford University Press.
Gazzaniga, M.S. (editor in chief). 2000. *The New Cognitive Neurosciences*, 2nd ed. Cambridge and London: MIT Press.
Gombrich, R.F. 1996. *How Buddhism Began: The Conditioned Genesis of the Early Teachings*. School of Oriental and African Studies, Jordan Lectures in Comparative Religion, no. XVII. London and Atlantic Highlands, NJ: Athlone.
Gyatso, K. 1995. *Ocean of Nectar: Wisdom and Compassion in Mahayana Buddhism*. London: Tharpa Publications.
Hacking, I. 1999. *The Social Construction of What?* Cambridge and London: Harvard University Press.
———. 2002. *Historical Ontology*. Cambridge and London: Harvard University Press.
Halbfass, W. 1988. *India and Europe: An Essay in Understanding*. Albany: State University of New York Press.
Hamilton, S. 2000. *Early Buddhism: A New Approach, The I of the Beholder*. Richmond, UK: Curzon Press.
Harvey, P. 1995. *The Selfless Mind: Personality, Consciousness and Nirvana in Early Buddhism*. Richmond, UK: Curzon Press.
Huntington, C.W., Jr. and Wangchen, N. 1989. *The Emptiness of Emptiness: An Introduction to Early Indian Mādhyamika*. Honolulu: University of Hawai'i Press.
Kaas, J.H. 2000. The reorganization of sensory and motor maps after injury in adult mammals, in Michael S. Gazzinga et al. (eds.), *The New Cognitive Neurosciences*, 2nd ed. Cambridge and London: MIT Press, pp 223–236.
Keown, D. 2003. *Dictionary of Buddhism*. Oxford: Oxford University Press.
Komito, D.R. 1987. *Nāgārjuna's "Seventy Stanzas": A Buddhist Psychology of Emptiness*. Ithaca, NY: Snow Lion Publications.
Lakoff, G. 1997. The internal structure of the self, in Ulric Neisser and David A. Jopling (eds.), *The Conceptual Self in Context: Culture, Experience, Self-Understanding*. Cambridge: Cambridge University Press.
LeDoux, J. 2002. *Synaptic Self: How Our Brains Become Who We Are*. New York: Viking Penguin.
Mageo, J.M. 1998. *Theorizing Self in Samoa: Emotions, Genders, and Sexualities*. Ann Arbor: The University of Michigan Press.
Merleau-Ponty, M. 1962. *Phenomenology of Perception*. Colin Smith (trans.). London and New York: Routledge and Kegan Paul.
Metzinger, T. 2003. *Being No One: The Self-Model Theory of Subjectivity*. Cambridge, MA: MIT Press.
Mitchell, D.W. 2002. *Buddhism: Introducing the Buddhist Experience*. New York and Oxford: Oxford University Press.
Padmakara Translation Group. 2002. *Introduction to the Middle Way: Chandrakirti's* Madhyamakavatara *with Commentary by Jamgön Mipham*. Boston and London: Shambhala Publications.
Reat, N.R. 1990. *The Origins of Indian Psychology*. Berkeley: Asian Humanities Press.
Roland, A. 1988. *In Search of Self in India and Japan: Toward a Cross-Cultural Psychology*. Princeton: Princeton University Press.
Ryle, G. 2002. *The Concept of Mind*. 1949. Reprint, with new introduction by Daniel C. Dennett. Chicago: University of Chicago Press.
Schutz, A. and Luckmann, T. 1973. *The Structures of the Life-World*, 2 vols. Richard M. Zaner and H. Tristram Engelhardt, Jr. (trans.). Evanston, IL: Northwestern University Press.
Schwartz, J.M. and Begley, S. 2002. *The Mind and the Brain: Neuroplasticity and the Power of Mental Force*. New York: Regan Books, HarperCollins.
Searle, J.R. 1997. *The Creation of Social Reality*. New York: Free Press.
Siewert, C.P. 1998. *The Significance of Consciousness*. Princeton: Princeton University Press.

Sweetser, E.E. 1990. *From Etymology to Pragmatics: Metaphorical and Cultural Aspects of Semantic Structure.* Cambridge Studies in Linguistics, no. 54. Cambridge: Cambridge University Press.

Wallace, B.A. 2000. *The Taboo of Subjectivity: Toward a New Science of Consciousness.* New York and Oxford: Oxford University Press.

Zhou, R. and Black, I.B. 2000. Development of neural maps: Molecular mechanisms, in Michael S. Grazzinga et al. (eds.), *The New Cognitive Neurosciences*, 2nd ed. Cambridge and London: MIT Press.

Zimmer, H. 1951. *Philosophies of India.* Joseph Campbell, (ed.). Bollingen Series, XXVI. Princeton: Princeton University Press.

CHAPTER TEN

Religion and Brain–Mind Science: Dreaming the Future

KELLY BULKELEY

> Our innermost being, our common ground, experiences dreams with profound delight and a joyous necessity.
> —Friedrich Nietzsche, *The Birth of Tragedy*

Introduction

Now that we are a few years beyond the "Decade of the Brain" (so proclaimed by the first President Bush in 1991), we can see how thoroughly the recent findings of brain–mind science have revolutionized our knowledge of human nature. Researchers have made astonishing discoveries about the workings of memory, language, vision, emotion, rationality, imagination, and many other basic features of psychological functioning. The implications of these findings are dramatic for many different fields of study, nowhere more so than in religious studies. Contemporary brain–mind science is giving us new insights into the evolved nature of our species, and this makes it directly relevant to the world's religious traditions insofar as they seek a deeper understanding of what it means to be human. The time has long since come when the abundant discoveries of brain–mind science and the extensive history of human religiosity should be compared, evaluated, and, where possible, integrated.

There are many obstacles to this integration, however, and many good reasons why religious studies scholars have been reluctant to make the effort of overcoming them. The research literature in brain–mind science can be extremely technical and all but incomprehensible to nonspecialists. New discoveries are being made with such dizzying rapidity and from so many different quarters that even professionals within the brain–mind science domain have difficulty keeping up. And most troubling of all, it seems that recent brain–mind science findings are overtly hostile toward religion, with some prominent figures arguing that all religious phenomena can be explained in naturalistic terms and reduced to the material operation of ordinary psychological processes. It is hard to blame religious studies scholars for being reluctant to engage with research that is apparently intent on eliminating their whole field of study.

In recent years, a significant number of attempts have been made to build new bridges between religion and brain–mind science, and these attempts have produced some promising results. But most of the projects are seriously flawed, in some cases flawed so badly that future growth in this area is threatened unless major conceptual and methodological changes are made. In this chapter, I will offer a preview of those necessary changes, as illustrated in current research on dreaming from the perspectives of both religious studies and brain–mind science. Although prayer and meditation have received more public attention in recent years, the study of dreams offers another opportunity for bridging new research findings from religious studies and scientific psychology. I will focus here on a study of what C.G. Jung called "big dreams" (Jung 1974). Since Jung's time, other researchers have explored the same territory using terms like "intensified dreams" (Hunt 1989), "impactful dreams" (Kuiken and Sikora 1993), "highly significant dreams" (Knudson 2001), "extraordinary dreams" (Krippner et al. 2002), and "apex dreaming" (Nielsen 2000). My contributions to this growing lexicon have been "root metaphor dreams" (Bulkeley 1999a, 1994) and "most memorable dreams" (Bulkeley 2000, 2004). Although differing somewhat in their conceptual emphases, these terms point to the existence of a cluster of relatively rare but widely experienced dream types involving (1) unusually intense emotions and physiological sensations, (2) striking visual images combining bizarreness, beauty, chaos, and symmetry, and (3) a high degree of memorability upon awakening. For people interested in religion and brain–mind science, the study of big dreams (I will use Jung's term for the remainder of this chapter) has several appealing features:

- Big dreams are frequently reported by "ordinary" people, not just by religious virtuosi or highly trained practitioners.
- Big dreams are reported in virtually every religious and cultural tradition throughout history, and are still reported by people in contemporary Western society.
- Big dreams are clearly grounded in the neural activity of the brain during sleep, a fact that opens the way to using sleep research technologies to investigate the neurological processes that produce such spiritually charged experiences.
- The cross-cultural occurrence of big dreams, combined with their rootedness in the brain, strongly suggest the possibility that such dreams serve powerful adaptive functions that can be explained and understood in evolutionary terms.

These are all features that favorably distinguish the study of big dreams from the study of meditation and prayer. Identifying these features is not, however, meant to diminish the significance of contemplative practices, nor to advocate big dreams as a new paradigm for uncovering the "essence" or "origin" of religion. What is being highlighted here is the cross-cultural and historical fact that extraordinary dreaming is a prominent religious phenomenon that must be accounted for by any theory aspiring to a comprehensive view of human religiosity. The evidence to support this view will be presented in the two main sections of the chapter. First will be an overview of the religious dimensions of dreaming as found in the traditions of Christianity, Islam, and Buddhism. Second will be an examination of the neuropsychological dimensions of dreaming as discovered by modern scientific researchers using a variety of experimental tools and methods. In the final section, I will draw upon both religious and neuropsychological approaches to propose a newly integrated understanding of the formation, function(s), and interpretation of big dreams.

Religious Dimensions of Dreaming

In the last few years, several major studies have been published about dreams in Buddhism, Islam, Judaism, Christianity, and the indigenous religious traditions of Africa, the Americas, and the South Pacific (Descola 1993; Ewing 1989; Gregor 1981; Harris 1994; Hermansen 2001; Irwin 1994; Jedrej and Shaw 1992; Jung 1974; Lohmann 2001; Mageo 2003; O'Flaherty 1984; Stephen 1995; Szpakowska 2001; Tedlock 1987, 2001; Trompf 1990; Young 1999). It would take much longer than space allows to describe the dream beliefs, practices, and experiences of all the world's religious traditions, so I will focus on three of the largest and most globally widespread ones—Christianity, Islam, and Buddhism.

Christianity

The founder of the religion, Jesus of Nazareth (died ca. 30 CE), is not reported in the New Testament as having experienced any dreams, possibly because he had none to report and possibly because the writers of the Gospels wanted to emphasize his uniquely immediate connection to God, even greater than the connection enjoyed by the Old Testament prophets Abraham, Jacob, and Joseph [though in the Hebrew Bible, Moses is said to have spoken to God "mouth to mouth, clearly, and not in dark speech" like dreams (Num. 12:8)]. One of the Gospels presents the story of Jesus' birth as directly shaped by several heaven-sent dreams warning his parents of threats against them and the life of their newborn child (Matt. 1:20–24, 2:12, 13:22). Dreams also serve as a source of divine guidance in the missionary work of Paul, directing him to visit certain regions, warning him of potential dangers, and reassuring him in times of anxiety (Acts 16:9, 18:9–11, 23:11, 27:23–25). Indeed, Paul's original conversion experience on the road to Damascus (Acts 9) became the preeminent model for conversion to the Christian faith, and over the centuries, wherever Christianity has spread, people have reported intense, revelatory dreams in which God appeared to them and compelled them to adopt the new religion. This is especially striking in the anthropological literature from Africa and the South Pacific, where dreams play a crucial role in the rejection of native religious beliefs and the adoption of Christian missionary teachings (Curley 1983; Fisher 1979; Jedrej and Shaw 1992; Kelsey 1991; Lanternari 1975; Lohmann 2001; Miller 1994; M'Timkulu 1977; Osborne 1970; Peel 1968).

One of the early converts to Christianity was Jerome (ca. 347–420), a well-educated young man who reported a startling dream in which he was brought before the judgment seat of God, accused of being "a follower of Cicero and not of Christ," and scourged by repeated strokes of the lash until he made a vow to reject all worldly books and devote himself exclusively to Christianity (Kelsey 1991). Despite the powerful impact of this dream on his own life ("thenceforth I read the books of God with a zeal greater than I had previously given to the books of men"), or perhaps precisely *because* of its transformative impact, Jerome, who went on to become a powerful church bishop, denounced the dream incubation practices that were still popular throughout the Mediterranean world and condemned people who "sit in the graves and the temples of idols where they are accustomed to stretch out on the skins of sacrificial animals in order to know the future by dreams, abominations which are still practiced today in the temples of Asklepius" (Miller 1994).

This basic theological tension—between a respect for the divine power of dreams and a deep fear of being deceived or misled by them—pervades the Christian tradition right into the present day. For many theologians, the key concern is the inescapable nature of sexuality in dreams. St. Augustine (354–430 CE) laments in Book X of his autobiographical *Confessions* (1991) that even though he has, like Jerome, converted from paganism to Christianity and has pledged himself to a life of chastity, he is still plagued by disturbingly vivid sexual dreams: "But in my memory of which I have spoken at length, there still live images of acts which were fixed there by my sexual habit. These images attack me. While I am awake they have no force, but in sleep they not only arouse pleasure but even elicit consent, and are very like the actual act. The illusory image within the soul has such force upon my flesh that false dreams have an effect on me when asleep, which the reality could not have when I am awake. During this time of sleep surely it is not my true self, Lord my God?" St. Thomas Aquinas (1225–1274) follows Aristotle in explaining many dreams (particularly sexual dreams) as nothing more than the unfortunate by-products of ordinary bodily processes, and he adds to Aristotle the Christian belief that dreams of sexual temptation are sent by the devil to lure the faithful astray. However, in the *Summa Theologica*, Aquinas admits that divine revelations in dreams are possible: "This can be seen in the fact that the more our soul is abstracted from corporeal things, the more it is capable of receiving abstract intelligible things. Hence in dreams and alienations of the bodily senses, divine revelations and foresight of future events are perceived the more clearly" (1.Q–12.11) (Kelsey 1991). Protestant reformer Martin Luther (1483–1546) refused to have anything to do with the seeming revelations of dreams: "I care nothing about visions and dreams. Although they seem to have meaning, yet I despise them and am content with the sure meaning and trustworthiness of Holy Scripture" (1945).

Despite these theological misgivings, Christian interest in dreams continued to exist at the popular level, although people had reason to be prudent about talking too openly about their dreams—the *Malleus Malificarum*, the fifteenth-century manual used by the inquisitors to detect and hunt down witches, lists dreams as one medium by which demons operate (Kramer and Sprenger 1971). For the most part, dream incubation practices continued unabated, and in many cases the rituals were performed in Asklepian temples that had been reconsecrated as shrines to Christian saints and martyrs (Achmet 1991). New manuals of dream interpretation were produced, including a text called the *Oneirocriticon* (1991), written by a Christian named Achmet in tenth-century Byzantium, who essentially copied the basic framework of Artemidorus's manual and filled it in with references to Christian scriptural and iconographic sources. These ritual and interpretive practices gave new theological expression to the same themes found in the earlier traditions regarding dreams as a source of revelation, guidance, warning, healing, and possible deception.

Looking at Christianity today in Western Europe and North America, the dream theory of C.G. Jung can be viewed as a psychological elaboration of those basic religious themes (Jung 1965, 1974), and his theory has in turn reignited Christian theological and popular interest in dreaming (Hall 1993; Kelsey 1991; Sanford 1982; Savary et al. 1984; Taylor 1983, 1992).

Islam

The Muslim faith emerged in seventh century CE Arabia as a profound revisioning of early Jewish and Christian traditions. One theme the religion's founder, the Prophet Muhammad (570–632), drew from the scriptures of those two traditions was a reverence for dreaming. In Islam's foundational text, the *Qur'an* (610–632), dreams serve as a vital medium by which God communicates with humans, offering divine guidance and comfort, warnings of impending danger, and prophetic glimpses of the future. Muhammad describes several of his own dreams in the *Qur'an* and in the *hadith* (the collected sayings of the Prophet). For example, he tells his followers (many of whom were battle-tested warriors), "I saw in a dream that I waved a sword and it broke in the middle, and behold, that symbolized the casualties the believers suffered on the Day [of the battle] of Uhud. Then I waved the sword again, and it became better than it had ever been before, and behold, that symbolized the Conquest [of Mecca] which Allah brought about" (Hermansen 2001). Muhammad also made a practice of asking his followers to share their dreams so he could interpret them, and he gave them instructions on how to purify themselves so they could have good, heaven-sent dreams [an incubation practice that came to be known as *istikhara*; see Hermansen (2001), Trimingham (1959), Von Grunebaum and Callois (1966)].

Inspired by these teachings from the *Qur'an* and the *hadith*, Muslim philosophers and theologians over the centuries developed new techniques and conceptual frameworks for the practice of dream interpretation (Lamoreaux 2002). The most famous of these interpreters was Ibn Sirin, whose name was reverently attached to dream interpretation manuals long after his death in 728 CE. The tradition of Ibn Sirin emphasized that the same dream image could have different meanings for different people, and many of his interpretations hinged on the identification of a special connection between a dream image and a passage from the *Qur'an*. The *Oneirocriticon* of Artemidorus was translated into Arabic in 877 CE, and it gave further stimulus to Muslim dream theory and practice. During an era of tremendous vitality in Islamic culture, the philosophers Ibn Arabi (1164–1240) and Ibn Khaldun (1332–1402) drew on both Muslim and Graeco-Roman dream traditions to devise a basic typology of dreams, which has shaped Muslim beliefs to the present day. Here is Ibn Khaldun's (1967) rendering of it:

> Real dream vision is an awareness on the part of the rational soul in its spiritual essence, of glimpses of the forms of events . . . This happens to the soul [by means of] glimpses through the agency of sleep, whereby it gains the knowledge of future events that it desires and regains the perceptions that belong to it. When this process is weak and indistinct, the soul applies to it allegory and imaginary pictures, in order to gain [the desired knowledge]. Such allegory, then, necessitates interpretation. When, on the other hand, this process is strong, it can dispense with allegory. Then, no interpretation is necessary, because the process is free from imaginary pictures . . . One of the greatest hindrances [to this process] is the external senses. God, therefore, created man in such a way that the veil of the senses could be lifted through sleep, which is a natural function of man. When that veil is lifted, the soul is ready to learn the things it desires to know in the world of Truth. At times, it catches a glimpse of what it seeks . . . Clear dream visions are from God. Allegorical dream visions, which call for interpretation, are from the angels. And "confused dreams" are from Satan, because they are altogether futile, as Satan is the source of futility.

The dream traditions of Islam have proven remarkably durable, as Muslims today in countries all around the world fully accept that basic typology and continue to look to their dreams for divine reassurance, future guidance, physical and emotional healing, and spiritual initiation [particularly with Sufi mystical practice—see Ewing (1989)]. Contemporary Muslims make use of an abundant literature of popular dream interpretation manuals, and they continue to practice dream incubation at shrines of deceased saints (Hermansen 2001; Hoffman 1997; Shaw and Jedrej 1992). Evidence of continued Muslim belief in the prophetic value of dreams during times of military conflict appears in the videotape publicly released in December 2001 in which Osama bin Laden and a group of followers discuss dreams that anticipated the September 11 attack on the World Trade Center and the Pentagon (see Bulkeley 2003).

Buddhism

This tradition emerged in fifth–sixth B.C. India, and its beginning is attributed to the decision of a prince named Gautama to renounce his privileged life and meditate under a Bo tree, which led him to become a Buddha, literally "one who has awakened" to the truth (Bowker 1997). Buddhist interest in dreaming starts with Gautama's conception. His mother, Queen Maya, is said to have had a dream by which she was impregnated with a child who was destined to become either a universal king or, if he renounced the world, a Buddha. Here is one version of Queen Maya's dream:

> At that time the Midsummer festival (Asalaha) was proclaimed in the city of Kapilasvatthu . . . During the seven days before the full moon Mahamaya had taken part in the festivities . . . On the seventh day she rose early, bathed in scented water, and distributed alms . . . Wearing splendid clothes and eating pure food, she performed the vows of holy day. Then she entered her bed chamber, fell asleep and saw the following dream: The four guardians of the world lifted her on her couch and carried her to the Himilaya mountains and placed her under a great sala tree . . . Then their queens bathed her . . . dressed her in heavenly garments, anointed her with perfumes and put garlands of heavenly flowers on her . . . They laid her on a heavenly couch, with its head towards the east. The Boddhisattva, wandering as a superb white elephant . . . approached her from the north. Holding a white lotus flower in his trunk, he circumambulated her three times. Then he gently struck her right side, and entered her womb. (Young 2001)

This story, which dramatically expands on the earlier Hindu theme of the cosmic creative power of dreaming, has been included in nearly every biographical text on the Buddha, and it has frequently been reproduced in painting, sculpture, and other forms of iconography. One of its noteworthy features is its ritual context, which reflects an idealized version of dream incubation. Indeed, there is abundant evidence throughout Buddhism's history of practices devoted to dream incubation, practices that in various ways model themselves on what Queen Maya did in terms of purifications, prayers, devotional activities, astronomical timing, and special sleeping conditions (Laufer 1931; O'Flaherty 1984; Ong 1985; Wayman 1967; Young 1999, 2001).

In keeping with the popular practice of dream incubation, several Buddhist medical texts were written to explain the origins and meanings of different kinds of dreams. Regarding origins, Buddhist explanations included disturbances of bodily humors, reflections of daily experience, personal fantasies, future prophecies, divine influences, strong emotions, sexual desire, and demonic attack (O'Flaherty 1984; Ong 1985; Wayman 1967; Young 1999). Other prominent texts spoke of dreams as a means by which Buddhist monks managed to convert Indian Kings to their faith. "Doubtless, these stories [of monks interpreting kings' dreams] reflect actual Buddhist practice, for other sources corroborate the tradition that Buddhists converted many Indian Kings by a combination of public debate, private counseling, ... and a kind of primitive psychoanalysis" (O'Flaherty 1984).

Buddhist monks, in the context of seeking to understand the illusory nature of all reality, took special interest in developing the ability to maintain awareness and intentionality within the dream state (Gillespie 1988; Norbu 1992; Rinpoche 1998; Young 1999). For example, the Tantric Buddhist Naropa (eleventh century C.E.), who left his home as a youth to seek out a teacher whose name he had been given in a dream, taught his followers a precise method for cultivating awareness in dreaming, including special visualizations, breathing exercises, and sleep postures (Gillespie 1988). In recent years, these Buddhist dream practices have become the subject of interest to Western researchers in "lucid dreaming," who have sought parallels between the conscious dream experiences of Buddhist monks and contemporary Westerners (Gackenbach and LaBerge 1988; LaBerge 1985). Their efforts have been met with the response that Buddhist dream experiences cannot be adequately understood outside the religious context in which they were generated (Doniger 2001; Lama 1997; Norbu 1992; Rinpoche 1998; Young 1999).

Summary

Christianity, Islam, and Buddhism are not the only religious traditions in the world, and of course there is much more to be said about the other traditions in relation to dreaming. But we have covered enough material to advance the following general observations about the dynamic interplay of dreaming and religion through history:

1. *Widespread public interest.* Although surviving dream reports from ancient times generally come from the social elite (prophets, kings, priests,

military leaders), there is ample evidence of strong popular interest in dreams among all segments of society in virtually every known culture—men and women, young and old, rich and poor. *Everybody dreams.*

2. *Different types of dreams.* All cultures make distinctions among different types of dreams. Some dreams are attributed to bodily processes, others to residual thoughts and feelings from the day, and still others to the influence of spiritual beings, powers, and realities. Although Jung's anthropology was clouded by Euro-centrism, he was essentially right that most cultures and religious traditions recognize that a rare but memorable percentage of dreams are *different* from other dreams—more powerful, more meaningful, more significant, with longer lasting effects on the individual's life.

3. *Qualities of dreaming.* Two qualities are especially prominent—the visual and the emotional. Most cultures emphasize the sense of sight in dreams; they speak of "seeing" dreams, and they connect dreaming at night to "visions" seen during the day. The strong emotional qualities of dreaming are also widely recognized, particularly the way dreams bring forth deep passions and powerful desires, often of a taboo nature.

4. *Origins of dreaming.* Although each culture has its own distinct understanding of the origins of dreaming, nearly all of them revolve around the paradox that dreaming is both passive and active, something people receive and create, something coming from outside and inside at the same time. The origins of dreaming are often interwoven with the origins of *everything*, for example, in myths of the cosmos as a divine dream.

5. *Functions of dreaming.* In keeping with the belief in different types of dreams, all cultures believe dreaming serves several different functions, among which the most important are anticipating the future, warning of danger, heralding new births, mourning death and other losses, envisioning sexual pleasure, healing illness, giving moral guidance, and providing divine reassurance in times of distress. In rare but widely reported cases, dreams are seen as serving the additional function of sparking religious conversions, that is, provoking radical transformations of personality and/or spiritual orientation.

6. *Dream content.* Mindful of the limited confidence we can have in the veracity of subjective dream reports, we can identify the following as the most prominent elements of dream content cross-culturally: characters personally known to the dreamer; characters with supernatural or divine qualities; animals, especially snakes; sexual activity; flying and falling; fighting, conflict, and aggression; bodily functioning (e.g., eating, excreting, illness).

7. *Dream interpretation.* In every known culture, there are people who have devoted extensive amounts of time and attention to the practice of dream interpretation. Most interpreters use a combination of personal details, common cultural symbols and metaphors, and linguistic analysis (e.g., wordplay, punning, parallels, and oppositions). These interpretive

practices are almost always accompanied by an awareness of the potential to be deceived or misled by dreams.

8. *Dream rituals*. Dream rituals are found all over the world, with the basic aims of fending off bad dreams and evoking good ones. The most prominent of these rituals are those intending to "incubate" a good dream by means of purifications, devotions, prayers, sleeping at a special time and in a special place (e.g., a temple, cave, mountain top, shrine, grave), and using a special body position or posture (Patton 2002).

This is what can be learned from the investigations of cross-cultural history. To sum it up in a phrase, *dreaming is a primal wellspring of religion*.

Neuropsychological Dimensions of Dreaming

The next question to address is how these findings from religious studies relate to the contemporary research of scientists who study sleep and dreaming. Unfortunately, most of these researchers pay no attention whatsoever to the history of religions, regarding the whole subject of religion as irrelevant to their work. One of my goals here is to show why that attitude is a foolish and self-defeating prejudice. We may begin by tracing the history of modern scientific studies of dreaming.

In the early 1950s, University of Chicago physiologist Nathaniel Kleitman and his student Eugene Aserinsky discovered quite by accident that during sleep humans experience regular periods of highly coordinated eye movements, intensified brain activity (as measured by the electroencephalograph, or EEG), irregular breathing, loss of muscle tone, increased heart rate, and increased blood flow to the genitals (leading to penile erections in men and clitoral swelling in women). William Dement, who soon joined the Aserinsky–Kleitman research team, coined the term "rapid eye movement" or "REM" for these stages of sleep, contrasting them with "non-rapid eye movement" or "NREM" stages of sleep. Of special interest to these researchers was the fact that subjects in the sleep lab who were awakened during a stage of REM sleep were much more likely to report a dream than people awakened during a state of NREM sleep. The REM dream reports frequently involved "strikingly vivid visual imagery," suggesting that "it is indeed highly probable that rapid eye movements are directly associated with visual imagery in dreaming" (Aserinsky and Kleitman 1955) (Aserinsky and Kleitman 1953; Dement 1972).

Following up on Aserinsky and Kleitman's discovery, researchers soon found that a typical night's sleep for an adult human follows a regular alteration of REM sleep and four distinct states of NREM sleep. On average, an adult human experiences each night four–five periods of REM sleep, of between 10 and 60 minutes each, for a total of approximately one-quarter of their total sleep time. Looking at sleep in other animals, researchers found that all mammals (with the apparent exception of the spiny anteater and the bottlenose dolphin) experience regular cycles of REM and NREM sleep; birds apparently have a kind of REM sleep, but reptiles do not. Furthermore, in most mammals (including humans), the percentage of REM sleep is higher among newborns than adults. Taken together, these findings suggest that REM sleep serves some adaptive function, developed over the course of evolutionary history, in the activation and maturation of the mammalian brain (Dement 1972; Flanagan 2000; Jones 1978; Jouvet 1999). Some of the adaptive functions that have been suggested include storing memories and newly learned skills (Smith 1993), processing information with a high emotional charge (Greenberg 1972), responding to waking life crises (Cartwright 1991), making wide-range connections in the mind (Hartmann 1995), and practicing responses to potential waking life threats (Revonsuo 2000).

Perhaps the most surprising finding in the early years of research on REM and NREM sleep was the intensity of activation in the brain during REM. Contrary to expectations that sleep was a time of quiescence, researchers found that in fact the brain is designed to engage in a cyclical pattern of highly complex and dynamic activities, whose intensity is often greater than that of brain functioning during wakefulness. "If we were not able to observe that a subject is behaviorally awake in the first case and sleeping in the second, the EEG alone would not be capable of indicating whether the subject is awake or [in REM sleep]" (Hobson 1988). Much of this intense neural activity can be attributed to the phasic discharge of PGO (ponto-geniculo-occipital) waves originating in the brain stem and spreading throughout the brain. "PGO waves represent unbridled brain-cell electricity. 'Brain-stem lightning bolts' is hyperbolic but to the point" (Hobson 1999).

One of the most influential theories regarding the relationship between REM sleep and dreaming comes from J. Allan Hobson, whose Activation–Synthesis model (Hobson 1988; Hobson and McCarley 1977) portrays an essentially unidirectional, bottom–up process of dream formation. In this view, dreaming is "activated" by the neurochemical processes of REM sleep that, as an accidental by-product, lead to the activation of random feelings, images, and memories in the sleeping mind. These arbitrary bits of mental content are "synthesized" by higher brain functions that struggle to make sense of them, leading to the imposition of meaning on what is fundamentally nonsensical material. The fact that this process starts in the brain stem and then moves to the forebrain explains why dreams are so bizarre, disjointed, and emotionally turbulent. As Hobson and McCarley say in their 1977 paper, "[T]he forebrain may be making the best of a bad job in producing even partially coherent dream imagery from the relatively noisy signals sent up to it from the brain stem" (p. 1347).

In more recent work, Hobson and his colleagues (1990) have examined in greater detail the neurochemical dimensions of REM sleep and dreaming. They have found that the transition from waking to sleeping is mediated by a shift in the relative balance of amine and acetylcholine molecules in the brain, with the aminergic system dominant in waking and the cholinergic system dominant in sleeping. Hobson says the loss of aminergic inhibition and the reciprocal gain in cholinergic excitation (a process initiated in the brain stem) offer a more precise explanation of why dreams are so filled with disorientation, visual hallucination, distractability, memory loss, and lack of self-awareness (pp. 43–44, 62). The reciprocal interactions of aminergic and cholinergic systems determine the mode of a person's brain/mind activity across a spectrum from "rational, logical, and self-aware (the traits of being awake), to delusional, illogical, and unreflective (the traits of dreaming)" (p. 66). Hobson also says dreaming, in both its psychological deficiencies and its neurochemical dominance by the cholinergic system, is akin to organic mental syndromes like Alzheimer's disease and Korsakoff's psychosis: "Dreaming, then, is not *like* delirium. It *is* delirium. Dreaming is not a *model* of psychosis. It *is* a psychosis. It's just a healthy one" (p. 44, italics in original). Hobson mentions his interest in a new class of anti-psychosis drugs, monoamine oxides inhibitors, whose therapeutic effectiveness involves strengthening the aminergic system and thereby improving the patient's mental functioning while awake, although at the cost of suppressing REM sleep. Noting that the people who take these drugs do not seem to suffer any obvious problems due to the cessation of their dreaming, Hobson speculates, "Maybe dreams really are an epiphenomenon, a secondary and unnecessary outcome of an underlying process. Maybe they are not important psychological experiences. This conclusion is already suggested by the near total amnesia that we all normally have for our dreams" (p. 282).

Even though Hobson's research is most directly aimed at refuting Freud's psychoanalytic theory of dream formation, he does grant that dreaming reflects an impressive degree of creative power in the human brain–mind system. Furthermore, he allows for the possibility that dreams can provide therapeutically valuable information in terms of revealing the primary emotional concerns of the dreamer, and he offers examples of his own dreams to illustrate this (Hobson 1988). If we set aside his barbed rhetorical flourishes against Freudian orthodoxy and perform a closer analysis of his work, we find it reveals a surprisingly high regard for the functional importance of sleep and dreaming. Consider the following list of functions he proposes in *Dreaming and Delirium* (1999):

1. By resting in REM sleep, the amines are better able to restrain the potentially disruptive effects of acetylcholine during the day. "[M]y theory is homeopathic: we have a sleep seizure in order to avoid a waking seizure. We dream so as not to hallucinate" (p. 135).
2. "During this unique brain–mind state of REM sleep, then, our circuits are being cleared and our battery is being recharged . . . REM, it seems, is some sort of supersleep" (p. 191).
3. REM sleep serves to restore the body's capacity for temperature regulation (p. 188), while NREM sleep enhances the immune system's ability to fight infection (p. 190).

4. The automatic cycling of waking, sleeping, and dreaming is directly involved in processing and storing new information (p. 30) and in forming, consolidating, and associating memories (pp. 114–117). "One function of my dream might thus be to associate my memories and so increase their versatility and redundancy" (p. 114).
5. REM sleep and dreaming function to exercise basic motor programs (especially the fight-or-flight system and the orienting response). "The motor programs in the brain are never more active than during REM sleep . . . [in order] to prevent their decay from disuse, to rehearse for their future actions when called on during waking, and to embed themselves in a rich matrix of meaning" (p. 117). This function of exercising basic motor programs extends to extremely positive dreams like those of flying: "our ecstatic dreams equip us for the elation and joy of life" (p. 152).
6. Dreaming reveals "the wondrous brain–mind with all of its creative power" (p. 56), a power that is rooted in the fundamentally random, chaotic, nonsensical nature of dream activation: "human creativity depends on a natural tension between the chaos and the self-organization of brain–mind states" (p. 25). Hobson says the brain–mind's drive toward self-organization begins *in utero* (pp. 141–142) and explains both the patent absurdity of most dreams and the occasionally startling creativity of a few of them. "A remarkable point is that some dreams do take up recurrent themes of undoubted significance to the life of the dreamer. Some seem to repeat almost exactly. How do we square this paradox? What part of dreaming is determined and what part is not? The only answer we have is chaos" (p. 216).
7. In light of current neuroscientific discoveries, Hobson is willing to agree that "Freud was right" in at least this sense: "dreams are trying to tell us something important about our instincts (sex, aggression), our feelings (fear, anger, affection), and our lives (places, persons, and times)" (p. 90).

Hobson's Activation–Synthesis model, with its emphasis on the unidirectional, bottom–up influence of brain stem processes on dream formation, has been challenged by Mark Solms, a clinical neurologist with training in psychoanalysis at London Hospital Medical College, who has carefully studied the dreams of 361 patients suffering a variety of brain lesions (Solms 1997). Solms found that almost all of the patients suffered one of four distinct "syndromes" or patterns of disrupted dreaming: "global anoneira," a total loss of dreaming; "visual anoneira," cessation or restriction of visual dream imagery; "anoneirgnosis," increased frequency and vivacity of dreaming, with confusion between dreaming and reality; and "recurring nightmares," an increase in the frequency and intensity of emotionally disturbing dreams (pp. 235–236). He analyzed these four syndromes in comparison with each other and in comparison with those patients (N = 24) who, despite suffering serious forms of brain damage, experienced no disruptions in their dreaming. On this basis, Solms has made several claims about what specific regions of the brain are responsible for the formation of dreams:

1. Both hemispheres—Solms says that global anoneira "can occur with strictly unilateral lesions in either hemisphere" (p. 42), suggesting that both left and right hemispheres make necessary contributions to normal dreaming. This refutes arguments that the right hemisphere has exclusive control over dreaming, and Solms says his findings support Doricchi and Violani's theory that both hemispheres play functional roles, with the right hemisphere providing the "perceptual 'hard grain' . . . which is probably indispensable for the sensorial vividness of the dream experience" and the left hemisphere providing the "cognitive decoding of the dream during its actual nocturnal development" (p. 217).
2. Global cessation of dreaming can occur after damage to several different regions of the brain, which Solms says supports a distributed, non-modular view of brain–mind functioning: "[C]omplex mental faculties such as reading and writing (and, we might add, dreaming) are not localized within circumscribed cortical centers . . . [They] are subserved by complex functional systems or networks, which consist of constellations of cortical and subcortical structures working together in a concerted fashion" (pp. 47–48).
3. Solms finds that language disorders such as aphasia were no more common among his dreaming patients than among his non-dreaming patients, and he comments "the high incidence of preserved dreaming among our aphasic patients . . . demonstrates that loss of the ability to generate language does not necessarily imply loss of the ability to generate dreams" (p. 161). Solms does, however, acknowledge that deeper semantic disorders related to language may be crucial to normal dreaming.
4. Solms finds a "double dissociation" between visual imagery in dreams and visual perception in waking: patients with visual problems had normal dreaming, and patients with nonvisual dreaming had normal visual abilities (p. 132). Specifically, he finds that brain areas V1 and V2, which are crucial for the processing of external visual signals, are not necessary for the generation and maintenance of normal dream imagery. This means, Solms says, that "dream perceptions are not perceptions; they are representations of perceptions" (p. 169), and he claims his findings refute theories that explain dreams as "internally generated images which are fed back into the cortex as if they were coming from the outside" (p. 135).

Taking these points together, Solms proposes a model of the normal dream process in which several particular brain regions make functional contributions: basal forebrain pathways, which contribute "a factor of appetitive interest" in terms of curiosity, exploration, and expectation; the medial occipito-temporal structures, which contribute visual representability; the inferior parietal region, which contributes spatial cognition; the frontal–limbic region, which adds "a factor of mental selectivity" in separating dreaming from waking (damage to this region leads to reality-monitoring problems in waking); and temporal–limbic structures, which contribute "a factor of affective arousal" and may, in their seizure-like behavior during sleep, be the ultimate source of dream generation (pp. 239–244). Solms claims his neuropsychological findings refute Hobson's Activation–Synthesis theory:

> [T]he neural mechanisms that produce REM are neither necessary nor sufficient for the conscious experience of dreaming . . . [N]ormal dreaming is impossible without the active contribution of some of the highest regulatory and inhibitory mechanisms of the mind. These conclusions cast doubt on the prevalent notion—based on simple generalizations from the mechanism of REM sleep—that "the primary motivating force for dreaming is not psychological but physiological" (Hobson and McCarley 1977). If *psychological forces* are equated with *higher cortical functions*, it is difficult to reconcile the notion that dreams are random physiological events generated by primitive brain stem mechanisms, with our observation that global anoneira is associated not with brain stem lesions resulting in basic arousal disorders, but rather with parietal and frontal lesions resulting in spatial–symbolic and motivational–inhibitory disorders. These observations suggest that dreams are both generated and represented by some of the highest mental mechanisms. (Solms 1997, pp. 153, 241–242, italics in original)

Unfortunately, the alternative offered by Solms to replace Activation–Synthesis turns out to be a pale restatement of Freud's sleep–protection theory (Freud 1965). Solms, unlike Freud, takes no interest whatsoever in the study of dream content (Solms 1997). However, he fully endorses Freud's sleep-protection model of dream formation, and the ultimate, though modestly stated, intention of Solm's 1997 book is to promote a neuropsychological revival of psychoanalytic theorizing.

In considering Hobson's work earlier, I suggested we distinguish his antipsychoanalytic polemics from his actual research findings. Here I suggest we make a similar distinction between Solms's pro-psychoanalytic theorizing and his clinico-anatomical data. From the perspective of our interest in the study of big dreams, one immediate point of interest in Solms's work is the syndrome of "excessive dreaming," or aneoneirognosis. This syndrome involves people experiencing intensely emotional and hyperrealistic dreams, often with unusual characters and other content features [although Solms takes no interest in content, his clinical descriptions of the 10 patients who had this syndrome include dream reports of meeting deceased loved ones (pp. 185–186), visiting the "pearly gates" (p. 178), visiting a very beautiful place

(p. 183), and having a black snake crawl into the dreamer's vagina (p. 192)]. Both in form and content, these anoneirognostic dreams are quite similar to historical and cross-cultural reports of big dreams. Although Solms's patients would probably be happy to give up their hyper-vivid dreams if they could just regain normal brain/mind functioning again, the experiential similarities of their dreams to the types of dreams most frequently reported in the world's religions is highly suggestive. If Solms is right that frontal–limbic lesions are the cause of this syndrome (pp. 197–200), this may be a key region of the brain to study in connection with the religiously oriented experience of big dreams.

Also of interest is Solms's account of the syndrome of recurrent nightmares. Patients who have damage to the brain's temporal–limbic areas are especially prone to an increase in nightmares, Solms says, because that particular kind of damage often produces intense neural discharges and seizure activities that overwhelm the ordinary functioning of the brain and inject the patients' dreams with a relentless sense of anxiety. Solms speculates that seizure activity anywhere in the brain can set the dreaming process in motion (p. 211). This suggests that a potentially fruitful way to study big dreams, some of which are terribly frightening and nightmarish, is to investigate their connection to states of heightened temporal–limbic activation, such as occurs in epilepsy and perhaps other kinds of religious experience. Many common features of big dreams—for example, the number and frequency of emotions, the activation of instinctive behavior patterns like fight-or-flight, the strong physiological arousal upon awakening—may be attributable in part to the extraordinary activation of certain neural processes in the temporal–limbic regions of the brain.

Recent neuroimaging studies [usefully summarized in Pace-Schott et al. (2003)] have supported many of Hobson's and Solms's basic insights by showing that both REM and NREM sleep involve wide-ranging chemical, electrical, and metabolic changes in the brain. This, I think, is a remarkable discovery that has not yet been fully appreciated: our lives involve a natural cycling of three very different states of brain–mind functioning—waking, NREM, and REM sleep. No theory of human nature and psychological functioning that fails to account for all three of these distinct modes of being can be accepted.

An emerging research consensus points to two basic changes in the brain during REM sleep, as compared with waking: a relative deactivation of the prefrontal cortex (especially the dorsolateral prefrontal cortex) and a relative activation of the anterior paralimbic REM activation area. The former region is central to the "executive" functions of focused attention, volition, and working memory, and the latter region is connected to emotion and instinct. This conveniently dovetails with the idea that dreaming is lacking in self-awareness, self-control, selective attention, and reality testing, and is pervaded by bizarre images and emotionally laden but meaningless events (the view of Hobson and Crick). That is a misleading conclusion to draw, however, for a number of reasons.

1. As David Foulkes (1962) showed in the earliest days of sleep laboratory research, dreaming occurs in NREM and REM, and attempts to distinguish sharply the content of REM dreams from NREM dreams have not been successful.

2. Later research by Foulkes (1999) on children's dreams has shown that the development of a capacity for dreaming tracks the development of high-order cognitive abilities for language, reasoning, and self-representation. This, combined with Solms's neuropsychological findings, indicates that dreaming has a much greater degree of high-order psychological structuring than is suggested by the neuroimaging data alone.

3. Several researchers [Tracey Kahan (Kahan 2000, 2001), Jayne Gackenbach (Gackenbach 1991), and Stephen LaBerge (LaBerge 1985)] have explored the phenomenon of lucidity, consciousness, and metacognition in dreaming, and their findings make clear that the "hypofrontality" of REM sleep does not necessarily preclude the development of some degree of self-awareness within the dream state. Whether or not this dreaming self-awareness is subserved by the same neural systems that operate in waking consciousness remains an open question.

4. The research literature on the content analysis of dreams (Domhoff 1996, 2001a, b; Hall 1966; Hall and Van de Castle 1966) has provided abundant empirical evidence of meaningful patterns in human dream experience. Based on the detailed, scientifically rigorous study of thousands of actual dream reports, we can see more clearly than ever that dreaming is largely continuous with our waking lives. We usually dream about the same people, places, and activities that are most important to us in our daily lives. On the whole, dreaming faithfully reflects the range and depth of our emotional concerns, so much so that accurate predictions about a person's waking life and identity can be made solely by means of analyzing the (manifest) content of his or her dreams. It is surprising how few neuroscientists take content analysis findings into account when they study dreaming—if they did, they would find that most dreams are not especially bizarre or emotionally intense, and also that dreaming usually preserves a high degree of linguistic ability, social intelligence, and sensory acuity.

5. Studies of chaos and nonlinear systems (Kahn and Hobson 1993; Kahn et al. 2000) suggest that what seems bizarre may in fact be a chaotic process unfolding along the edge of order and disorder. In his more generous moments, Hobson will grant the potential of chaos theory to help our understanding of the tremendous creativity that occasionally bursts forth from the dreaming imagination. Whether or not chaos theory as it now stands is the best way to explain this spontaneous creative power, the important point is that future dream research cannot proceed if we continue to privilege stable, rational waking consciousness (the Cartesian *cogito*) as the highest state of being, a prejudice that automatically defines any *other* mode of awareness as deviant, immature, deficient, pathological, and/or dangerous.

Conclusion: The Formation, Function, and Interpretation of Big Dreams

Let me conclude by drawing these various strands of research together and weaving them into a provisional whole.

The study of religion and brain–mind science in the early years of the twenty-first century is proceeding vigorously on at least three different fronts: the evolutionary function of religion, the neuroanatomy of religious experience, and the study of sleep and dreaming. Of these three areas, I would argue, the greatest potential for future growth lies with sleep and dreaming. Although research in the other two areas will undoubtedly produce important new knowledge in the years to come, their limitations are such that no adequate theory of religion can be developed with reference to them alone. Their findings should increasingly be supplemented with the latest advances in dream research, particularly the investigation of big dreams. This is where religious studies and brain–mind science can come together in an especially fruitful way, with a mutually advantageous deepening of knowledge about the cross-cultural phenomenon of highly significant and memorable dreams.

In a previous work (Bulkeley 1997), I characterized the major dream theories of the twentieth century (Freud, Jung, Perls, Hall, etc.) in terms of their claims about the formation, function(s), and interpretation of dreams. Here I will use that same tripartite framework to integrate the foregoing discussion of religious and neuroscientific studies of dreaming in the twenty-first century, and to suggest directions for research in coming years.

Formation

Big dreams emerge from brain/mind processes that are intimately related to, but not strictly dependent on, the neural activities of REM sleep. Big dreams involve unusually intense feedback loops between higher and lower regions of the brain/mind system, which points to the likelihood that big dreams are shaped by the same kinds of nonlinear dynamics and self-organizing tendencies that researchers have identified in other chaotic systems (e.g., the weather, natural ecologies, water currents, star formation). Although much more research is needed on this point, current neuroscientific findings suggest that several different brain regions contribute to the formation of big dreams, especially the brain stem (responsible for generating PGO waves), the frontal–limbic area (responsible for reality-monitoring), the temporal–limbic area (responsible for emotional arousal), and posterior occipital lobe (responsible for visual imagery). The ultimate generative source of a big dream is usually felt by the dreamer to lie outside his or her ordinary waking self. This is where the world's religious traditions come in, offering explanations that refer to the soul of the dreamer and/or the influence of transpersonal beings and powers. Neither the religious nor the scientific explanation can be decisively proven or disproven, and perhaps the wisest course to follow is that of William James (1958) when he suggests that religious experiences involve, on the nearer side, the subconscious workings of the individual's mind, and, on the hither side, the realm of transpersonal powers.

Function

Evidence from both religious history and brain–mind science show that big dreams serve a variety of functions: anticipating the future, warning of danger, heralding new birth, mourning death and other losses, envisioning sexual pleasure, healing illness, and stimulating creativity. Each of these functions can be understood (though not proven) in evolutionary terms as contributing to the adaptive fitness and reproductive success of *Homo sapiens*. These functions can also be understood in religious terms as the means by which divine beings and powers interact with humans and benevolently guide them through their lives. The religions of the world have also affirmed that big dreams serve other functions: enabling otherworldly journeys, teaching esoteric knowledge, offering divine reassurance, and sparking deep personality transformation. Brain–mind science has yet to examine these more specifically religious functions in any detail, but it seems reasonable to view them as further extensions of what neuroscience already knows, that is, that dreaming generates a tremendous degree of creative power, which is directly involved in promoting, preserving, and sustaining the individual's healthy brain/mind functioning.

To put my theory in a phrase, *dreaming provokes greater consciousness*. Dreaming calls forth new voices and possibilities into the dreamer's conscious awareness. Especially in the experience of big dreams, the supreme function is to stimulate the dreamer to grow in consciousness—to consider new thoughts, acknowledge new feelings, remember past experiences, and take new actions in waking life. By provoking greater consciousness, big dreams fundamentally reorient people's views of themselves and the world, and this deep reorientation inevitably extends to their relationship to ultimate reality, however they personally and culturally conceive it.

Interpretation

At a certain level, it is impossible to give a satisfying interpretation of a big dream. Such dreams almost always involve images, feelings, and sensations that exceed ordinary categories and defy translation into waking language. Nevertheless, people throughout history have tried to interpret their dreams, and the most frequently used methods are seeking associations to the dreamer's personal life, identifying connections with common cultural symbols and metaphors, and making a linguistic analysis of wordplay, punning, parallels, and oppositions (all the while remaining vigilant toward possible deception). Other people can help in the interpretive process; most cultures recognize certain individuals as specialists in discerning the meanings of dreams, just as we have Freudian and Jungian therapists today. Few neuroscientists take any interest in the interpretation of actual dreams, and Hobson is especially harsh on Freudian psychoanalysts for reading their own preconceived ideas into their patients' dreams. Although I disagree with Hobson's claim that whatever meaning a dream may have is simple and obvious, his concern about being deceived by dream interpretation is legitimate, and in fact echoes the Biblical prophet Jeremiah's warning against false dreamers who mistake their own fantasies for divine revelations. Here we confront the ultimate paradox of self-awareness: we have learned so much about the unconscious workings of our brain–mind system that we can never be sure we are not, at some level, deceiving ourselves. To make that confession is not, however, to abandon all efforts at dream interpretation. What it means is that dream interpretations involve complex aesthetic responses rather than simple mathematical calculations. The interpretation of a dream can never be proven right or wrong, only better or worse relative to the interests of the interpreter and/or the dreamer.

I close as I began, with a quote from Nietzsche—who, through his profound influence on Freud and Jung, is an unacknowledged progenitor of Western psychological dream research:

> The beautiful illusion of the dream worlds, in the creation of which every man is truly an artist, is the prerequisite of all plastic art, and an important part of poetry also . . . The aesthetically sensitive man stands in the same relation to the reality of dreams as the philosopher does to the reality of existence; he is a close and willing observer, for these images afford him an interpretation of life, and by reflecting on these processes he trains himself for life. (1967)

References

Achmet. 1991. *Oneirocriticon*. S.M. Oberhelman (trans.). Lubbock: Texas Tech University Press.
Aserinsky, E. and Kleitman, N. 1953. Regularly occurring periods of eye motility, and concomitant phenomena, during sleep. *Science*, 118, 273–274.
———. 1955. Two types of ocular motility occurring in sleep. *Journal of Applied Physiology*, 8, 1–10.
Augustine, A. 1991. *Confessions*. H. Chadwick (trans.). Oxford: Oxford University Press.
Bowker, J. ed. 1997. *The Oxford Dictionary of World Religions*. Oxford: Oxford University Press.
Bulkeley, K. 1994. *The Wilderness of Dreams: Exploring the Religious Meanings of Dreams in Modern Western Culture*. Albany: State University of New York Press.
———. 1997. *An Introduction to the Psychology of Dreaming*. Westport: Praeger.
———. 1999a. *Visions of the Night: Dreams, Religion, and Psychology*. Albany: State University of New York Press.
———. 2000. *Transforming Dreams: Learning Spiritual Lessons from the Dreams You Never Forget*. New York: John Wiley & Sons.
———. 2003. *Dreams of Healing: Transforming Nightmares into Visions of Hope*. Mahwah: Paulist Press.

———. 2004. *Revision of the Good Fortune Scale: A New Tool for the Study of "Big Dreams."* Paper read at International Association for the Study of Dreams, Copenhagen.
Cartwright, R. 1991. Dreams that work: The relation of dream incorporation to adaptation to stressful events. *Dreaming*, 1(1), 3–10.
Curley, R.T. 1983. Dreams of power: Social process in a West African religious movement. *Africa: Journal of the International African Institute*, 53(3), 20–37.
Dement, W. 1972. *Some Must Watch While Some Must Sleep: Exploring the World of Sleep*. New York: W.W. Norton.
Descola, P. 1993. *The Spears of Twilight: Life and Death in the Amazon Jungle*. New York: The New Press.
Domhoff, G.W. 1996. *Finding Meaning in Dreams: A Quantitative Approach*. New York: Plenum.
———. 2001a. A new neurocognitive theory of dreams, *Dreaming*, 11(1), 13–33.
———. 2001b. Using content analysis to study dreams: Applications and implications for the humanities, in K. Bulkeley (ed.), *Dreams: A Reader on the Religious, Cultural, and Psychological Dimensions of Dreaming*. New York: Palgrave, pp. 307–320.
Doniger, W. 2001. Western dreams about Eastern dreams, in K. Bulkeley (ed.), *Dreams: A Reader on the Religious, Cultural, and Psychological Dimensions of Dreaming*. New York: Palgrave, pp. 233–238.
Ewing, K. 1989. The dream of spiritual initiation and the organization of self representations among Pakistani Sufis. *American Ethnologist*, 16, 56–74.
Fisher, H.J. 1979. Dreams and conversion in Black Africa, in N. Levtzion (ed.), *Conversion to Islam*. New York: Holmes and Meier, pp. 217–235.
Flanagan, O. 2000. *Dreaming Souls: Sleep, Dreams, and the Evolution of the Conscious Mind*. Oxford: Oxford University Press.
Foulkes, D. 1962. Dream reports from different states of sleep. *Journal of Abnormal and Social Psychology*, 65, 14–25.
———. 1999. *Children's Dreaming and the Development of Consciousness*. Cambridge: Harvard University Press.
Freud, S. 1965. *The Interpretation of Dreams*. J. Strachey (trans.). New York: Avon Books.
Gackenbach, J. 1991. Frameworks for understanding lucid dreaming: A review. *Dreaming*, 1(2), 109–128.
Gackenbach, J. and LaBerge, S. (eds.) 1988. *Conscious Mind, Sleeping Brain: Perspectives on Lucid Dreaming*. New York: Plenum Press.
Gillespie, G. 1988. Lucid dreams in Tibetan Buddhism, in J. Gackenbach and S. LaBerge (eds.), *Conscious Mind, Sleeping Brain: Perspectives on Lucid Dreaming*, New York: Plenum Press, pp. 27–36.
Greenberg, R., Pillard, R., and Pearlman, C. 1972. The effect of dream (REM) deprivation on adaptation to stress. *Psychosomatic Medicine*, 34, 257–262.
Gregor, T. 1981. "Far, far away my shadow wandered . . . ": The dream symbolism and dream theories of the Mehinaku Indians of Brazil. *American Ethnologist*, 8(4), 709–729.
Hall, C. 1966. *The Meaning of Dreams*. New York: McGraw Hill.
Hall, C. and Van de Castle, R. 1966. *The Content Analysis of Dreams*. New York: Appleton-Century-Crofts.
Hall, J.A. 1993. *The Unconscious Christian: Images of God in Dreams*. Mahwah: Paulist Press.
Harris, M. 1994. *Studies in Jewish Dream Interpretation*. Northvale: Jason Aronson.
Hartmann, E. 1995. Making connections in a safe place: Is dreaming psychotherapy? *Dreaming*, 5(4), 213–228.
Hermansen, M. 2001. Dreams and dreaming in Islam, in K. Bulkeley (ed.), *Dreams: A Reader on the Religious, Cultural, and Psychological Dimensions of Dreaming*. New York: Palgrave, pp. 73–92.
Hobson, J.A. 1988. *The Dreaming Brain*. New York: Basic Books.
———. 1999. *Dreaming as Delirium: How the Brain Goes Out of Its Mind*. Cambridge, MA: MIT Press.
Hobson, J.A. and McCarley, R. 1977. The brain as a dream state generator: An activation–synthesis hypothesis of the dream process. *American Journal of Psychiatry*, 134, 1335–1348.
Hoffman, V. 1997. The role of visions in contemporary Egyptian religious life. *Religion*, 27(1), 45–64.
Hunt, H. 1989. *The Multiplicity of Dreams: Memory, Imagination, and Consciousness*. New Haven: Yale University Press.
Irwin, L. 1994. *The Dream Seekers: Native American Visionary Traditions of the Great Plains*. Norman: University of Oklahoma Press.
James, W. 1958. *The Varieties of Religious Experience*. New York: Mentor.
Jedrej, M.C. and Shaw, R. (eds.) 1992. *Dreaming, Religion, and Society in Africa*. Leiden: E.J. Brill.

Jones, R.M. 1978. *The New Psychology of Dreaming*. New York: Penguin Books.
Jouvet, M. 1999. *The Paradox of Sleep: The Story of Dreaming*. L. Garey (trans.). Cambridge, MA: MIT Press.
Jung, C.G. 1965. *Memories, Dreams, Reflections*. R. Winston and C. Winston (trans.). New York: Vintage Books.
———. 1974. On the nature of dreams, in *Dreams*. Princeton: Princeton University Press (Original work published in 1948), pp. 67–84.
Kahan, T.L. 2000. The "problem" of dreaming in NREM sleep continues to challenge reductionist (2-Gen) models of dream generation (commentary). *Behavioral and Brain Sciences*, 23(6), 956–958.
———. 2001. Consciousness in dreaming: A metacognitive approach, in K. Bulkeley (ed.), *Dreams: A Reader on the Religious, Cultural, and Psychological Dimensions of Dreaming*. New York: Palgrave, pp. 333–360.
Kahn, D. and Hobson, J.A. 1993. Self-organization theory and dreaming. *Dreaming*, 3(3), 151–178.
Kahn, D. Krippner, S., and Combs, A. 2000. Dreaming and the self-organizing brain. *Journal of Consciousness Studies*, 7(7), 4–11.
Kelsey, M. 1991. *God, Dreams, and Revelation: A Christian Interpretation of Dreams*. Minneapolis: Augsburg Publishing.
Khaldun, I. 1967. *The Muqaddimah*. F. Rosenthal (trans.). Princeton: Princeton University Press.
Knudson, R. 2001. Significant dreams: Bizarre or beautiful? *Dreaming*, 11(4), 167–178.
Kramer, H. and Sprenger, J. 1971. *The Malleus Maleficarum*. M. Summers (trans.). New York: Dover.
Krippner, S. Bogzaran, F. and de Carvalho, A.P. 2002. *Extraordinary Dreams and How to Work with Them*. Albany: State University of New York Press.
Kuiken, D. and Sikora, S. 1993. The impact of dreams on waking thoughts and feelings, in A. Moffitt, M. Kramer, and R. Hoffmann (eds.), *The Functions of Dreaming*. Albany: State University of New York Press, pp. 419–476.
LaBerge, S. 1985. *Lucid dreaming: The Power of Being Awake and Aware in Your Dreams*. Los Angeles: Jeremy Tarcher.
Lama, The Dalai. 1997. *Sleeping, Dreaming, and Dying*. Boston: Wisdom Publications.
Lamoreaux, J.C. 2002. *The Early Muslim Tradition of Dream Interpretation*. Albany: State University of New York Press.
Lanternari, V. 1975. Dreams as charismatic significants: Their bearing on the rise of new religious movements, in T.R. Williams (ed.), *Psychological Anthropology*. Paris: Mouton, pp. 221–235.
Laufer, B. 1931. Inspirational dreams in East Asia. *Journal of American Folk-Lore*, 44, 208–216.
Lohmann, R. 2001. The role of dreams in religious enculturation among the Asabano of Papua New Guineam, in K. Bulkeley (ed.), *Dreams: A Reader on the Religious, Cultural, and Psychological Dimensions of Dreaming*. New York: Palgrave, pp. 111–132.
Luther, M. 1945. *Luther's Works*. J. Pelikan (trans.). St. Louis: Concordia Publishing House.
Mageo, J.M. ed. 2003. *Dreaming and the Self: New Perspectives on Subjectivity, Identity, and Emotion*. Albany: State University of New York Press.
Miller, P.C. 1994. *Dreams in Late Antiquity: Studies in the Imagination of a Culture*. Princeton: Princeton University Press.
M'Timkulu, D. 1977. Some aspects of Zulu religion, in J. Newell and S. Booth (eds.), *African Religions: A Symposium*. New York: NOK Publications, pp. 13–30.
Nielsen, T. 2000. Cognition in REM and NREM sleep: A review and possible reconciliation of two models of sleep mentation. *Behavioral and Brain Sciences*, 23(6), 851–866.
Nietzsche, F. 1967. *The Birth of Tragedy*. W. Kaufmann (trans.). New York: Vintage (original work published in 1872).
Norbu, N. 1992. *Dream Yoga and the Practice of Natural Light*. Ithaca: Snow Lions Publications.
O'Flaherty, W.D. 1984. *Dreams, Illusion, and Other Realities*. Chicago: University of Chicago Press.
Ong, R.K. 1985. *The Interpretation of Dreams in Ancient China*. Bochum: Studienverlag Brockmeyer.
Osborne, K.E. 1970. A Christian graveyard cult in the New Guinea highlands. *Practical Anthropologist*, 46(3), 10–15.
Pace-Schott, E., Solms, M., Blagrove, M., and Harnad, S. (eds.) 2003. *Sleep and Dreaming: Scientific Advances and Reconsiderations*. Cambridge: Cambridge University Press.

Patton, K. 2002. *Dream Incubation: Theology and Topography*. Paper read at 19th International Conference of the Association for the Study of Dreams, June 19, Boston, Massachussetts.

Peel, J.D.Y. 1968. *Aladura: A Religious Movement among the Yoruba*. London: Oxford University Press.

Revonsuo, A. 2000. The reinterpretation of dreams: An evolutionary hypothesis of the function of dreaming. *Behavioral and Brain Sciences*, 23(6), 877–901.

Rinpoche, T.W. 1998. *The Tibetan Yogas of Dream and Sleep*. Ithaca: Snow Lions Publications.

Sanford, J. 1982. *Dreams: God's Forgotten Language*. New York: Crossroads.

Savary, L.M., Berne, P.H., and Williams, S.K. 1984. *Dreams and Spiritual Growth: A Christian Approach to Dreamwork*. Mahwah: Paulist Press.

Smith, C. 1993. REM sleep and learning: Some recent findings, in A. Moffitt, M. Kramer, and R. Hoffmann (eds.), *The Functions of Dreaming*. Albany: State University of New York Press.

Solms, M. 1997. *The Neuropsychology of Dreams: A Clinico-Anatomical Study*. Mahway: Lawrence Erlbaum.

Stephen, M. 1995. *A'Aisa's Gifts: A Study of Magic and the Self*. Berkeley: University of California Press.

Szpakowska, K. 2001. Through the looking glass: Dreams in ancient Egypt, in K. Bulkeley (ed.), *Dreams: A Reader on the Religious, Cultural, and Psychological Dimensions of Dreaming*. New York: Palgrave, pp. 29–44.

Taylor, J. 1983. *Dream Work*. Mahwah: Paulist Press.

———. 1992. *Where People Fly and Water Runs Uphill*. New York: Warner Books.

Tedlock, B. 2001. The new anthropology of dreaming, in K. Bulkeley (ed.), *Dreams: A Reader on the Religious, Cultural, and Psychological Dimensions of Dreaming*. New York: Palgrave, pp. 249–264.

———. (ed.) 1987. *Dreaming: Anthropological and Psychological Interpretations*. New York: Cambridge University Press.

Trimingham, S. 1959. *Islam in West Africa*. Oxford: Clarendon Press.

Trompf, G.W. 1990. *Melanesian Religion*. Cambridge: Cambridge University Press.

Von Grunebaum, G.E. and Callois, R. (eds.) 1966. *The Dream and Human Societies*. Berkeley: University of California Press.

Wayman, A. 1967. Significance of dreams in India and Tibet. *History of Religions*, 7, 1–12.

Young, S. 1999. *Dreaming in the Lotus: Buddhist Dream Narrative, Imagery, and Practice*. Boston: Wisdom Publications.

———. 2001. Buddhist dream experience: The role of interpretation, ritual, and gender, in K. Bulkeley (ed.), *Dreams: A Reader on the Religious, Cultural, and Psychological Dimensions of Dreaming*. New York: Palgrave, pp. 9–28.

CHAPTER ELEVEN

Religion Out of Mind: The Ideology of Cognitive Science and Religion

JEREMY CARRETTE

> As important as what a scientific programme emphasizes is what it leaves out.
> —John Dupré, *Human Nature and the Limits of Science*

Cognitive theory and neuroscience have provided many useful models for medical and therapeutic understanding, but the problem with any theory is when it oversteps the limits of its domain or operates on unexamined conceptual foundations. Classification and abstract theorizing have always been a part of the academic study of religion and there is no reason to assume that the insights of theoretical psychology cannot offer some useful frameworks of analysis. The use of complex models to represent mental processes that escape physical laws has a value in scientific discourse, as the use of complex models to understand social and economic behavior is a well-established practice. For example, psychologists of religion Robert McCauley and Thomas Lawson believe in theory and its scientific value and embrace a cognitive approach to study sociocultural systems. They have generated, along with others, some intriguing debates in the field of religion. Indeed, they have openly tried to "make trouble," because of their "frustration with the timidity that characterizes so many scholars' discussions of religious behaviour" (1990, p. 1). Just over 10 years later, they still felt frustration at the "disdainful proclamations" that discouraged "theoretical precision and empirical responsibility" (2002, p. ix).

Following on from Lawson and McCauley's innovative modeling, Jensine Andresen, in an extremely useful edited collection of essays, suggests considering the key works in the field, such as those by Sperber, Boyer, and Lawson and McCauley, "to determine if there are good reasons to doubt the validity of their postulates" (2001, p. 276). The scope of this essay does not permit a detailed study of all of these thinkers, but by taking some of the insights of the latter two thinkers as a starting point to explore the wider sociopolitics of the field, we can open some space to take theory seriously. In "frustration with the timidity" of thinking in theoretical psychology and religion, I want to celebrate theory by offering some contexts to test its use and value and, thus, prevent it from being lost in the uncritical celebration of the thesis and the inevitable uncritical antithesis.

Theory provides the representational politic of our world and it is for this reason that it requires careful scrutiny. As Heinrich Klüver notes in his introduction to a study of theoretical psychology by the economist Hayek: "It is one thing . . . to develop a theory based on the detailed experimental analysis of a particular problem; it is another thing to examine the conceptual tools of theoretical psychology itself" (1952, p. xvi). What is at stake in cognitive theory is the uncritical use of theory and the lack of understanding of how such theory operates in the sociopolitical world. Cognitive theory of religion emerges at a certain point in history and finds support because its models of the mind have explanatory use-value for the society, but what is not clear is how it is used and to whom it is of value.

In this essay, I want to explore cognitive theory in its encounter with religion by examining the conditions of knowledge that make such thought possible. By tracing key epistemological problems and theoretical lacunae, I want to raise some ideological issues at the heart of the project. In this process, I am not disputing the value of the science as such, but questioning the assumption that this knowledge takes place outside the social, political, and economics realities, which are hidden in the conceptual camouflage of the scientific theory. Theory is always in the abstract space between mental processes and the physical world and it is therefore open to the flux of conditions of how we think about the problem of thinking. At the very least, I am questioning how a cognitive theory of religion operates and at most I am suggesting that our models of being human are reflections of our contemporary sociopolitical worlds. The "reflective" space between mind and the physical world allows ideology to emerge (Pyysiäinen 2002, p. 322). The problem is not so much that human beings might have some innate, universal cognitive functions, but how human beings socially and politically *represent* the mind and its functions.

Until we appreciate the link between politics and models of ourselves, we will never appreciate how our scientific ideas are formed by metaphors and the imagination. In order to understand this crossover of ideas and appreciate its challenge to scientific knowledge, we have to question the reification of knowledge and the assumption that knowledge and politics are somehow separate. I am not simply restating a debate between rationalism and empiricism or a debate between biological determinists and social constructivists. Such framing of the debates only clouds the issues to enable exponents on either side to remain closed and reinforce narrow and entrenched positions. The reality of some universal biological facts and, equally, the reality of some cultural determinants is not under question. What is in question here is how we seek to represent the "truth" of ourselves and why the models we create emerge at certain points of history.

Howard Gardner's excellent history of the cognitive revolution admits that he does not explore the links between the fields of sociology and economics (1987, p. 7). While he was implying the possible examination of these subjects from the safety of cognitive theory, his omission reflects the way fields of study are, in part, colonized by cognitive theory as an all encompassing theoretical backdrop. What Gardner may not have considered is that, rather than assuming cognitive theory as the meta-discourse, it might be possible to identify the socioeconomic and political nature of cognitive theory itself. Such an idea is difficult, even aggressively denied as relevant to cognitive theory, because it is seen as a science in search of truth and, in such a position, truth and economics are not related! However, it is a naive and fearful science that seeks to deny the socioeconomic history of scientific knowledge, and we suffer from "disciplinary amnesia" (Carrette 2001) if we assume the conceptual framework or representational models of a science do not reflect the social world. We only have to look at history to show how the modeling of the mind changes and how such changes mirror the social world. As Kurt Danziger indicates in his study of the history of psychological theory:

> In talking about a field like scientific psychology we are talking about a domain of constructions. The sentences in its textbooks, the tables and figures in its research reports, the patterned activity in its laboratories, these are first of all products of human construction, whatever else they may be as well. If the world of scientific psychology is a constructed world, then the key to understanding its historical development would seem to lie in those constructive activities that produced it. (1990, p. 2)

There is no reason to assume we have arrived at any final truth, and some critical reflection from historical and political thinking is much required. These issues at least need answering or exploring at some level. It is, of course, an epistemological "mind-field" for the representational categories of "cognition" to be examining the representational category of "religion." The basic fallacy of the cognitive theory of religion, and related neuroscientific models, is that there is no such singular object called religion or "God" in the material world, but rather a semantic space, which is caught in the politics of representation (Fitzgerald 2000; King 1999). Although it would be grossly unfair to accuse all writers in the field of the cognitive science of religion of committing this fallacy, the fact is that few researchers advance the socioeconomic or sociopolitical issues, even as they apply for funding from key supporting agencies. It is, however, time to ask some difficult questions of the cognitive science of religion. While there is a legitimate sociological exercise to be done in examining the financial and political issues behind the science, I want to explore these from within the discourse by examining the "conceptual" politics of cognitive theory and religion. But before opening the wider issues of ideology, let us return to the more prosaic issues of religion and cognitive science and, in particular, to the work of Lawson and McCauley, in order to show how some of the theoretical problems at work in their writing open up the ideological question. I will focus my attention on their two main monographs, *Rethinking Religion* (1990) and *Bringing Ritual to Mind* (2002). My project will be twofold. I will first reveal the epistemological errors of their cognitive science of ritual and then show how these errors conceal a wider ideological position, irrespective of whether it is consciously adopted or not. In this twofold procedure, I will identify four areas of critique: (1) the social codification of the cognitive science of religion; (2) the recycling of old ideas, or disciplinary amnesia; (3) metaphorical mutation and reification; and, (4) mechanistic modeling of the mind. The politics of representation become apparent in this critical reading of the theory.

The History of Psychology, Structuralism, and the Emperor's New Clothes

At the beginning of James's *Principles of Psychology* (1890/1983), he reminisces about his graduation day (p. 16) and argues "no mechanical cause can explain this process" of imaginative re-creation of experience. His suggestion is recognition of the complexity of a single event and appreciation of the plurality of interpretation required. The public ritual of graduation holds all sorts of different registers and emotions that defy easy organization. Cognitive theory has attempted to model some of these complexities of rituals and provided useful insights in terms of "regularity" (Whitehouse 1995), "form" (Lawson and McCauley 2002), and "modes of religiosity" (Whitehouse 2002). There is no doubting the fact that some illuminating ideas have been generated in these studies, but the disturbing question that needs to be asked is *whether cognitive theory as such was required at all to offer the insights usefully generated*. If we take out the cognitive science rhetoric from the work of Lawson and McCauley, we can see that in large measure they are performing an academic ritual in concept formation or what we might call, following Jones and Elcock (2001, p. 174), "knowledge production." Lawson and McCauley acknowledge that their thinking goes back to the French tradition of anthropology in such writers as Claude Levi-Strauss and Dan Sperber, and this returns us to the question of whether much of their work is really a new version of structural anthropology (1990, pp. 10, 68). While the metaphorical language of "structure" is very strong in the work of Lawson and McCauley (2002, pp. 117, 118), it is not my aim to establish a methodological link with structural anthropology [see Greenwood (1997)], but rather to suggest that insights can be generated in the same frame of earlier studies without the language of cognitive science. As Durkheim (1912/1995) recognized, models of classification are central to cultural life and this includes the cultural representation of the mind. Whitehouse (1995, p. 220) and Lawson and McCauley (2002, p.105) note the importance of the "style of codification" for religious ritual and such ideas need to be extended to questions about the codification of mind in theory. Why is it that "cognitive processing" rather than "mental processing" is the key conceptual space? We might also ask why the metaphor of "processing" is so celebrated, but that is to step ahead of ourselves by opening questions about the politics of metaphors in science.

My allusion to Durkheim is no accident, but rather a question of the politics of disciplinary memory. Andresen (2001) usefully asks cognitive theory to return to classical thinkers in the study of religion to consider whether anything had been left out. We might also add that we might return to classical thinkers to ask whether much of what we presently see has *already been said or could be said in the terms of past discussions*, especially in relation to emotion and ritual. Could we establish the insights about religion in terms of regularity, form, and mode in the language of sociology and anthropology, without recourse to "cognitive science"? Does cognitive theory of religion say anything more than other studies on memory and repetition in religion, as in Halbwachs's (1992) idea of "collective memory" and, more recently, in Hervieu-Léger's (1993/2003) idea of "chain memory"? Is, for example, the sophisticated language of cognitive science required to make the following insight? "Whitehouse and we agree that the comparative emotional arousal these rituals produce contributes to enhanced memory for aspects of these events" (Lawson and MaCauley 2002, p. 123).

We do not require the discursive space of cognitive science to establish the relation between emotion and memory. The same conclusions have been made in previous anthropological and sociological studies of religious behavior without the politic of the "mind-stuff." Even if we introduce the question of mind into anthropology, which has long been an epistemological rather than a cognitive theory question (something I will examine later), we can find fascinating correlations between old truths and the repackaged cognitive version. As, for example, Alfred Gell indicates, Lévy-Bruhl has exercised "a powerful influence over the development of cognitive anthropology" (2001, p. 56), and arguably offered as much insight without the technical wizardry of cognitive science. Although tainted with major ethnocentric and postcolonial errors, Lucien Lévy-Bruhl's study in 1910, *Les functions mentales dans les sociétés inférieures* (translated in 1926 as *How Natives Think*), holds all the conceptual tools of mental functioning without the modern discourse of cognition. While Lévy-Bruhl's earlier work is highly problematic, Fiona Bowie (2000, pp. 221–222, 242) has rightly argued that we should always read Lévy-Bruhl in terms of the later *Notebooks on Primitive Mentality* (1949/1975), where he corrects many of the problems of his earlier work in response to critics. Lévy-Bruhl tried to identify laws and patterns of mental functioning without the rhetoric and empirical modeling of psychology; he even foresaw ideas in cognitive archeology with his sense of a "pre-logical" thought, something that relates to fascinating discussions in contemporary cognitive philosophy about nonlinguistic creatures (see Bermúdez 2002). However, what is most striking, as Scott Littleton notes in the introduction to the 1985 edition of Lévy-Bruhl's work, is the way he recognized the importance of "cognitive relativity." This is the idea "that the logic we bring to bear in our descriptions of the world is *not* universal, but rather a function of our immediate techno-environmental circumstances and our particular linguistic and ideological heritage, and that no one logic is necessarily superior to any other logic" (1985, p. vi).

This exercise in disciplinary memory is important when we see how the cognitive science of religion believes its insights are somehow unprecedented. Why reinvent the wheel? Or more precisely, what is being offered in contemporary cognitive theory of religion that is categorically different from earlier sociological and anthropological material other than the discursive reordering? Does the inclusion of the category of the cognition (in its later sense) offer any substantive insight to religious studies? It is my view that much of the cognitive theory of religion is an academic ritual of amnesia and reinvention, made more complex by the wider cultural and political energy that drives such ideas.

In order to unravel this problem, let me explore the question of whether it is possible to have a cognitive theory of the cognitive theory of religion. I follow this line of argument, provocatively of course, to unearth a wider set of problems. I have elsewhere (Carrette 2002) taken this strategy to expose the errors of Persinger's (1987) neuroscience of religion and return to it here, not simply to ask a question about the power relations between science and religion (that is now well developed), but to reveal how the language of cognitive science can easily evaporate under its own internal logic. Let us then explore the issue further in Lawson and McCauley.

Lawson and McCauley point out that concepts "earn our allegiance because of the achievements of the theories that inspire them" (2002, p. 9). They outline "predictive and problem-solving power, explanatory suggestiveness, generality, and empirical accountability" as the grounds for support. However, each of these lines of support depends on the acceptance of the cognitive model in a circular logic, such that where Ritual Act "A" is abstractly equated with Cognition "B," $B = A$, irrespective of the politics of theory construction X, Y and Z. Such logic supports Shotter's sense that an "isolation of everything from everything typifies cognitivism" (1997, p. 322), especially in the study of religion, which holds so many abstract epistemological and metaphysical entities not captured so neatly under Lawson and McCauley's (2002, p. 8) CPS (culturally postulated superhuman) agents. Thus, the predictive register of cognition, the problem solving of cognition, the explanatory nature of cognition, and the discovery of behavior are based on associated cognitive concepts. Underlying this criteria is the question of "empirically useful insights about religious ritual," but to whom is it "useful" and what is the criteria of "use-value," in Marx's (1867/1895/1995, p. 13ff) economic sense or not. As I will explore later, this "use-value" of cognitive theory is not unrelated to the political climate that creates the conditions for financial and political advancement of cognitive science.

Supporters of Lawson and McCauley may feel that this is too much of a theoretical abstraction from their own empirically grounded theoretical abstractions, and a closer reading of their work is required. Recognizing the textual nature of the theory that seeks to "empirically" ground itself, we do well to return to the texts of Lawson and McCauley. Establishing a critique by questioning the category of religion has already frustrated Lawson and McCauley and we will not pursue such a primary quagmire of critical analysis save we have no shared object of study to make cognitive theory of religion a possible subject (Lawson and McCauley 2002, pp. 9, 25). We, of course, already recognize that as believers in theory, they do not "desire to engage in debates about definitions," as long as we accept definitions of cognition, "CPS agents," "scripts," "the tedium effect," and the like (pp. 49, 184). They would have done well to remember that James had already marked out the fact that psychological processes could never be distinct processes separate from any other human mental function, because religion is the fabric of life (1902/2002, pp. 26–28). However, as the history of psychological theory is forgotten, or never read if more than 20 years old, old debates are replayed, and eventually Lawson and McCauley arrive at the well-established premise that, actually, the cognitive science of religion is not about religion, but about mental functioning generally. And so we read: "Instead, it is *only a theory* about actions that individuals and groups perform within *organized communities of people* who possess *conceptual schemes* that include presumptions about those actions' connections with the actions of agents who exhibit various *counter-intuitive properties*" (2002, p. 9, emphasis added).

It becomes clear from their account of religion that Lawson and McCauley's study applies as much to the academic institutions of Western society as any other ritual performance. It also seems that the procedures and tasks in academic practice have an uncanny resemblance to the so-called accounts of "religious" ritual. We can therefore infer that if certain academic rituals serve to establish the permanency of cognitive theory, they must, as Lawson and McCauley say of religious ritual, "*convince* participants that something profound is going on" (p. 122). Academic rituals, such as papers, conferences, journal articles, subdivisions of organizations, and large external grants, have a function to maintain the "belief" in cognitive theory as a powerful social discourse. Danziger (1990) has already identified how academic psychology depends on these external rituals to "convince" people and the community of psychologists that something important is going on. In religion, Lawson and McCauley believe there must be evidence that CPS agents are responsible for the actions that are carried out through "special agent rituals" (2002, p. 122). In academic cognitive theory rituals, CPS agents are not required because the metaphysical system is principally reductive materialist in nature and therefore only requires a "culturally postulated scientific system (with economic support)," a CPSS(E) agent. The use of intellectual parody at this point illustrates the dangers of reification in psychology and should not distract us from recognizing the power of academic agency, especially when the rituals of truth and concept formation are disassociated from their economic base. The move, as Danziger (1990, p. 185) notes, from ordinary language to jargon, and we might add abbreviation, is a part of convincing and energizing the participants and creating a new knowledge-market. The study of religion can find investment at those points where the dominant discourse in a society, in this case scientific materialism, can be seen to verify, or uphold, ideologically useful pictures of religious truth through the emergence of a new set of conceptual ideas.

Rituals of Cognitive Theory

The problem of cognitive theory relates to the totalitarian impulse in some forms of contemporary psychology to find a meta-theory to explain everything. In this situation, the power of cognitive theory is the shift of the register of cognition into scientific modeling and the aura that subsequently surrounds such science and its utterances. Such an aura demands that it be exempt from turning its own analysis on itself. In this respect, it is fascinating to note how Lawson and McCauley unnecessarily embellish their fascinating insights in the discursive codes of a science. This can be seen in the contributions of Barrett and Keil (1996) to their work. In the preface to their later study, Lawson and McCauley argue that "one of the most important" things they have been taught from Barrett and Keil is that "*bona fide* experimental *cognitive* psychologists" are working on religion. The key here is "bona fide," because it reveals an important issue about who has the authority and power to speak and shows the rituals of academic practice that maintain and structure such truth–power statements. We learn also that Barrett suggested the use of figures or diagrams in Lawson and McCauley's later study. These diagrams (Lawson and McCauley 2002, p. 185ff) service the cognitive science of religion by presenting an apparent mathematical modeling behind the observed and metaphorical inventions. This modeling is another step in convincing the participants of the importance and power of the theory. It also overcomes the "tedium" effect of academic study and adds to the aura of belief. What does it add to the narrative of Lawson and McCauley to include the diagrams, other than as a part of a ritual of persuasion? Why does mathematical modeling offer over the narrative? We could, for instance, add a diagram here with a vertical axis showing increased abstract modeling (p) and a horizontal axis showing increase in persuasion, or capital investment (q), and a diagonal line across the picture showing that an increase in p leads to an increase in q. This would overcome the cognitive tedium located in the area where p and q show the lowest register. It is not that the diagrams used by Lawson and McCauley offer any sense of mystique or hidden authority to the trained eye; it is rather that their explanatory power and worth is caught inside an ideology of representation. The Cambridge mathematician Timothy Gowers, in a short introduction to his subject, points out the nature of abstract modeling. He writes:

> When devising a model, one tries to ignore as much as possible about the phenomenon under consideration, abstracting from it only those features that are essential to understanding its behaviour . . . A graph theorist can leave behind the real world entirely and enter the realm of pure abstraction. (2002, p. 16)

Graphic modeling has its place and importance in our world, but questions as to the location and use of this science in the field of religion arise. What are the aims and importance of the graphic modeling of religion and what "features" are considered "essential" in the modeling? Who does it benefit to explain cultural or social traditions of human meaning inside such measurement? Or, we might add, what is the "use-value" of such an exercise? It is important to make clear that I am not antiscientific or devaluing the usefulness of cognitive thinking. Rather, I am trying to expose the limits of psychology and show that at the edges of cognitive science is an uncomfortable question about the politics of representation itself. It is my contention that the cognitive science of religion in Lawson and McCauley goes beyond the limits of its methodological tools, reveals ideas and insights already mapped or identified in existing discourses about religion, operates on hidden rituals of persuasion, and clouds its insights in a discourse that holds highly problematic ideological structures.

Religion, Cognition, and Representation

> [T]he representational approach is crucial, but it is not the whole story.
> —Jensine Andresen, *Religion in Mind*

Jensine Andresen concludes her edited collection of essays on the topic of cognitive theory and religion by identifying some central issues and problems for the new field of study. She recognizes the need for scholars of religion to "familiarize themselves with cognitive science's key orienting concepts and methodologies" and shows appreciation for its rapidly expanding literature (2001, p. 275). As well as appealing for much needed experimental data, she opens useful discussions on basic theories. Andresen's caution and open attitude is important for establishing some linguistic hygiene in a world that generates abstract concepts, such as the confusions she identifies between "cognitive modules" and "cognitive operators." There is also much that remains "vague" and "troubling" not just, as Andresen discusses, in specific theories such as that of Newberg and d'Aquili, but in the entire epistemological space of cognitive theory and neuroscience. Following again Andresen's recommendation to revisit classical theories of religion "to determine if any of the propositions raised by these scholars have been overlooked" (p. 276), it might be worth considering something of William James's insights into psychological terminology before any final assessment is made of the cognitive revolution in religion. James at least still held to the philosophical scrutiny that is often lacking in the broader literature on religion and cognitive science and recognized how terms can develop a life of their own even if problems persist.

> The fundamental conceptions of psychology are practically very clear to us, but theoretically they are very confused, and one easily makes the obscurest assumptions in this science without realizing, until challenged, what internal difficulties they involve. When these assumptions have once established themselves (as they have a way of doing in our very descriptions of phenomenal facts) it is almost impossible to get rid of them afterwards or to make anyone see that they are not essential features of the subject. (1890/1983, p. 148)

James's insight into the problems of psychological terminology is very much applicable to cognitive theory and neuroscience. The abstractions within these fields have a dangerous tendency to embed themselves and the problems of reification abound. Cognitive theories of religion, as we have seen, easily become lost in the rhetoric and social currency of the language, to the extent that they never ask primary questions about the nature of the theory or examine the basic assumptions behind the framing of religion within this meta-discourse.

Gardner rightly recognizes one of the key features of cognitive science as relating to the issue of representation. As he states: "Cognitive science is predicated on the belief that is it legitimate—in fact, necessary—to posit a separate level of analysis which can be called the 'level of representation' " (1987, p. 38).

At one level, all of life is determined by models of representation and there is nothing unique or different about cognitive theory, but whereas we might recognize the nature of the representational dynamic in poetry, religious symbols, or even quantum mechanics, there is an extraordinary amnesia in cognitive theory, because it either assumes the modeling is a direct representation of thought or "that there are neural mechanisms by which cognitive tasks are performed" (Harré 2002, p. 6). This reflects a problem of slippage in cognitive theory (the language of mental operations) into the language of cognitive neuroscience (the language of brain tissue). It is therefore vital, as Rom Harré indicates, to "keep in mind that a computational model is an *abstract* representation of whatever process it models" (p. 127). Once we recognize that there are other ways of modeling the mind, we can see that the use of cognitive science is not neutral or somehow more accurate than any other viewpoint (even if we concede that it may be most useful for the present political world or ideology). Lakoff and Johnson (1999, p. 129) are correct when they recognize that conceptual metaphors are one of the "greatest of our intellectual gifts," but perhaps even greater is our ability to recognize critically how our metaphors conceal lived political realities.

Cognitive Illusions: Why Humans are not Machines?

> [Computational Theory of Mind] is, in my view, by far the best theory of cognition that we've got; indeed, the only one we've got that's worth the bother of a serious discussion . . . But it hadn't occurred to me that anyone could think that it's a very *large* part of the truth; still less that it's within miles of being the whole story about how the mind works.
>
> —Jerry Fodor, *The Mind Doesn't Work that Way*

In two insightful works by Hubert Dreyfus, *Mind over Machine* (1986), with his brother Stuart Dreyfus, and, his revised 1972 work, *What Computers Still Can't Do* (1992), there is an attempt to show those areas of life and human creativity that escape the modeling of the mind as a computer. These works raise questions about Artificial Intelligence and Cognitive Simulation by highlighting issues related to intuition, common sense, skill acquisition, and learning. While not rejecting the value of computers, there is an important qualification of the value of machines and the danger of human beings servicing machines. What these works show is that computers require a certain formalism to operate and that human experience escapes this process. We might also add, with John Dupré, that, unlike machines, humans are self-conscious, exist within communities of language users, and "are fundamentally different from machines in that they have no controls" (2001, pp. 34–35, 175). In its attempt to establish control over life, like a machine, cognitive science appears to forget humans do not operate like machines. But despite the logic against representing human mental functioning as a machine, it continues because the conditions for such thinking persist. It is for this reason that we can see that cognitive science has the shape of an ideology rather than a scientific theory. It holds its own closed rationality, based on a "monistic and reductive" metaphysics (p. 8).

The advantages and disadvantages of the comparison of the mind to the computer have been well-documented—see, for example, Valentine's (1992, pp. 130–146) useful summary—and we should never assume that the analogy is ever to simple machines. However, the persistence of the machine and computational metaphors in psychology relates, as Lakoff and Johnson (1999, p. 124) make clear, to the "conceptual mapping" of a term, which remains either alive or dead. Metaphors have a currency in their ability to carry a "conceptual map" from the social world and the very fact that metaphors live and die reflects the shifts of social understanding rather than *a priori* truth. Metaphors are not so much about similarity as the politics of social discourse. In this respect, it is important to make clear that the idea of *cognition is not synonymous with computational theory* and that the metaphorical modeling of cognition in this frame can be extremely restrictive. This can be illustrated best by considering Alan Turing's famous essay on machines and intelligence, where he argues that machines could not go through the same learning process as a normal child. As he states:

> It will not be possible to apply exactly the same teaching process to the machine as to the normal child. It will not, for instance, be provided with legs, so that it could not be asked to go out and fill the coal scuttle. Possibly it might not have eyes. But however well these deficiencies might be overcome by clever engineering, one could not send the creature to school without the other children making excessive fun of it. (1950/1960, p. 27)

The socialization process is important in recognizing not only the limits of Artificial Intelligence, but also the limits of the cognitive metaphor for sociocultural processes. It is striking and revealing of the metaphorical limits that on this point, Lawson and McCauley (1990, p. 183) depart company with Chomsky's cognitive linguistic model, who they otherwise use to support their fledgling project. The shift from the cognitive to the cultural opens up some key epistemological problems, which allow us to see precisely the political motivations.

Chomsky: Epistemic States and Cognitive Processes

Lawson and McCauley appeal to Chomsky to support their introduction of cognition into cultural anthropology. Following Chomsky, they see abstraction as appropriate in being able "to discover and formulate explanatory principles" [Chomsky (1986, p. 211), quoted in Lawson and McCauley (1990, p. 68)]. They welcome the description of Chomsky's work as cognitive (Lawson and McCauley 1999, p. 183), but go beyond Chomsky in seeing a link between the cognitive and the cultural. Chomsky does not form part of Lawson and McCauley's later study and with this there is also a simultaneous, and easy, move to discussing "cognitive machinery," "cognitive foundations," "cognitive apparatus," "cognitive processes," and "cognitive systems." The question is of how the adjective is being employed in the later work and how we can distinguish this from Chomsky's own understanding.

Lawson and McCauley's study is at one level transparent (2002, p. 36). They acknowledge that they hold a valid "general explanatory strategy" (insights possible within the traditions of anthropology and sociology) to explore ritual, but they then try to locate this within "the framework of emerging cognitive science." This second move is redundant in so far that it offers very little to the study other than a resonance with a second-order discourse. What does the adjective cognitive add to the study? Take, for example, a representative sentence: "Psychologists construe the *cognitive* foundations of memory for such actions in terms of scripts. A script is a *cognitive* representation for 'a predetermined, stereotyped sequence of actions that defines a well-known situation' " (p. 49, emphasis added). By definition, the adjective cognitive is superfluous to the meaning, because the terms memory and representation sufficiently indicate the subject of the sentence. Lawson and McCauley elsewhere (pp. 39, 41) refer to Dan Sperber's *mental representations* and we begin to see that the use of the word cognitive is not simply implying the "mind," but reflecting a whole set of ideals about the mental faculties in terms of universal structure, the physical sciences, and totalized explanation. Chomsky, in his 1999 University of Siena lectures, is cautious about developments in cognitive theory and the science of the brain. He relates such issues back to the history of science. "The Galilean model of intelligibility has a corollary: when mechanism fails, understanding fails. The apparent inadequacies of mechanical explanation for cohesion, attraction, and other phenomena led Galileo finally to reject 'the vain presumption of understanding everything' " (2002, p. 51).

Chomsky is also of the opinion that the early modern scientific revolution is "the first cognitive revolution" and, perhaps, the only one deserving the name of "revolution" (p. 69). It is, as Chomsky indicates, Descartes' philosophy of mind that provides the groundwork for the later developments of neurophysiology and this usefully positions Chomsky's thinking, as well as offering a critical context for reading Lawson and McCauley (p. 69; see also Chomsky 1966). We can illuminate this issue further by considering Jerry Fodor's (2001) critical essay on computational psychology and revealing the shifts of meaning in the notion of cognition from Chomsky to later cognitive theorists.

Fodor in critique of Steven Pinker's (1997) work, highlights the difference between Pinker and Chomsky. He argues that Chomsky establishes his thesis of nativism based on knowledge and belief; it is principally a theory of knowledge that frames the linguistic issues in terms of epistemology (2001, p. 11). This position is distinguished from Pinker's position, or what Fodor calls "New Synthesis psychological theories." New Synthesis psychology builds its models of cognition from computational analogy; it is the language of cognitive processes. A mental process in this position is "a formal operation on syntactically structured mental representations" (p. 11). As Fodor notes, there is a reliance on syntax and an assumption that "thoughts themselves have syntactic structure" (p. 13). This shift from "epistemic states" in Chomsky to cognitive process is central in appreciating the long history of cognition in philosophy and anthropology and how the framing of this idea does not have to be located within the computational structure. Lawson and McCauley appear to make this move from *Rethinking Religion* to *Bringing Ritual to Mind* and, in consequence, distort much of their original project. This is an unfortunate and precarious move, riding as it does on the wave of what we might call "ideological cognition," or what Prochiantz calls in a critical appreciation of the subject "the new ideology of nature" (2001, p. 171). When the language shifts to New Synthesis models, the richness of their work, in echoing Lévis-Strauss and cultural anthropology, is lost and converted into a different political register. The shift is also witnessed in the movement away from Sperber's (1975, p. 87) epistemic understanding of cognition to more concrete operations. Sperber, influenced by Fodor's earlier work, *The Language of Thought* (1975), at least recognizes that cognitive psychology is not "without problems nor immune from revisions" (Sperber 1985, p. 50). Cognition and anthropology need to have a vital relationship, as the French tradition of thinking illustrates so helpfully, but Lawson and McCauley distort this perception in their ideological move toward computation and the essentialism of brain functions. Valuable insights are now lost in the rhetoric of a new ideology.

Fodor's distinction between Pinker and Chomsky saves us from the confusion that may arise in assuming Chomsky's cognitive theory has any metaphorical connection with the ideological models of computational analysis. John Lyons identifies this in an early study of Chomsky:

> It is Chomsky's conviction that human beings are different from animals and machines and that this difference should be respected both in science and in government; and it is this conviction which underlies and unifies his politics, his linguistics and his philosophy. (1970, p. 14)

The link between politics and computational psychology at this point might be testing to the narrow logic of syntactical structures, but it is the guiding metaphorical structure that bridges the different domains of knowledge and opens up questions about the wider epistemology that governs thinking. This can be seen in the work of the economist Philip Mirowski, who shows the way contemporary economics is shaped by the same analogical patterns as cognitive psychology. In his work *Machine Dreams*, we can see how there is a link between the dominant analogies of economics and the current market system of neoliberal ideology. It is not just minds but markets that "resemble computers" (2002, p. 539). The New Synthesis psychologists could see this analogy as further evidence that cognitive processes form the basis of all human life, but this form of intellectual imperialism operates on an amnesia about metaphors and the changing frames of analysis in the history of Western thought. The problem of the cognitive science of religion is that it is part of a wider cultural shift in the signification of cognition to the computer, and many of the traditions it seeks to study offer alternative models of human cognition or mental representation. When Lawson and McCauley use the notion of cognition, we must interrogate its use not just in terms of the science of psychology but also in terms of the wider ideology it supports. Cognition is not some neutral conceptual space; it has evolved since Descartes and has taken on "conceptual patterning" that renders it not simply an analysis of mind, but a desire for mind to be controlled in the metaphorical and ideological space of the machine. As Lawson and McCauley conclude their 1990 study: "Human cognitive systems have generated a stunning array of cultural forms. Any method of controlling this splendour theoretically offers a glimpse of the structures of the imagination. Dionysus dances not in the heaven but in our heads" (1990, p. 184).

The cognitive psychology of religion requires the powers of the imagination not only to govern that which exceeds representation, but also to suppress the imagination that thinks beyond the cognitive model. It is not Dionysus but Apollo who commands the rituals of cognitive science, and in that it distorts the dance of the mind.

Cognitive Theory as Ideology

[C]ognitive Science is a product of the post-industrial present and, like many intellectual movements, it has used analytical tools and ruling metaphors that reflect its own immediate cultural environment.
—M. Donald, *The Future of Cognitive Revolution*

The fault lines within Lawson and McCauley need answering even if we do not follow the next part of my argument, but I would suggest that the fault lines themselves reflect an ideological smoke screen that appeals to cognitive science. The ideological influences on cognitive theory relate to the sociopolitical spaces of conceptual mapping, which determine the metaphors of science. As I have already indicated, cognitive science, in its computational processing form, relates to the ideology of the machine and political totalitarianism, insofar as it seeks to find a *stabilizing structure* for being human. This is not simply a desire for universals and a common biological ground, but the location of mental activity in a one-dimensional metaphorical pattern. Such regimes of knowledge seek to establish control over the human mind and have a use-value for the state or corporate apparatus (Fitzgerald 2003) insofar as they increase the mechanization of society and its commercial output. This might seem a difficult line of argument for those working on a presupposition of science as truth, rather than one shaped by the early modern scientific understanding and more recent ideas of metaphorical modeling. Those who see cognition as a stable referent will also find it difficult to imagine that contemporary uses of the term reflect specific sociopolitic forces or historical evolutions of the idea from Descartes. The idea of cognition is shifting in its meaning away from mind to machine, from epistemic state to cognitive process, and the results of this move can only distort the imaginative possibilities of being human.

The question of ideology and cognitive theory has long plagued the anthropology of ritual. In Maurice Bloch's 1976 Malinowski's Memorial Lecture, there is a critique of Durkheimian and relativist notions of cognition.

Bloch unravels the danger of assuming certain cognitive universals from studying ritual discourse and, in his Marxist reading, questions anthropology as reflecting ideology (in the sense of a legitimating practice), rather than typical modes of cognition. As he sharply concludes his critique of anthropology: "[T]hey have confounded the systems by which we know the world with the systems by which we hide it" (1977, p. 290). If Bloch hoped that cognitive science would correct the relativist Durkeimian position, he was mistaken in limiting ideology to religion. We now face the reverse position. *It is no longer the problem of seeing ideology (religious ritual) as shaping the understanding of cognition in different cultures, but of seeing cognitive theory as an ideology (dominant politic regime) in the representation of our mind.*

The problem of cognitive theory, as John Shotter notes, is that as a "unifying theory" it makes it very difficult for "alternative voices" and "vocabularies" to be heard (1997, p. 318). The financial support and politic climate allow cognitive science to flourish, not only because cognitive science supports a mechanistic and reductionistic worldview (Dupré 2001, p. 15), but also because it orders human identity for a use-value in the present political system. As Shotter goes on to indicate, cognitivism parallels certain social structures: "Cognitivism achieves the main aims of bureaucratic organisation—efficiency, predictability, quantification and control—first, by the substitution of non-decision-making human functionaries for decision-making persons, and then by substituting non-human for human functionaries" (1997, p. 323).

It would be naive to assume that governments are deliberately setting up a way of talking about mental processes for their own services in some kind of conspiracy theory; only those wishing to dismiss arguments about metaphorical modeling would read my work in this way. The emergence of metaphors is caught in the wider social currency and the dominant registers of a society—something the history of science underscores. In the information age, our models of self and of mind will emerge from those areas of life where power is most heavily invested. The investments in the economy of the machine and its military power make us all cyborgs, but whether this metaphorical structure of psychological language is the best one to describe and model human beings and understand religion is another matter.

Howard Gardner makes it clear that war made an important contribution to the development of cognitive science (1987, p. 7). The key issue is that "the war effort demanded calculating machines" (p. 15). Unfortunately, the calculation was not just, as Gardner indicates, a matter of dealing with large sets of numbers for decoding; it was also the calculation of the efficiency of killing. *Machines are necessary for efficiency, but human beings defy efficiency.* The cognitive science of religion absorbs a cultural ideology of efficiency and measurement to explore and exploit those areas of life that seek to offer an alternative vision to the ideologies of control. If my critical readings of Lawson and McCauley are correct, it is perhaps the dream of the cognitive science of religion to extend the rational project in such a way as to hide the ideology of its own endeavor. Cognitive theory desires a human being who can be "programmed" and reduced to a unit of human resource, where hopes, dreams, and aspirations can become frequencies, structures, and acts of measurement. This in turn generates successive rituals of academic performance for the production of such discourses. The problem with cognitive theory and religion is that the former is a model of mental processes that operates in a closed system, unlike its own metaphors. Religion, on the other hand, is shaped by an open-ended social force, captured in the interactions about mind and cognition. The desire to stabilize the truth of cognition through the machine is to limit human beings to the machine. In such a context, we need is to bring cognition back to mind, to the mind that imaginatively exceeds the machine and the logic of "cognitive theory."

Conclusion: Religion within the Limits of Mind Alone

The study of religion in the last 20 years has been enjoying a confused liaison with the insights of cognitive theory and its related area of neuroscience. It has in turn silenced many of its problems and errors inside the fetish of the new. Scholars have been asserting their authority and power through the physicalist imperialism of the brain, attempting to locate the truth of religion in a reductive, jargon-deluding narrative, which throws out philosophical scrutiny with dazzling, technical pictures of the cerebral cortex or common sense ideas dressed in the fanciful language of cognitive theory. There is no arguing that we have before us a fascinating model for reading religious behavior built from post–War computational analysis, but the tragedy of this encounter is that it lacks a critical reading of epistemology, the politics of metaphor and the structure of psychological language. Indeed, its exponents undermine the very value of the model by turning it into a totalizing explanation, even presenting it—in delusions of grandeur—as a solution to world violence and the way to understand the "sociopolitical history of our species" (Whitehouse 2002, p. 312) . In this respect, much of the literature points toward a totalizing answer to all problems on the basis of finding the mechanisms to control the human species through knowledge of cognitive processes. It does not take a cognitive "science" of sociopolitical history to realize the politics of knowledge involved in this kind of cognitive theory. We can still have common biological structures and speak different languages and, above all, think differently. The future success of cognitive theory will be to rescue it from its computational and ideological frames. It is time to rescue cognition from recent cognitive science of religion and develop a counter-epistemological discourse, where religion is given back to the lived realities of people's minds in the empirical findings of sociology, anthropology, and psychology.

The creativity of Lawson and McCauley's work is to offer imaginative perspectives on the patterns of ritual; the problem is that they seek to locate these insights inside the rhetoric of a computational cognition. The love affair between religion and cognitive theory will in time fade and the next generation of religion scholars will fall for yet another romance, with yet more scholars who desperately run toward the "will to truth," dangerously enhanced by the popular science market (Sampson 1997, p. 12). Time and history reveal the flux of ideas, the temporary excitements and the desire of the masses for an easy solution to complex situations and the yearning for new gurus to offer complete answers to truth, which ensures political control and market success. Such waves of ideas are hidden behind professional advancements in the air-conditioned worlds of academia, rather than revealing the financial institutions and the political realities of global exchange, which seek to dictate the long-held problems of the nature of being human. The tragedy of psychological theory is that it forgets its history and in so doing hides the fault lines of the project behind the bright lights of the power of scientific advancement. What is left out of the debate between religion and cognitive theory is the history of psychology, the history of philosophy, and the economics of science, all of which can be seen by examining the fault lines of Lawson and McCauley's cognitive discourses on religion. The reason so much is open to debate, and so much is at stake, is that the ideas about being human and the subject of religion are trapped inside representational categories and the politics of representation. The categories of knowledge are not stable fixed entities, which allow us to present the truth of ourselves, and in this respect we are looking, at the very least, at the limits of cognitive theory in the study of religion. We have to recognize what Lawson and McCauley are *not* saying as much as what they are saying and uncover the complex interweaving of ideas and the shifts in metaphorical patterns. Theory is always political; it is a register of the time and, in the end, the lived realities behind the ideas will be the base from which to judge the results. "Such are the ways of the gods" (Lawson and McCauley 2002, p. 212).

References

Andresen, J. 2001. *Religion in Mind: Cognitive Perspectives on Religious Belief, Ritual and Experience.* Cambridge: Cambridge University Press.

Barrett, J. and Keil, F. 1996. Conceptualizing a non-natural entity: Anthropomorphism in God concepts. *Cognitive Psychology*, 31, 219–247.

Bermúdez, J.-L. 2002. Rationality and psychological explanation without language, in J.-L. Bermúdez and A. Millar (eds.), *Reason and Nature: Essays in the Theory of Rationality*. Oxford: Oxford University Press, pp. 233–264.

Bloch, M. 1977. The past and the present in the present. *Man: The Journal of the Royal Anthropological Institute*, 12(2), 278–292.

Bowie, F. 2000. *The Anthropology of Religion*. Oxford: Blackwell.

Carrette, J.R. 2001. Post-structuralism and the psychology of religion: The challenge of critical psychology, in D. Jonte-Pace and W.B. Parsons (eds.), *Religion and Psychology: Mapping the Terrain*. Londong: Routledge.

Carrette, J.R. 2002. The return to James: Psychology, religion and the amnesia of neuroscience, in W. James (ed.), *The Varieties of Religious Experience*, London: Routledge, pp. xxxix–lxiii.
Chomsky, N. 1966. *Cartesian Linguistics: A Chapter in the History of Rationalist Thought*. New York: Harper and Row.
———. 1986. *Knowledge of Language: Its Nature, Origin and Use*. New York: Praeger.
———. 2002. *On Nature and Language*. Cambridge: Cambridge University Press.
Danziger, K. 1990. *Constructing The Subject: Historical Origins of Psychological Research*. Cambridge: Cambridge University Press.
Donald, M. 1997. The mind considered from a historical perspective: Human cognition phylogenesis and the possibility of continuing cognitive evolution, in D.M. Johnson and C.E. Erneling (eds.), *The Future of The Cognitive Revolution*. Oxford and New York: Oxford University Press, pp. 355–365.
Dreyfus, H. 1992. *What Computers Still Can't Do*. Cambridge: MIT Press.
Dreyfus, H.L. and Dreyfus, S.E. 1986. *Mind Over Machine: The Power of Human Intuition and Expertise in the Era of the Computer*. Oxford: Basil Blackwell.
Dupré, J. 2001. *Human Nature and the Limits of Science*. Oxford: Clarendon.
Durkheim, E. 1995. *The Elementary Forms of Religious Life*. New York: Free Press (original work published in 1912).
Fitzgerald, T. 2000. *The Ideology of Religious Studies*. Oxford: Oxford University Press.
———. 2003. Playing language games and performing rituals: Religious studies as ideological state apparatus. *Method and Theory in the Study of Religion*, 15 (3), 209–254.
Fodor, J. 2001. *The Mind Doesn't Work That Way*. Cambridge, MA: MIT Press.
Gardner, H. 1987. *The Mind's New Science: A History of the Cognitive Revolution*. New York: Basic Books.
Gell, A. 2001. *The Anthropology of Time: Cultural Constructions of Temporal Maps and Images*. Oxford: Berg.
Gowers, T. 2002. *Mathematics: A Very Short Introduction*. Oxford: Oxford University Press.
Greenwood, J.D. 1997. Understanding the "Cognitive Revolution" in Psychology. *Journal of the History of the Behavioural Sciences*, 35(1), 1–22.
Halbwachs, M. 1992. *On Collective Memory*. Chicago: University of Chicago Press.
Harré, R. 2002. *Cognitive Science: A Philosophical Introduction*. London: Sage.
Hayek, F.A. 1976. *The Sensory Order: An Inquiry into the Foundations of Theoretical Psychology*. Chicago: University of Chicago Press (original work published in 1952).
Hervieu-Léger, D. 2000. *Religion as a Chain of Memory*. London: Polity Press (original work published in 1993).
James, W. 1983. *Principles of Psychology*. Cambridge, MA: Harvard University Press (original work published in 1890).
———. 2002. *The Varieties of Religious Experience*. London: Routledge (original work published in 1902).
Johnson, D.M. and Erneling, C.E. (eds.) 1997. *The Future of the Cognitive Revolution*. Oxford and New York: Oxford University Press.
Jones, D. and Elcock, J. 2001. *History and Theories of Psychology: A Critical Perspective*. London: Arnold.
King, R. 1999. *Orientalism and Religion*. London: Routledge.
Klüver, H. 1952. Introduction, in F.A. Hayek (ed.), *The Sensory Order: An Inquiry into the Foundations of Theoretical Psychology*. Chicago: University of Chicago Press, pp. xv–xxii.
Lakoff, G. and Johnson, M. 1999. *Philosophy in the Flesh: The Embodied Mind and its Challenge to Western Thought*. New York: Basic Books.
Lawson, E.T. and McCauley, R.N. 1990. *Rethinking Religion: Connecting Cognition and Culture*. Cambridge: Cambridge University Press.
———. 2002. *Bringing Ritual to Mind: Psychological Foundations of Cultural Forms*. Cambridge: Cambridge University Press.
Lévy-Bruhl, L. 1975. *The Notebooks on Primitive Mentality*. Oxford: Blackwell (original work published in 1949).
———. 1985. *How Natives Think*. Princeton: Princeton University Press (original work published in 1910).
Littleton, C.S. 1985. Lucien Lévy-Bruhl and the concept of cognitive relativity, in L. Lévy-Bruhl (ed.), *How Natives Think*. Princeton: Princeton University Press, pp. v–viii (original work published in 1910).

Lyons, J. 1970. *Chomsky*. London: Fontana.
Marx, K. 1995. *Capital*. Oxford: Oxford University Press [original work published in 1895 (1967)].
Mirowski, P. 2002. *Machine Dreams: Economics Becomes a Cyborg Science*. Cambridge: Cambridge University Press.
Persinger, M.A. 1987. *Neuropsychological Bases of God Beliefs*. New York: Praeger.
Pinker, S. 1997. *How the Mind Works*. New York: Norton.
Prochiantz, A. 2001. *Machine-Esprit*. Paris: Editions Odile Jacob.
Pyysiäinen, I. 2002. A theory of ideology: Implications for religion and science. *Method and Theory in the Study of Religion*, 14, 316–333.
Sampson, G. 1997. *Educating Eve: The "Language Instinct" Debate*. London: Cassell.
Shotter, J. 1997. Cognition as a social practice: From computer power to word power in D.M. Johnson and C.E. Erneling (eds.), *The Future of the Cognitive Revolution*. Oxford and New York: Oxford University Press, pp. 317–314.
Sperber, D. 1975. *Rethinking Symbolism*. Cambridge: Cambridge University Press.
———. 1985. *On Anthropological Knowledge*. Cambridge: Cambridge University Press.
Turing, A.M. 1964. Computing machinery and intelligence, in A.R. Anderson (ed.), *Minds and Machines*. Englemood Cliffs, NJ: Prentice-Hall, pp. 4–30 (original work published in 1950).
Valentine, E. 1992. *Conceptual Issues in Psychology*. London: Routledge.
Whitehouse, H. 1995. *Inside the Cult: Religious Innovation and Transmission in Papua New Guinea*. Oxford: Clarendon Press.
———. 2002. Modes of religiosity: Towards a cognitive explanation of the sociopolitical dynamics of religion. *Method and Theory in the Study of Religion*, 14(3/4), 293–315.

CHAPTER TWELVE

Brain Science on Ethics: The Neurobiology of Making Choices

WALTER J. FREEMAN

I was invited to address the graduating students of the high school in Villa San Giovanni during my visit to Reggio Calabria, Italy, to receive the Premió Calabria 2002. My comments are summarized as follows.

I am a brain scientist. I have studied brains for half a century. I have taught thousands of students about brains. I have taught future doctors and nurses what they would need to know in order to diagnose and treat diseases of brain and mind. I have taught future engineers what they need to build better computers and robots for the information age. Most important, I have taught the future leaders of my state and country the principles of brain function that relate to ethics.

The greatest power of brains is their capacity to create their own futures. We call this property "self-determination." In order to create themselves, brains must learn about themselves and the world around them. They learn by taking action into the world. They create within themselves their visions of what they can do, what the world holds, what they want from the world, and what course of action will get them what they want. Brains thirst for the knowledge that enables them to act effectively.

Brains are not like sponges. They do not passively absorb whatever the world brings to them. They boldly, or timidly, thrust their bodies into the world and suffer the consequences of their actions. They learn from their experiences through their own bodies. Without bodies, there is no action. Without action, there is no learning.

Most scientists, philosophers, and sociologists preach determinism. Behaviorists claim environmental determinism, by which our parents, schools, friends, and enemies make us who we are. Sociobiologists and geneticists claim genetic determinism, by which our genes have the first and final say.

But brains are not passive machines. They have the freedom to act within the constraints set by their genes, and to do the best they can with whatever their world brings to them, for better or worse. They assemble whatever knowledge they can get, and they choose the courses of action that seem most likely to bring rewards.

Here is the first principle of freedom. Each of us is born free. We are born with varying degrees of freedom. Those of us who are born into modern democracies, or into the ruling classes of despotic states, have the most freedom. Others born in bondage, or with tyrannical parents, have much less freedom. That is the luck of the draw. Some of us start with more, others with less, but everyone can and must choose something of the future.

Here is the second principle of freedom. Brains seek optimal degrees of freedom. We can choose by our actions to increase our freedom, or we can choose to give away what we have. Too little freedom prevents brains from achieving full self-realization. Too much freedom brings confusion and disorientation. That experience is frightening, as well as demanding and disagreeable.

Here is the third principle of freedom. The choices we make affect not only our personal freedom; our choices also affect the freedom of others with whom we live. Brains do not grow and flourish in isolation. They repeatedly dissolve their internal structures and reform them. They reform themselves to act in concert with others. Our brains realize their full potential only through the accompanying growth of others and of the societies in which we live. When we act and study alone, we isolate ourselves and weaken our freedom. When we act in concert with others, we enhance it.

We know these three principles from millennia of human experience. Now we also know the hard biological facts of brain growth and development, which tell us how brains operate. We know the chaotic dynamics by which brains create themselves. We understand the branch points that mathematicians call bifurcations, at which brains make choices. We know the chemical systems by which brains get rewarded and punished by the consequences of their own actions, and by which we experience emotions.

And we know how brains dissolve their structures by regressing to earlier stages of development, in order to make way for new growth. We experience dissolution most painfully in the catastrophe of falling in love. We cannot choose to fall in love, though we can choose not to, in pursuit of a lesser freedom that carries the price of loneliness. Paradoxically, here is the gift of a path to true freedom, if it is taken correctly. Falling in love. The risks are high, the path is arduous, and the rewards are immense. The greatest freedom is found in commitment to others.

Here is where ethics enter. We must predict and evaluate the consequences of our actions. Prediction and evaluation are the prime functions of brains. The reason we study is to acquire knowledge by which to strengthen these functions. We must first choose whether to increase or decrease our degrees of freedom by our actions. We must next choose whether to act for the costs and benefits of our own futures, or those of others.

On the one hand, those who have too much freedom may decrease it by accepting arbitrary limits. They can accept guidance from civil, religious, military, or neighborhood organizations, which make decisions for them. On the other hand, those who have too little freedom can only choose to act either rapidly and efficiently, or grudgingly and resentfully, but they have always the choice to create plans and fantasies for liberation.

You students here are among the elite of your generation. You are receiving the best education that your state can give. You are the future leaders of your state and country. We expect that you will choose to enhance your freedom. You already know that if you choose unwisely to lie, cheat, steal, take drugs, or neglect your studies, you will diminish your freedom. You know that if you choose wisely to work hard at your studies, cooperate with your families, and take good care of your bodies, you will enhance your freedom.

True leaders have deep understanding of themselves, and they have the self-control and command of situations that express true freedom. Here is your opportunity—today, tomorrow, and throughout your lives. You can choose and keep on choosing, wisely, whenever you are ready to do so. With talent, hard work, luck, and help from your family and friends, you may get what you choose, and, above all, find that it is worth having.

INDEX

Abdullaev, Y.G., 123
Absolute Unitary Being (AUB), 177
    vs. God, 179
Achmet, 223
activation–synthesis model of dreaming, 229–31
Advaita Hinduism, 83
Aitken, Robert, 113, 114, 115, 127, 128, 132
Albright, Carol Rausch, 65, 176–7, 178, 181, 194
Allen, D., 95
Allport, G.W., 99, 107
Alper, Matthew, 151, 155
Alvarado, Carlos, 80
Alzheimer's disease, 149, 193, 230
anatman (no self), 198–200; see also sunyata (emptiness)
Andresen, Jensine, 38, 109, 242, 246, 250–1
animal behavior, 11–14
    chimpanzees, 12, 191–2
    gorillas, 14
    monkeys, 14, 213
    see also termite behavior
anomalous experience, 62–3
Anterior Attention Network, 122, 123, 124, 131
Anttonen, V., 109
Apollo, 256
Arabi, Ibn, 224
Aristotle, 222
Arnold, M.B., 97
Aronson, Harvey, 215
Artemidorus, 223
Aruni, 199–200
Aserinsky, Eugene, 228
Ashbrook, James, 176, 178, 181, 194
Ashby, F.G., 106

Asklepius, temples of, 222, 223
attention, 152
    cognitive-behavioral research on, 117–21
    definition of, 115
    and executive functions, 121–2
    neural mechanisms of, 121–3
    regulation of, 121
    selective, 116, 117–20, 124
    sustained, 117–18, 120–1, 124
    sustained selective, 114
    in zazen meditation, 113–15, 126–30
Attentional Processing, 116
AUB (Absolute Unitary Being), 177
aurora borealis (Northern Lights), 71
Austin, James, 6, 39, 113, 114, 122, 123, 130, 131, 133, 152, 215
autocatalysis, 143
awareness, 48–50, 57
awe, 99–101; see also gratitude; hope; reverence; wonder
ayahuasca brew, 61
Azari, N.P., 151

Baars, B., 121, 124, 128, 133
Baddeley, A., 121
Baer, R.A., 102
Balkin, T.J., 145
Banks, R., 168
Barber, J., 37
Barber, T.X., 79
Barbour, Ian, 180, 181
Bargh, J., 128
Barrett, J., 249
Barrie, J.M., 155
Barrow, J.D., 25
Basmajian, J.V., 37
Bear, M.F., 27, 28, 30
Beatty, J., 37

Bechara, A., 148
Bechert, H., 215
Begley, Sharon, 211–14, 215–16
Behavioral and Brain Sciences, 12
Being No One, 204
Bellah, R., 193
Belvedere, E., 71
Benard instability, 141
Ben-Yehuda, N., 68
Ben Ze'ev, A., 96
Berger, Peter, 210
Bermúdez, J.-L., 247
Bertocci, P.A., 98
Biblical passages
  Acts, 167, 222
  Deuteronomy, 95, 187
  Exodus, 187
  Ezekiel, 187
  Genesis, 187
  Hebrews, 101
  Isiah, 167
  Jeremiah, 167, 187
  Luke, 169
  Matthew, 169, 221
  Numbers, 221
  1 Peter, 167
  Pslams, 167
  Timothy, 190
Bielfeldt, Dennis, 48, 56, 58
"big dreams," 220–1, 236–8
bin Laden, Osama, 225
biofeedback, 37–8, 51
Birth of Tragedy, the, 219
Black, I.B., 212, 213
Blackmore, Susan, 80, 83, 130, 132–3
Blakeslee, S., 65, 163
Blanke, Olaf, 150
Bloch, Maurice, 256–7
Bohm, David, 58
Boies, S.J., 116
Bootzin, R.R., 80
bottom-up causation, 27, 36, 40; *compare* top-down causation
Bowie, Fiona, 246
Bowker, J., 225
Bowler, Peter J., 212
Boyd, Robert, 11, 15, 17
Boyer, Pascal, 4, 109, 151, 175, 242
Bracken, J.A., 194
Braddon-Mitchell, D., 212

brain activity and states of mind, 144–52
brain damage, effects of, 147–9
  on dreaming, 231–4
brain formation and genetics, 20–6
brain functioning
  and ethics, 263–4
  executive, 145, 147
  and freedom, 263
  trauma to, 147–8
brain imaging studies, 146
brain–mind functioning, 144
  and dreaming, 144–7
  and morality, 150
  and religious experience, 150–1
brain–mind problem, 175–6
brain–mind research, social implications of, 262–4
brains, animal vs. human, 22–5
brain stimulation
  and out of body experience, 150
  and "sensed presence," 176
Braitenberg, V., 22, 24
Brand, W., 101
Bratus, Boris, 63
Braud, L.W., 78
Braud, W.G., 71, 75, 78
Braun, A.R., 145, 147
Braver, T.S., 122, 123
Bressler, S.L., 140
Bringing Ritual to Mind, 245, 254
British Society for Psychical Research, 74
Britton, W.B., 80
Broadbent, D., 116
"broaden-and-build" theory, 105–6, 109
Broughton, R.S., 69
Brouwer, W.H., 116
Brown, D., 36, 37
Brown, K.W., 104–5, 130
Brown, Muriel, 115
Brown, W.S., 181, 182, 183, 186, 189, 192, 194
Buddha, Gautama, 225
Buddhism
  dreams in, 225–6
  Dzogchen, 83
Budzynski, T.H., 78
Bulkeley, Kelly, 101, 220, 225, 236
Bullard, P.L., 102
Burns, C.P.E., 194

Bush, George H.W., 221
Buss, D., 18

Cabeza, R., 155
Callois, R., 224
Campbell, K., 86
Candrakirti, 197–8, 206–9, 214, 215, 216
Capgrass syndrome, 149
Cardena, E., 62, 64, 68
Carlson, L, 38
Carrette, J.R., 244, 247
Cartwright, Rosalind, 229
causality, circular, 154
causation, mental, 39, 42, 45–6, 47–8, 50, 52, 56
  bottom-up, 27, 36, 40
  downward, 39, 41, 42, 44, 45, 46, 47–8, 51, 52, 53
  top-down, 27, 36, 39, 40, 46, 51–2
Center for Theology and the Natural Sciences, 183
central-state materialism, 85–6
Chalmers, David, 43, 49, 53, 54, 56, 57, 175, 215
chaos in religious conversion, 163–4
chaos theory, 86–7
Chartrand, T., 128
Cheney, Dorothy L., 13, 14
Chesterson, G.K., 107
Child, I.L., 70
Chimpanzee Cultures, 12
chimpanzees, *see* animal behavior
Chomsky, N., 13, 253–5
Christianity, dreams in, 221–3
Christiansen, M., 15
Christian soul, *see* soul in Christianity
circular causality, 154
City of God, 190
clairvoyance, 69, 85
  in dreams, 70–3
Clark, W.H., 110
Clayton, Philip, 40–2, 48, 51, 52, 56
Cleary, T., 113, 114, 127, 128
clinical interventions, effect on neurophysiology, 38
cognitive linguistics, 203–4, 211
cognitive neuroscience of religious beliefs, 151
cognitive relativity, 247

cognitive science and the soul, 183–5
cognitive theory of religion
  critique of, 243–59
  as ideology, 256–8
  and mathematical modeling, 249–50
  politics of representation in, 250–2, 259
Cohen, R.M., 123, 131
collective behavior, 143
Collins, A.M., 126
Collins, Francis, 23
Collins, H.M., 85
Collins, Steven, 199, 205
Concept of Mind, the, 210
conciliationists, 183, 185, 186, 192
conditioning, 47
Confessions of St. Augustine, 222
consciousness, 40, 47, 49, 52, 175–6, 214–15
  as cause, 51–2
  as emergent property, 42–4
  incompleteness of theories of, 53–6
  and religion, 56–7
Consciousness, 130
conversion, Christian, *see* conversion, religious
conversion, religious
  chaos, role in, 163–4
  Christian language of, 166–72
  cognitive science and, 163–4
  emotion in, 107–8
  metaphor in, 168–72
  metaphor theory and, 165, 166
  theories of, 160–2
Cooper, John, 182, 187, 188–9, 194
Copernicus, 174
Corbetta, M., 122
Cosmides, Leda, 17–18, 33
Cox, S., 118, 127
Coyle, C.T, 104
Crick, Francis, 175, 182, 234
Crossan, John Dominic, 160, 171, 172
Cullman, O., 182, 188
culture
  and the Christian soul, 191–4
  cumulative transmissible, 11, 13, 14, 20, 25, 26, 29
  definitions of, 11
  vs. genetics, 19–20
  language and, 13–14
  vs. nature, 11–15

cumulative transmissible cultures, *see under* culture
Curley, R.T., 222

Dalton, Kathy, 71
Damasio, A.R., 148, 175
Danzinger, Kurt, 244, 248
d'Aquili, Eugene, 65, 67, 122, 123, 130, 131, 133, 177–8, 179, 251
Darwin, Charles
  influence on brain–mind research, 4
  personal views of religion, 3
  theory of evolution, 3–4
Davidson, Donald, 45, 48
Davidson, R.J., 39, 130, 152
"Decade of the Brain," 219
de Duve, Christian, 24, 25
Degas, Edgar, 72
degrees of intentionality, 13
Deikman, A.J., 39, 120
Delbrück, Max, 32
Delio, Ilia, 180
Dement, William, 117, 228, 229
Deneubourg, J.L., 143
Dennett, D.C., 13, 36, 54
Dennis, S.P., 71, 75
dependent arising, 198, 201–2
Derr, J.S., 66
DesCamp, Mary Therese, 166–7
Descartes, Rene, 6, 190–1, 212, 254, 255, 256
Descola, P., 221
Desimone, R., 117, 122
determinism, 26, 262
de Waal, F.B.M., 11
Dickson, Barry J., 22
DiGirolamo, G.J, 122, 123, 124, 131
Dillard, J., 74
Dionysus, 256
DiScenna, P., 74
distributed cognition, 210
Djikic, M., 107
Doblin, R., 65
Dobzhansky, Theodosius, 17, 22
Dogen Kigen Zenji, 114
Domhoff, G.W., 235
Donald, M., 192, 256
Doniger, Wendy, 226; *see also* O'Flaherty, Wendy D.
Donovan, W., 65

Doob, Penelope, 170
Doricchi, F., 232
downward causation, 39, 41, 42, 44, 45, 46, 47–8, 51, 52, 53
Draganski, B., 27
dreaming
  activation synthesis model of, 229–31
  and brain damage, 231–4
  in Buddhism, 225–6
  in Christianity, 221–3
  as delirium, 230
  functions of, 22, 230–2, 237
  in Islam, 223–5
  neurophysiology of, 147
  neuropsychology and, 228–35
Dreaming and Delirium, 230
dreaming brain, 144–7, 228–35
dreams
  "big," 220–1, 236–8
  clairvoyant, 70–3
  nightmares, recurrent, 231, 234
  precognitive, 73
  telepathic, 70–2
Drewry, M.D.J., 79
Dreyfus, Hubert, 252
Dreyfus, Stuart, 252
Droit, R.P., 215
dualism, 40, 47, 182, 185, 188–9, 190–1, 206, 212, 215
  interactive, 50
  compare monism
DuBreuil, S., 37
Duerlinger, James, 198, 208
Duffy, E., 117
Duncan, J., 117, 122
Dupré, John, 242, 252, 257
Durham, W., 15
Durkheim, E., 245–6

Eccles, John., 40, 50, 182
Edelman, G.M., 114, 140
Edelstyn, N.M., 149
Edwards, Jonathan, 96
Edwards, P., 175, 176
EEG studies, 130, 145, 146
Ehrenwald, Jan, 77
Einstein, Albert, 23
Elbert, T., 27
Elcock, J., 245
Ellis, G.F., 48, 184

emergence, 40–1, 42, 44, 48, 50, 51, 144, 183–4
emergent behavior, neurons and, 144
Emmons, Robert, 93
emotional regulation, role of religion in, 100–5
Emotion and Personality, 97
emotion in religious conversion, 107–8
emotions, religious, *see* sacred emotions
emotions, sacred, *see* sacred emotions
emptiness (sunyata), 198, 201–2, 207; *see also* no self (anatman)
Enard, W., 23
Engel, A.K., 155
Enright, R.D., 104
entheogens, *see* ayahuasca brew; hoasca brew; LSD; psilocybin; yage brew
Entry into the Middle Way, 206–7
epilepsy, limbic temporal lobe, 76
epiphenomenalism, 85–6, 181–2, 183
epiphenomenal mental activity, 33, 39, 40
    dreaming as, 230
Epstein, M., 130
Erdi, P., 144
Erickson, T.C., 150
Erwin, William, 72, 74
ethics and brain function, 263–4
evolution of intelligence, 24–5
Ewing, K., 221, 225
exceptional human experiences (EHEs), 83–5
executive function, neural mechanisms of, 121–3

Feinberg, T.E., 149
Feit, J.S, 178
Feldman, J., 37
Fierro, B., 155
Findlay, J., 54
Fisher, H.J., 222
Fitzgerald, T., 244, 256
Flanagan, Owen, 24, 183, 217, 229
Flavell, J.H., 121
Flournoy, Theodore, 62
fMRI studies, 152
Fodor, Jerry, 252, 254, 255
Foote, S.L., 123
forgiveness, 103–4
Forstl, H., 149
Foucault, Michel, 85

Foulkes, David, 71, 234
Fox, E., 116, 118, 119, 127, 128
Franks, B., 109
Franks, N., 143
Frederickson, B.L., 93, 94, 105, 109
Freedman, S.R., 104
freedom
    and brain function, 263
    principles of, 263
Freeland, Matthew, 133
Freeman, Walter, 139, 154, 155, 163–4
Fregoli syndrome, 149
Freud, Sigmund, 2, 6, 176, 230, 231, 233, 236
Frith, C., 155
Fromm, E., 37
Fuster, Joaquin M., 123, 213
Future of Cognitive Revolution, the, 256
Future of the Body, the, 87

Gackenbach, Jayne, 226, 235
Gage, Phineas, 148
Galef, Jr., Bennett G., 12
Galileo, G., 174, 254
Gardner, Howard, 244, 251, 257
Garfield, J.L., 202
Geertz, Clifford, 204
Gelade, G., 117, 119, 120
Gell, Alfred, 246
Genes, Genesis and God, 4
genetics
    and brain formation, 20–6
    vs. culture, 19–20
geomagnetic activity and psi experiences, 71–3, 75
geomagnetic field, 71
George, L., 79
Geschwind, N., 66
Gibbons, A., 23
Gillani, N.B., 102
Gillespie, G., 226
Gillman, N., 187
Globus, Gordon, 71
God vs. AUB (Absolute Unitary Being), 179
Goedel, Kurt, 53–4
Goldapple, K., 38
Goldman-Rakic, P.S., 123, 147
Goleman, D., 39
Gombrich, Richard, 200, 201

Goodenough, U., 98
gorillas, see animal behavior
Gorsky, J., 187
Gowers, Timothy, 250
Grace, G.M., 149
gratitude, 98–9; see also awe; hope; reverence; wonder
Gray, C.M., 155
Greeley, A.M., 63, 68
Green, A., 37, 38
Green, E., 37, 38
Green, Joel, 189
Greenberg, R., 229
Greene, J., 150
Greenwood, J.D., 245
Gregor, T., 221
Greyson, Bruce, 79, 80
Griffin, David, 50, 56
Griffith, R.M., 108
Griffiths, P.E., 94
Grof, Stanislav, 67, 68
Gross, J.J., 102
"Guernica," 55
Gurney, E., 74

Hacking, Ian, 215
hadith, 223, 224
Haidt, J., 100, 101
Haken, Hermann, 139, 140, 141, 144, 154
Halbfass, W., 215
Halbwachs, M., 246
Hall, Calvin, 71, 235, 236
Hall, G. Stanley, 2
Hall, J.A., 223
Hallett, B., 38
Halligan, P.W., 149
hallucinogens, see ayahuasca brew; hoasca brew; LSD; psilocybin; yage brew
Hamer, Dean, 27–8
Hamilton, Sue, 201
Hanh, T.N., 113, 127, 128, 132, 202
Harcourt, A., 14
Hardy, A., 100
Harran, Marilyn, 168
Harré, Rom, 251
Harris, M., 221
Hartmann, Ernest, 229
Harvey, Peter, 200
Hastings, A., 83
Hauser, M., 14

Haviland-Jones, J.M., 93
Hayed, F.A., 243
health benefit of sacred emotions, 102–5
Hearne, Keith, 71
Hegel, Georg, 194
Heil, J., 48, 58
Heinz, S.P., 119
Heisenberg uncertainty principle, 54
Helminiak, Daniel, 63
Hermansen, M., 221, 224, 225
Hervieu-Léger, D., 246
Hill, P.C., 97
Hirst, W., 116, 119
Hitt, J., 67
hoasca brew, 61
Hobson, J. Allan, 74, 86, 139, 144, 145–6, 229–31, 233, 234, 235, 238
Hoffman, V., 225
Hofstadter, D.R., 54
Hood, R.W., 97, 100
hope, 101–2; see also awe; gratitude; reverence; wonder
Houston, Jean, 64
How Natives Think, 246
How Religion Works, 4
Huag, S.T., 104
Huber, Sherry, 115
Hudson, A.J., 149
human genome project, 26
Humanizing Brain, the, 176–7
Human Nature and the Limits of Science, 242
Humphreys, Christmas, 126, 127, 128, 132
Humphreys, M.S., 117
Hunsberger, B., 110
Hunt, Harry, 113, 121, 133, 220
Huntington, C.W., 207, 208
Hurst, L.A., 78
Hutch, R.A., 96
Hutto, D., 49, 53
hypnosis, 37

impermanence in Buddhism, 199
intelligence, evolution of, 24–5
intentionality, 47
degrees of, 13
interdependence (pratiyasamutpada), 198, 201–2
"Internal Structure of the Self, the," 202
International Journal of Transpersonal Studies, 84

Irwin, Lee, 221
Irwin, W., 152
Isen, A.M., 106
Islam, dreams in, 223–5

Jackson, F., 212
James, William, 2, 6, 62, 116, 117, 237, 245, 248, 251
    on attention, 115
    on mysticism, 64
    on religious emotions, 106–7
Jasper, H., 80
Jedrej, M.C., 221, 222, 225
Jehovah's Witnesses, 67
Jeremiah, prophet, 238
Jesus of Nazareth, 221
Jewett, P.K., 185
Jilek-Aall, L., 77
John, E.R., 133
Johnson, Chris, 165
Johnson, Mark, 164, 165–6, 252
Johnston, W.A., 119
Jones, D., 245
Jones, J., 47, 54–5, 56, 57, 58
Jones, R.M., 229
Joseph, A.B., 149
Joseph, Rhawn, 66, 67
Journal of Consciousness Studies 2001, 40
Jouvet, M., 229
Judaism, 95, 108
Jung, Carl Gustav, 2, 6, 8, 62–3, 220, 221, 223, 227, 236

Kaas, J.H., 213
Kabat-Zinn, J., 38
Kahan, Tracey L., 235
Kahn, David, 139, 144, 146, 235
Kahneman, D., 116
Kant, Immanuel, 6
Kaplan, S., 116
Kapleau, P., 113, 127
Kaufman, P., 114
Keil, F., 249
Kelsey, M., 222, 223
Kelso, J.A.S., 139, 140
Keltner, D., 100, 101
Kennedy, Robert, 73
Keown, D., 215
Khaldun, Ibn, 224
Khilstrom, J., 56

Kim, J., 58, 183
"kindling," 66
King, R., 244
Kirby, S., 15
Kleitman, Nathaniel, 228
Klimo, J., 79
Kluckhohn, C., 33
Klüver, Heinrich, 243
Knight, R., 123
Knudson, Roger., 220
Koch, Christof, 28, 116
Koenig, H.K, 109
Kolb, B., 123
Komito, D.R., 199
Korsakoff's psychosis, 230
Kosslyn, S., 38
Kramer, H., 223
Krippner, Stanley, 70, 72, 73, 74, 79, 81, 84, 86, 220
Kristeller, J.L., 38
Kroeber, A.L., 33
Kruger, Ann Cale, 12
Kuiken, Don, 220
kundalini meditation, 152; *see also* meditation

LaBerge, David, 114, 117, 119, 120–1, 126, 128, 132
LaBerge, Stephen, 226, 235
Lakoff, George, 7, 164, 165–6, 172, 203–4, 252
Lama, the Dalai, 226
Lamoreaux, J.C., 224
Langer, Ellen, 126, 128
language
    and culture, 10, 13–14
    human vs. animal, 13–14
Language of Thought, the, 255
Lanternari, V., 222
Lather, P., 85
Laughlin, Charles, 86
Laurent, Giles, 28
Lavie, Nilli, 116, 118, 119, 126, 127, 128
Lawson, Thomas, 4, 8, 242, 245–50, 253–7, 259
Lazer, S.W., 151–2
Leahy, G.E., 85
Leahy, T.H., 85
LeDoux, Joseph, 204, 213
Leon-Defour, 167

Leonid meteor showers, 67
Les functions mentales dans les societes inferieures, 246
Levi-Strauss, Claude, 245, 255
Lévy-Bruhl, Lucien, 246–7
Lewis, J., 108
Lewis, M., 93
Lewontin, Richard, 20
Libet, B., 53, 54
Linehan, M., 39
Littleton, Scott, 247
Llinas, R.R., 145, 155
Loftus, Elizabeth, 74, 126
Lohmann, R., 221, 222
Lou, H.C., 151–2
Lowe, L.M., 77
LSD, 67, 68
Lucas, J., 54
Luckmann, Thomas, 210
Luna, Beatriz, 20
Luther, Martin, 223
Lynn, S.J., 80
Lyons, John, 255

MAAS (Mindful Attention Awareness Scale), 104
MacDonald, W.L., 63
Machine Dreams, 255
Mackworth, N.H., 120
Mackworth's Clock Test, 120
MacLean, Paul, 194
MacLeod, C., 120
Madhyamaka (Middle Way) thought, 201–2, 206–7
Mageo, Jeannette, 205, 221
Maguire, E.A., 27
Mahoney, A., 97, 98
Maimonides Medical Center, 70, 71, 74
Makarec, K., 65
"Malinowski's Memorial Lecture," 256
Malleus Malificarum, 223
Mandel, A., 66
Maquet, P., 145, 147
Marks, J., 23
Martlett, G.A., 38
Martyr, Justin, 189–90
Marx, Karl., 247–8
Maslow, A., 100
Masters, Robert, 64

materialism, 176
  central-state, 85–6
Matthews, G., 117
Max-Planck Institutes, Germany, 23
May, E.C., 79
Mayer, J.D., 108, 109
Maynard Smith, John, 10
Mayr, E., 15, 25
mazeway reformulation, 160
McCarley, R., 229, 233
McCauley, Robert, 4, 8, 95, 109, 242, 245–50, 253–7, 259
McClenon, James, 63
McCraty, R., 102
McCullough, M.E., 98, 99, 103
McFague, Sally, 161, 164, 171, 172
McGinn, C., 53
Mead, Margaret, 11
Mechelli, A., 27
meditation, 57, 65, 76
  attention in zazen, 113–15, 126–30
  and the brain, 151–2
  fMRI study of, 152
  kundalini, 152
  PET study of, 152
  and psychophysiological functions, 39
  research, 38–9, 122
  SPECT study, 177
  zazen, 102, 114, 126–30
MEG studies, 145, 146
Mele, A., 48, 58
memory, 147
memory retrieval, 74
mental imagery, power of, 37
Merleau-Ponty, Maurice, 210
Merzenich, Michael, 26
Mesulam, M.M., 78, 147
metacognition, 121, 124
  in zazen meditation, 126–30
metaphoric structure of language, 203
metaphors
  of Christian conversion, 168–72
  primary and complex, 164–6
metaphor theory and Christian conversion, 165, 166
metastability, 140
Metcalfe, J., 121, 127
Metzinger, Thomas, 204
Metzner, Ralph, 161, 170
Meyering, T., 46, 47

Millard, R.M., 98
Miller, E.K., 122
Miller, P.C., 222
Mills, A., 80
mind
  in evolutionary context, 15
  as machine, critique of, 252–3
  role in shaping the brain, 26–9
Mind and the Brain, the, 211
mind–body problem, 175–6
Mind Doesn't Work that Way, the, 252
Mindful Attention Awareness Scale (MAAS), 104
mindfulness, 104–5
Minding God, 189
Mind Over Machine, 252
Mipham, Jamgon, 207, 208, 215
Mirowski, Philip, 255
Mitchell, D.W., 215
Molman, Jurgen, 192
monism, 41–2, 182, 189
  structural, 86
  compare dualism
Monk, T.H., 117
monkeys, *see* animal behavior
Moody, Raymond, 82
Mormonism, 67
Morowitz, H.J., 55
Morrison, J.H., 123
Morrison, Karl, 162
Motter, B.C., 122, 123
M'Timkulu, D., 222
Muhammad, Prophet, 66, 223
Mulamadhyamakakarika, 202
Murphy, Michael, 65, 87
Murphy, Nancey, 36, 44, 45–8, 51, 52, 57, 58, 183, 184, 186
mutable maps, 26
mystical experience, 64–8, 179–80
  psychoneurological correlates of, 65–6
Mystical Mind, the, 177

Nagarjuna, 199, 202, 206, 216
Nagel, T., 43, 49
Narens, L., 121, 127
Naropa, 226
National Human Genome Research Institute, 23
natural theology, 180, 186
nature and culture, 11–15

Nature of Consciousness, On the, 113
near-death experiences (NDEs), 79–81
Neher, Andrew, 69, 85
Neibur, E., 116
Nelson, T.O., 121, 127
Neppe, Vernon, 76, 77, 78
Neppe Temporal Lobe Questionnaire, 77
Neuhaus, O.W., 33
neural mechanisms of attention, 121–3
neurons
  and emergent behavior, 144
  and self-organization, 139–40
neuroplasticity, 212–13
neuropsychology and theology, *see* neurotheology
neuropsychology of dreaming, 228–35
neuroscience and the Christian soul, 181–3
neurotheology, 67, 176–81
Newberg, A.B., 67, 151–2, 177–8, 179, 122, 123, 130, 131, 133, 251
new epiphenomenalism, 85–6
Newman, J., 133
New Synthesis psychology, 254, 255
New Testament, 167, 168, 188–9
Nicolis, G., 140, 141
Nielsen, T., 220
Nietzsche, Friedrich, 219, 238
nightmares, recurrent, 231, 234
nirvana, 179
Nofzinger, Eric, 145
nonreductive physicalism, 36, 39–42, 48, 50–3, 55–6, 57, 183–5
Norbu, N., 226
Northern Lights (aurora borealis), 71
no self (anatman), 198–200; *see also* emptiness (sunyata)
Notebooks on Primitive Mentality, 246
Novak, P., 113, 130
NREM sleep, 228–9, 234, 235; *compare* REM sleep
Nyberg, L., 145, 155

Oatley, K, 107
OBEs, *see* out-of-body experiences
obsessive-compulsive disorder (OCD), 38
O'Flaherty, Wendy D., 221, 226; *see also* Doniger, Wendy
Okumura, S., 113, 126–7, 128
Old Testament, 167, 221
O'Leary, D.H., 149

Oneirocriticon, 223, 224
Ong, R.K., 226
Orientation Association Area, 122
Orten, J.M, 33
Osborne, K.E., 222
Otto, R., 96, 97, 99, 100
out-of-body experiences (OBEs), 80–1
  via brain stimulation, 150
Oyebode, F., 149

Pace-Schott, Edward, 234
Padmakara, 207, 208
Pahnke, Walter, 64
Palla, Shabana, 133
Palmer, J., 68
Paloutzian, R.E., 93, 107
Pannenberg, W., 190
panpsychism, 50, 51
Pantev, C., 27
Parapsychological Association, 68, 82, 87
parapsychology, 85; see also psi-related experiences
Parasuraman, R., 116, 117, 120, 121, 127
Pare, D., 145
Pargament, K.I., 97, 103
Park, C.L., 102
"past-life" experiences (PLEs), 80–1
Patton, Kimberly, 228
Peel, J.D.Y., 222
Penfield, W., 40, 54, 80, 150
Pentecostal movement, 108
Perls, Fritz, 236
Persinger, Michael, 65–6, 67, 71–2, 73, 75, 76–7, 78, 80, 163, 176, 247
Peters, Ted, 182
Peterson, G., 174, 176, 189
Peterson, S.E., 122, 123, 130, 131
PET studies, 122, 123, 131, 145, 151, 152
physicalism, nonreductive, 36, 39–42, 48, 50–3, 55–6, 57, 183–5
Picasso, Pablo, 55
Pinch, T.J., 85
Pinker, Steven, 254, 255
PLEs ("past-life" experiences), 80–1
Pool, R., 86
Popper, K., 40, 50
positive emotions, 93, 105–6, 109; see also religious emotions
Posner, M.I., 113, 116, 122, 123, 124, 130, 131

Posterior Network, 122, 123, 124, 130
Potts, R, 23
Poulton, E.P., 43
Prasangika, 206
pratiyasamutpada (interdependence), 198, 201–2
precognition, 69, 85
  in dreams, 73
Premack, David, 13
Pribram, K.H., 74
Prigogine, Ilya, 140, 141
Principles of Psychology, 245
Process Theology, 180
Prochiantz, A., 255
Pruyser, P.W., 95, 96, 97
psilocybin, 64–5
psi-related experiences, 68–70
  and the brain, 74–9
  and geomagnetic activity, 71–3, 75
  implication for religious belief and spiritual practice, 81–5
  and religious belief, 68–70
  and the temporal lobes, 76–8
psi research and postmodern science, 85–7
psychokinesis, 69, 71, 85
Psychology of Religion: Classic and Contemporary, 2
psychoneuroimmunology, 36, 39, 50
psychophysical aggregates (skandhas), 198, 201
Pyysiäinen, Ilkka, 4, 109, 243

Queen Maya, 225, 226
Quider, Robert F., 78
Qur'an, 223, 224

Radin, D., 69
Rafal, R.D., 122
Ramachandran, V.S., 65, 163
Rambo, L., 162
Rani, N.J., 113, 114, 130
Rao, K.R., 68, 83
Rao, P.V.K., 113, 114, 130
Ratner, Hiliary Horn, 12
Reat, N.R., 200
reductionism, 26
reductive materialism, 176
Rees, G., 133
Reiser, Morton, 74

relationship in Christian theology, 193
religion
    and consciousness, 56–7
    and dreams, 221–6
    generation of emotions in, 96–7
    role of in regulating emotion, 100–5
Religion Explained, 4
Religion in Mind, 250
religious belief
    neuroscience of, 151
    and psi-related experiences, 68–70
religious emotions, *see* sacred emotions
religious feelings, brain basis for, 151
REM sleep, 145, 146, 147, 228–35
    compare NREM sleep
Rendell, L., 11
"Report on the Census of Hallucinations," 69
Rethinking Religion, 4, 245, 254
Revelle, W., 117
reverence, 99–101; *see also* awe; gratitude; hope; wonder
Revonsuo, A., 229
Rhine, J.B., 82, 83
Rhine, Louisa, 69, 70
Ribary, U., 155
Richerson, Peter J., 11, 15, 17
Rinpoche, T.W., 226
rituals of cognitive theory, 249–50
Robbins, T., 117
Roberts, R.C., 94, 99–100, 100–101
Robertson, E.M., 155
Robinson, H.M., 43
Roe, C.A., 71
Roland, Alan, 205
Roll, William, 77, 82–3
Rosch, Eleanor, 164, 172
Rosenberg, E.L., 94, 95, 130
Ross, E.D., 78
Rothbart, M.K., 114, 122, 123, 124, 131
Runciman, W.G., 33
Rurzyla-Smith, P., 37
Russell, Charles Taze, 67
Russell, R.J., 40, 179
Ryan, Chris, 61
Ryan, R.M., 104–5, 130
Ryff, C., 106
Ryken, L., 168
Ryle, Gilbert, 210–11, 214

sacred emotions, 98–102
    effect on health, 102–5, 109
    functions of, 105–6
    historical studies of, 108–9
    and spiritual experience, 96
    and spiritual transformation, 107–8
    uniqueness of, 106–7
Sampson, G., 259
Sanford, J., 223
Sargeant, R., 149
Savary, L.M., 223
Schaut, G.B., 73, 77
Schimmel, S., 102
Schleiermacher, F., 96
Schmeidler, Gertrude, 75, 78, 79
Schneider, W., 119, 120
"School of the Dance," 72
Schopenhauer, Arthur, 215
Schutz, Alfred, 210
Schüz, A., 22, 24
Schwartz, G., 37
Schwartz, Jeffrey, 38, 211–14, 215–16
science, cognitive, *see* cognitive science
Scott, A., 53, 140
Searle, J.R., 215
Segal, Z., 38
selective attention, 114, 116, 117–20, 124
self, concept of
    Candrakirti's refutation of, 206–9
    contemporary vs. Buddhist views, 206
    monistic vs. dualistic, 209
    objective, subjective, intersubjective, 209–11
    social construction of, 205–6
    Western views on, 203–5
self-determination, 262
self-organizing systems, 139–43
self-transcendence, human, 29–33
Seligman, M.E.P., 102
Sethi, S., 102
Seventy Stanzas Explaining How Phenomena are Empty of Inherent Existence, 199
Seyfarth, Robert M., 13, 14
Shapiro, S., 38, 130
Shaw, R., 221, 222, 225
Sheikh, A., 37
Sherwood, S.J., 71
Shiffrin, R.M., 119, 120
Shimamura, A.P., 121, 127
Shotter, John, 247, 257

Sidgwick, H., 69
Siewert, C.P., 215
Sikora, S., 220
Silberman, I., 95
Silberstein, M., 53, 58
Singer, B., 106
Singer, W., 155
Sirin, Ibn, 224
skandhas (psychophysical aggregates), 198, 201
sleep and dreaming, 143–7, 228–34
Slingerland, Edward, 166
Smith, Curtis, 161, 229
Smith, Huston, 184
Smith, J.C., 102
Smith, Joseph, 67
Snyder, C.R., 101
Social Construction of Reality, the, 210
social construction of the self, 205–6
social insects, 143
Society for Psychical Research, 69, 79
Solmes, Mark, 133, 231–4
Solomon, R.C., 98, 100
Solso, R.L., 116, 120
soul in Christianity
    and cognitive science, 183–5
    culture as guardian of, 193
    embodied, 187–91
    as emergent property, 184
    and neuroscience, 181–3
Spanos, N., 37, 63
SPECT study, 66, 177
Spelke, E., 116, 119
Sperber, Dan, 242, 245, 254, 255
Sperry, Roger, 36, 40–2, 46, 48, 51, 52, 57, 58, 182
Spottiswoode, James, 73
Sprenger, J., 222
Stace, W.T., 64
Stanovich, K., 68
St. Augustine, 168, 190, 222
Stearns, P.N., 108
Stephen, M., 221
Sterelny, K., 19
Stevenson, Ian, 70, 73, 81
Stewart, K., 14
St. Francis of Assisi, 66, 183–4
Stokes, D., 69

Stroop, J.R., 120
Stroop Effect, 120
structural monism, 86
St. Theresa of Avila, 170–1
St. Thomas Aquinas, 190, 192, 222
Subrahmanyam, S., 73
Summa Theologica, 222
sunyata (emptiness), 198, 201–2; *see also* anatman (no self)
supervenience, 40, 44–8, 50, 51, 52, 183–4, 185
sustained attention, 114, 117–18, 120–1, 124
Suzuki, S., 114, 115, 127
Sweet, P.I., 128–9
Sweetser, Eve, 165, 166–7, 172, 211
Symons, D., 18
synesthesia, 62
Szathmáry, Eörs, 10
Szectman, H., 38
Szentagothai, J., 144
Szpakowska, K., 221

Tarakeshwar, N., 98
Targ, E., 69
Targ, Russell, 83
Tart, Charles, 82
Taylor, D.C., 78
Taylor, Jeremy, 223
Tedlock, Barbara, 221
Teichner, W.H., 120
telepathy, 62, 69, 85
    in dreams, 70–2
temporal lobes
    and psi-related experience, 76
    and religious experience, 163
    and "sensed presence," 176
termite behavior, 143
Tertullian, 190
Teske, J., 194
Teyler, T.J., 74
Thatcher, Adrian, 189
Thayer, R.E., 103
theological anthropology, 187–8
theology and neuropsychology, *see* neurotheology
theology of nature, 180, 186
Theraulaz, G., 143
Thompson, Evan, 164, 172
Thouless, Robert, 83

Tillich, Paul, 179
Tipler, F.J., 25
Tipper, Steven, 117, 119
TMS studies, 151
Tomasello, Michael, 12, 192
Tomita, H., 122
Tong, F., 150
Tononi, G., 140
Tooby, John, 17–18
top-down causation, 27, 36, 39, 40, 46, 51–2; *compare* bottom-up causation
Toulmin, S., 55
traducianism, 190
transmissible cultures, cumulative, *see under* culture
trauma and brain functioning, 147–8
Treatise Concerning Religious Affections, 96
Treatise on the Soul, 190
Treisman, A., 117, 119, 120
Trimingham, S., 224
Trompf, G.W., 221
Truzzi, Marcello, 62, 85
Turing, Alan, 253
Turnbull, O., 133
Tylor, Edward B., 11

Ullman, C., 107
Ullman, Montague, 70, 72, 73
Umitla, C., 121, 133
uncertainty principle, Heisenberg, 54
Underwood, T., 121
Uttal, W.R., 80
Utts, J., 69

Valentine, E.R., 128–9, 252
Van de Castle, Robert, 71, 235
Van Gulick, R., 46, 47
Van Zomeren, A.H., 116
Varela, Francisco, 114, 132, 133, 164, 172
Varieties of Anomalous Experience, 68
Varieties of Religious Experience, 62
Vasubandhu, 197
Vaughan, Alan, 73
Vaughn, F., 113
Vedanta, 215
Velmans, M., 43, 49, 53, 56, 57
Venter, J. Craig, 21, 26, 30
Vigilance Network, 123, 124, 131

Violani, 232
vitalism, 212
von Grunebaum, G.E., 224
von Simson, G., 215

Wacome, Donald, 183, 185
Wade, Jenny, 67
Wade, N., 21
Walker, M.P., 147
Wallace, Alan, 114, 130, 131
Wallace, Anthony F.C., 160, 171
Wallace, B.A., 215, 216
Walsh, R., 113
Wangchen, N., 207, 208
Wapogoro tribe, 77
Ware, K., 188
Warren, C.A., 79
Washburn, S.L., 20
Watkins, J., 43
Watkins, P.C., 99
Watts, Fraser, 95, 102, 179
Wayman, A., 226
Western Approach to Zen, A, 126, 128
Wettstein, H., 99
What Computers Still Can't Do, 252
Whishaw, I.Q., 123
White, Rhea A., 63, 83–4
Whitehead, H., 11
Whitehouse, H., 245–6, 258
"Why the Mental Matters," 51
Wiener, Norbert, 30
Wigner, Eugene P., 32, 54
Wildman, D.E., 23
Wilkins, A., 123
Williams, M., 102
Wilson, D.S., 17
Wilson, E.O., 23
Wilson, Sheryl, 79
Winkelman, Michael, 67, 77
Witvliet, C.V., 103–4
Woerlee, G.M., 79
Wolfson, H.A., 190, 194
wonder, 101; *see also* awe; gratitude; hope; reverence
Wood, G., 37
Woodruff, P., 100, 100–1
Woods, D., 123
Woody, E., 38
Wrangham, R., 12
Wulff, David, 2, 64, 65

yage brew, 61
Yoga Nidra, 152
Young, Serinity, 221, 225, 226

Zaehner, R., 199
Zaleski, P., 114
zazen, *see under* meditation

Zen and the Brain, 6, 130
Zhabotinsky chemical reaction, 140
Zhou, R., 213
Zimmer, H., 200
Zizioulas, J.D., 191, 194
Zygon, 57